MANY-BODY THEORY OF SOLIDS

An Introduction

MANY-BODY THEORY OF SOLIDS

An Introduction

JOHN C. INKSON

Cavendish Laboratory
University of Cambridge
Cambridge, England

PLENUM PRESS · NEW YORK AND LONDON

Library of Congress Cataloging in Publication Data

Inkson, John C., date–
Many–body theory of solids.

Includes bibliographies and index.
1. Solid state physics. 2. Many–body problem. 3. Green's functions. I. Title.
QC176.I44 1983 530.4′1 83-17771
ISBN 0-306-41326-4

A Division of Plenum Publishing Corporation
233 Spring Street, New York, N.Y. 10013

Printed in the United States of America

PREFACE

There exists a gap in the present literature on quantum mechanics and its application to solids. It has been difficult to find an introductory textbook which could take a student from the elementary quantum mechanical ideas of the single-particle Schrödinger equations, through the formalism and new physical concepts of many-body theory, to the level where the student would be equipped to read the scientific literature and specialized books on specific topics. The present book, which I believe fills this gap, grew out of two courses which I have given for a number of years at the University of Cambridge: "Advanced Quantum Mechanics," covering the quantization of fields, representations, and creation and annihilation operators, and "Many Body Theory," on the application of quantum field theory to solids. The first course is a final-year undergraduate physics course while the second is a joint first-year postgraduate physics course and fourth-year undergraduate mathematics course. In an American context this would closely correspond to a graduate course at the masters level.

In writing this book I have tried to stress the physical aspects of the mathematics preferring where possible to introduce a technique by using a simple illustrative example rather than develop a purely formal treatment. In order to do this I have assumed a certain familiarity with solid-state physics on the level of a normal undergraduate course, but the book should also be useful to those without such a background.

The book is divided into three basic parts. The first four chapters introduce the basic tools and concepts of advanced quantum mechanics in a simple context. In this way the reader will have already been introduced to the Green's functions, diagrams, etc., before the complicated formalism is developed. The second part introduces the Green's function self-energy techniques. These form the backbone of the application of quantum mechanics to interacting systems. The method used is the so-called "equation of motion" method which I believe is more physically understandable than the use of Wick's theorem (which leads directly to

Feynman diagrams) but the use of the diagrammatic interpretation of the resulting iterative solution is stressed. The final section consists of the application of the techniques to a number of examples. These have been chosen to bring out particular physical aspects rather than for their inherent importance. A number of examples and a bibliography have been provided at the end of each chapter for readers wishing to look further into the areas treated.

I'd like to express my appreciation to my wife Pam for all her patient hours of typing, and to the staff of Plenum Press, especially Ken Derham and Steve Pisano, for their help in the preparation of this book. Finally, I'd like to dedicate this book to my family, Pam, Andrea, Beverley, and Jonathan, for making it all worthwhile.

JOHN INKSON

CONTENTS

CHAPTER 1. THE INTERACTING SYSTEM

CHAPTER 2. GREEN'S FUNCTIONS OF THE SINGLE-PARTICLE SCHRÖDINGER EQUATION

CHAPTER 3. QUANTIZATION OF WAVES (SECOND QUANTIZATION)

CHAPTER 4. REPRESENTATIONS OF QUANTUM MECHANICS

CHAPTER 5. INTERACTING SYSTEMS AND QUASIPARTICLES

CHAPTER 6. MANY-BODY GREEN'S FUNCTIONS

CHAPTER 7. THE SELF-ENERGY AND PERTURBATION SERIES

CHAPTER ONE

THE INTERACTING SYSTEM

1.1. THE BASIC PROBLEM

The calculation of the motion of one particle is, under classical or quantum mechanics, a well-defined problem with a well-defined solution. True, in the latter case we must then be careful as to how we interpret our solution and remember that we work with probabilities, but in principle the solution can be achieved to the Schrödinger equation for that particle. In the usual form we have

$$H\psi = E\psi \qquad [1.1]$$

with

$$H = -\frac{\hbar^2}{2m}\nabla^2 + V(\mathbf{r}) \qquad [1.2]$$

and, with its boundary conditions, it forms a well-defined computational problem. If we consider two or three particles, we may consider the problem soluble whether they have a mutual interaction or not. Once we get into the realm of large numbers of particles, however, sheer computational difficulty prevents a solution. The only exception is the case of a *noninteracting set* of particles, for then the total Hamiltonian is of the form

$$H = \sum_i -\frac{\hbar^2}{2m_i}\nabla_i^2 + V(\mathbf{r}_i) \qquad [1.3]$$

and splits into a set of independent single-particle Hamiltonians

$$H = \sum_i H_i \qquad [1.4]$$

The independent solutions ψ_n^i for each of the H_i

$$H_i\psi_n^i = E_n\psi_n^i \qquad [1.5]$$

can be combined to form a total wave function for the system according to some prescribed rules. This separation is, of course, due to the fact that, if they are noninteracting, there will be no way in which the behavior of one particle could be influenced by the others. Once we turn on an *interaction* term, no matter how weak, the situation must change. The motion of any particle now becomes a function of all other particles so that the very concepts of single-particle motion and properties become much less useful. This is rather similar to the situation one has in statistical physics, where the motions of "noninteracting" particles in a gas are randomized by collisions with each other and the walls of the container. Even though these collisions may play very little part in determining the gross properties, they have a drastic effect on the individual motions. In fact, there is a very strong formalistic and conceptual connection between the statistical physics and the many-body theory of solids, which we will exploit in later chapters. But first let us study exactly what the problems are if we are to study the physics of a solid body in terms of a collection of interacting particles.

A *solid* consists of an array of atoms in close proximity. This array may be periodic, as in a crystal lattice, but this is not essential. Each atom consists of a positive nucleus surrounded by a neutralizing set of electrons. The Hamiltonian is then

$$H = \sum_i -\frac{\hbar^2\nabla_i^2}{2M_i} + \frac{1}{2}\sum_{i,j}{}' \frac{Z^2e^2}{|\mathbf{R}_i - \mathbf{R}_j|} + \sum_k -\frac{\hbar^2}{2m}\nabla_k^2 + \frac{1}{2}\sum_{k,l}{}' \frac{e^2}{|\mathbf{r}_k - \mathbf{r}_l|} - \sum_{k,i} \frac{Ze^2}{|\mathbf{r}_k - \mathbf{R}_i|} \qquad [1.6]$$

where the nuclear positions are given by $(\mathbf{R}_i, \mathbf{R}_j)$ and the electron positions by $(\mathbf{r}_k, \mathbf{r}_l)$, Z is the nuclear charge, and (M_i, m) are the masses of the nuclei and electron, respectively.

The primes on the summations mean that we must exclude the terms $i = j, \ldots,$ to prevent self-interactions. We have three interaction terms between ions, electrons, and the electron–ion coupling term, as well as the kinetic energy of the ions and electrons. Thus, the solid is a coupled system, consisting of two species interacting with themselves and with each other. There are a number of simplifications which must be made in order to set up a soluble problem:

1. Of the electrons in the solid, only a fraction will determine many of the properties, that is, the valence electrons in the outermost shells. We treat these separately while the other, inner electrons we consider as moving rigidly with the nuclei to form the ion core. Thus, in Eq. (1.6) we lump these electrons into the mass of the nucleus (M_i), and the charge Z is altered appropriately; otherwise they disappear for all intents and purposes from the equation. For specific problems they may, of course, be necessary, but in such situations they can be treated appropriately.
2. Because the ion cores are so much more massive than the electrons, their characteristic velocity is much lower. The electrons generally react so quickly that the ion cores can be considered stationary. The first two terms in the Hamiltonian are therefore irrelevant if we are considering the electronic motion and so can be ignored for the present. This is the famous Born–Oppenheimer approximation.

Both of these approximations can and will be relaxed during the course of this text; but of the two, the second is by far the more difficult to justify and, in general, gives rise to careful considerations in special circumstances. It is a common approximation, however, and for most of the cases of interest fully justified.

This leaves us with a Hamiltonian for the electrons

$$H = \sum_k -\frac{\hbar^2}{2m}\nabla_k^2 + \frac{1}{2}\sum_{k,l}{}' \frac{e^2}{|\mathbf{r}_k - \mathbf{r}_l|} - \sum_{k,l} \frac{Ze^2}{|\mathbf{r}_k - \mathbf{R}_i|} \qquad [1.7]$$

This Hamiltonian forms the basic starting point for an investigation of the effect of electron interaction on the properties of a solid. If we ignored the second term we would, of course, have the independent particle model of a solid, which in itself has been the subject of much work. The interested reader is referred to one of the many standard texts on solid-state physics.

1.2. THE JELLIUM SOLID

Solids come in various forms. They range from the rare-gas solids, in which the atomic nature of the constituents has hardly changed, to the simple metals, in which the outer electrons of the atoms are free to move throughout the entire crystal and cannot be considered to be associated with any one atom. In between these extremes, we have the ionic and covalent solids, which in many ways have the characteristics of giant molecules. In all cases, however, the electrons occupying the outer shells of the atom are dominant in determining the form of the solid and the properties it will have. The core of the atom in most cases takes very little part in the solid's properties. In a first step of simplification we can ignore it and treat it as a source of positive charge, which neutralizes the negative charge of the outer electrons. One further step is to neglect the fact that the charge is concentrated in the core regions and, metaphorically, hammer it flat so that it is smeared out uniformly over the region taken up by the solid. Thus, we obtain the *jellium model* of a solid. Neglecting surface effects, the solid is now translationally invariant and we have, at a stroke, lost all the properties that make a diamond different from lead. The only parameter we have retained is the overall density of the electrons. This must be, however, our basic starting system, for only if we can treat the effect of the electron interaction in this highly simplified system can we have any hope of understanding its effects on the behavior of real solids and liquids. It concentrates all our attention on the single question "If we have a large number of charged particles interacting with one another, how do their properties differ from those of an equal number of noninteracting particles?" The purpose of the rigid positive background is simply to cancel the mean charge so that the electron system is stable and does not explode under the influence of its combined coulomb repulsion.

For the present this then forms our basic system—a set of interacting electrons whose total charge is balanced by a uniform rigid positive background that takes no active part in the dynamics of the system. The way we treat the interaction gives rise then to the vast range of techniques and results that are the main subject of this text. The applications to real systems are by way of special methods and approximations to this ideal case, but those we will treat as they come.

1.3. HARTREE THEORY—THE SOMMERFELD MODEL

As we saw in the first section, the main obstacle to the use of single-particle wave functions is the interaction term, since this prevents the separation of Eq. (1.4). This means that the potential seen by an electron is dependent upon the positions of all of the other particles. The equation to be solved is then a true many-body one. If, however, we can in some way make the interaction a function only of the individual positions of the particles, the total problem will divide immediately into a set of single-particle equations, which are more tractable than the original equation. Thus, we wish to replace the interaction term

$$V_{\text{INT}}(\mathbf{r}_1 \cdots \mathbf{r}_N) = \frac{1}{2} \sum_{i,j}{}' \frac{e^2}{|\mathbf{r}_i - \mathbf{r}_j|} \qquad [1.8]$$

by an approximate form

$$\tilde{V}_{\text{INT}}(\mathbf{r}_1 \cdots \mathbf{r}_N) = \sum_i V_i(\mathbf{r}_i) \qquad [1.9]$$

The most physically satisfying way to do this is to appeal to our ideas of electrostatics. If we take the ith particle at a position $\mathbf{r}_i$, then the potential seen by it will be given by the contribution from all of the other electrons, i.e.,

$$V_i(\mathbf{r}_i) = \sum_j{}' \frac{e^2}{|\mathbf{r}_i - \mathbf{r}_j|} \qquad [1.10]$$

If we then fixed the position of all of the other electrons, we would obtain an equation for the motion of the ith particle of the form

$$(H_0(\mathbf{r}_i) + V_i(\mathbf{r}_i) - E_i)\psi_i(\mathbf{r}_i) = 0 \qquad [1.11]$$

H_0 is that part of the Hamiltonian which separates naturally, i.e., kinetic energy and electron–ion interactions. In the same way we could obtain a set of wave functions for all of the electrons corresponding to a particular distribution of particles. As the system develops, however, these wave functions change and the whole procedure quickly becomes meaningless. The picture of the electron in a particular single-particle wave function is, however, useful because, if we postulate the set of wave functions which are occupied, we can make the whole procedure self-consistent by writing for $V_i(\mathbf{r}_i)$

$$V_i(\mathbf{r}_i) = \sum_{\substack{j \neq i \\ j \text{ occupied}}} \int \frac{e^2 |\psi_j(\mathbf{r}_j)|^2}{|\mathbf{r}_i - \mathbf{r}_j|} d\mathbf{r}_j \qquad [1.12]$$

This is the *Hartree approximation,* for now we have replaced the actual interaction by an approximate form, which postulates each electron moving in a field due to a distributed charge (through $|\psi|^2$) of all the other electrons. The Schödinger equation becomes

$$\left[E - \sum_{i=1}^{N} (H_0(\mathbf{r}_i) + V_i(\mathbf{r}_i)) \right] \Psi(\mathbf{r}_1, \mathbf{r}_2, \ldots, \mathbf{r}_j, \ldots, \mathbf{r}_N) = 0 \qquad [1.13]$$

and splits into a set of N one-electron equations on the assumption of a product wave function of the form

$$\Psi(\mathbf{r}_1, \mathbf{r}_2 \cdots \mathbf{r}_N) = \psi_1(\mathbf{r}_1)\psi_2(\mathbf{r}_2) \cdots \psi_N(\mathbf{r}_N) \qquad [1.14]$$

so that we have N equations

$$(H_0(\mathbf{r}) + V_i(\mathbf{r}) - E_i)\psi_i(\mathbf{r}) = 0 \qquad [1.15]$$

Starting from an assumed set of ψ, the equations are solved self-consistently. It is important to realize that this is not a trivial approximation. Besides the internal self-consistency, through the definition of $V_i(\mathbf{r})$, each of the N Hartree equations has, itself, a complete set of solutions. These are *not* the excited states of that particle, however. The Hartree solution also does not incorporate the antisymmetric nature of the true solution; instead it attempts to include the Pauli principle simply through the requirement that no more than two particles can be in each electronic state.

The Hartree equation only exists to produce that one state for each particle. As we shall see, the other solutions are strictly meaningless. Let us take, for instance, a simple example: the lithium atom. Here we have a central core potential in which three interacting electrons move. If there were no electronic interactions, two would occupy the $1s$ state and one either a $2s$ or $2p$ orbital. We can take these as the starting wave functions for a Hartree solution. Suppose, for instance, we take the initial states as $1s$, $1s$, and $2s$, then the potentials for the three particles become

$$V_{1s}(\mathbf{r}) = e^2 \int \frac{|\psi_{1s}(\mathbf{r}')|^2}{|\mathbf{r} - \mathbf{r}'|}\, d^3r' + e^2 \int \frac{|\psi_{2s}(\mathbf{r}')|^2}{|\mathbf{r} - \mathbf{r}'|}\, d^3r' \qquad [1.16a]$$

$$V_{2s}(\mathbf{r}) = 2e^2 \int \frac{|\psi_{1s}(\mathbf{r}')|^2}{|\mathbf{r} - \mathbf{r}'|}\, d^3r' \qquad [1.16b]$$

From these we obtain new Schrödinger equations to solve and give new solutions corresponding to the old $1s$ and $2s$ states. These can be resubstituted into expressions for the potential. The whole procedure is repeated until we obtain a self-consistent solution in the sense that the potentials and wave functions change infinitesmally from one iteration to the next. It is now possible to investigate whether the solution is the correct ground-state solution by repeating the whole procedure with some other combination (say $1s$, $1s$, $2p$). The total energies can be calculated in both cases as the sum of the energies of the single-particle Hartree energies; the one with lowest energy will be the ground state.

In practice we find that the $1s$, $1s$, $2s$ has the lowest energy. Writing the solutions as $\Phi_{1s}(\mathbf{r})$, $\Phi_{1s}(\mathbf{r})$, $\Phi_{2s}(\mathbf{r})$, it is meaningful to ask whether the other solutions of the equation for the $\Phi_{2s}(\mathbf{r})$ state

$$\left\{ H_0(\mathbf{r}) + 2e^2 \int \frac{|\Phi_{1s}(\mathbf{r}')|^2}{|\mathbf{r} - \mathbf{r}'|}\, d^3r' - E_k \right\} \Phi_k(\mathbf{r}) = 0 \qquad [1.17]$$

have any meaning.

There will be a complete set of solutions $\Phi_k(\mathbf{r})$ corresponding approximately to the hydrogenic series, and it is very tempting to identify these, and their energies, with the excited states of the lithium atom. Thus, we would write

$$E_0 = 2E_{1s} + E_{2s} \qquad [1.18]$$

for the ground state and

$$E_{\text{exc}} = 2E_{1s} + E_k \qquad [1.19]$$

for the "excited" state with the particle raised to the kth state of Eq. (1.17). This is *not* true, however, for in taking the higher state of Eq. (1.17) we have *neglected* the change in the Hartree potential due to the change in the 1s wave functions brought about by the movement of the electron from 2s orbital to the kth orbital (Fig. 1.1). This phenomenon of *relaxation* is quite general; it simply means that if the electrons form an interacting system, the higher energy states will be determined not only by the change in a single-particle state but also by the reaction of the system as a whole to that change. The eigenvalues of the Hartree equations are only useful if they are one of the occupied ones, and calculations of the excited states must go by way of a separate calculation of both states. The amount to which the energies of the excited states change due to relaxation is, fortunately, sometimes negligible. In particular, in the case of large systems, the relaxation effect tends to become quite small, since the relaxation perturbation on each state is so small. For this reason the actual Hartree solutions can often be used as if they referred to the real state of the system. This simplifies both the calculation and interpretation enormously. It must not be forgotten, however, that this is an approximation, even within the Hartree model, and each case must be judged on its merits.

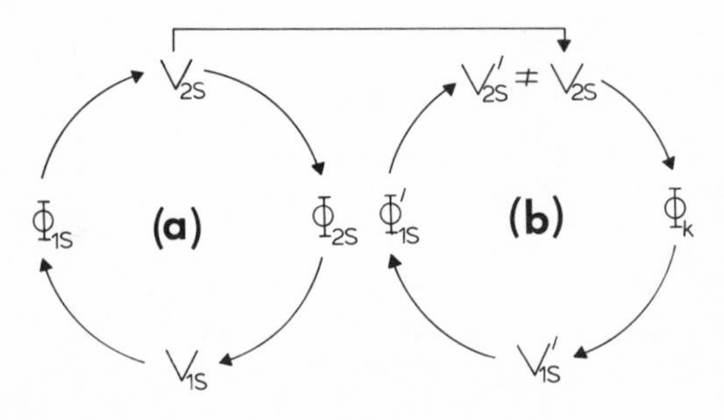

FIGURE 1.1. (a) Self-consistency loop for the ground-state wave functions and Hartree potentials. (b) "Excited state" calculation. Since $V'_{2S} \neq V_{2S}$, the excited state Φ_k is *not* an eigenfunction.

We can now apply the Hartree approximation to the jellium solid. Since everything is uniform, we take as our first-order approximation to the ground state a product wave function with the single-particle wave functions being the plane-wave set of states

$$\psi_{\mathbf{k}}(\mathbf{r}) = \frac{1}{\sqrt{\Omega}} e^{i\mathbf{k}\cdot\mathbf{r}} \qquad [1.20]$$

normalized to the volume of the system (Ω). Each $\mathbf{k}$ is then doubly occupied up to the Fermi wave vector. The Hartree potential becomes

$$V(\mathbf{r}) = \frac{2e^2}{\Omega} \sum_{\mathbf{k}=0}^{\mathbf{k}_F} \int \frac{d^3r'}{|\mathbf{r} - \mathbf{r}'|} - \frac{e^2}{\Omega} \int \frac{d^3r'}{|\mathbf{r} - \mathbf{r}'|} \qquad [1.21]$$

The first term is canceled by the positive background, since the system is neutral, while the second is the self-interaction term (i.e., the contribution from the state considered). There are two things to note. First, the self-interaction term is zero in the limit of a large system. Taking $\mathbf{r}$ as the origin

$$\begin{aligned} \frac{e^2}{\Omega} \int \frac{d^3r'}{|\mathbf{r}'|} &= \frac{4\pi e^2}{\Omega} \int \frac{(r')^2}{r'} dr' \\ &= \frac{4\pi e^2}{\Omega} \int r' \, dr' \end{aligned} \qquad [1.22]$$

The integral is proportional to the area of the system so the whole term will go to zero as the total volume increases. Second, the potential is the same for all states. The Hartree solution thus degenerates to one differential equation, all the solutions are exact states, and there is no relaxation.

The Hartree equation for the jellium model is thus

$$\left[\frac{\hbar^2}{2m} \nabla^2 + E(\mathbf{k}) \right] \psi_{\mathbf{k}}(\mathbf{r}) = 0 \qquad [1.23]$$

It has plane-wave solutions, and within this model, the system is described exactly by the complete set of single-particle wave functions. Thus, the Hartree model for the jellium solid has degenerated into the noninteracting electron gas, the energy of interaction simply going to cancel the effect of the positive background charge. This is commonly known as the *Sommerfeld model* of a solid. It works quite well for many of the properties of simple metals such as sodium or potassium.

In the same way the Hartree solution of real solids simply returns us to the one-electron picture and provides for the replacement of the original electron–ion potential by a new one-electron potential evaluated self-consistently through the relationships of Eqs. (1.12) and (1.15). To obtain interesting interaction effects we must go beyond the Hartree approximation.

1.4. HARTREE–FOCK

The way we derived the Hartree approximation was by physical reasoning; there is, however, a more mathematical method which lends itself directly to improvements, that is, the variational method. If the total wave function Ψ for the system is known, the total energy E is given by

$$E = \frac{\langle \Psi | H | \Psi \rangle}{\langle \Psi | \Psi \rangle} \qquad [1.24]$$

where the denominator is unnecessary if the wave function is properly normalized. Normally, however, we do not know Ψ, but a given approximation to the wave function will give us, however badly, an estimate for the energy. The *variation principle* simple states that the best approximation to the energy is given when

$$\frac{\delta E}{\delta \Psi} = 0; \qquad \frac{\delta^2 E}{\delta^2 \Psi} > 0 \qquad [1.25]$$

Obviously, the better our original approximation for Ψ the better the resulting energy, but it must be remembered that to a large extent our initial conditions govern the final result. The Hartree approximation comes from the assumption

$$\Psi(\mathbf{r}_1 \cdots \mathbf{r}_N) = \psi_1(\mathbf{r}_1)\psi_2(\mathbf{r}_2)\psi_3(\mathbf{r}_3) \cdots \psi_N(\mathbf{r}_N) \qquad [1.26]$$

The operation of varying Ψ is simply the variation of each of the one-electron states. This results in the set of Hartree equations to give the best single-particle wave functions and energy consistent with the initial approximation of a product total wave function and need not have any relation whatsoever to reality.

The next step is to improve our first estimate of the wave function. We do this by introducing one extra piece of physics, the *Pauli exclusion principle,* and take an antisymmetric total wave function composed of one-electron states. There is no *a priori* reason, however, for believing this to be better than the Hartree approximation in any sense other than including the Pauli principle. It can only be justified by appeal to experimental results. Our wave function now becomes

$$\Psi(\mathbf{r}_1 \cdots \mathbf{r}_N) = \frac{1}{\sqrt{N!}} (-1)^P P \{\psi_\alpha(\mathbf{r}_1)\psi_\beta(\mathbf{r}_2) \cdots \psi_\eta(\mathbf{r}_N)\} \qquad [1.27]$$

where P is a permutation operator which interchanges the labels α, β, ... Thus, for example, for two particles

$$\begin{aligned} \Psi(\mathbf{r}_1,\mathbf{r}_2) &= \frac{1}{\sqrt{2}} (-1)^P P \{\psi_\alpha(\mathbf{r}_1)\psi_\beta(\mathbf{r}_2)\} \\ &= \frac{1}{\sqrt{2}} (\psi_\alpha(\mathbf{r}_1)\psi_\beta(\mathbf{r}_2) - \psi_\beta(\mathbf{r}_1)\psi_\alpha(\mathbf{r}_2)) \end{aligned} \qquad [1.28]$$

while for N particles we can express Ψ as a "Slater" determinant.

$$\Psi(\mathbf{r}_1 \cdots \mathbf{r}_N) = \frac{1}{\sqrt{N!}} \begin{vmatrix} \psi_\alpha(\mathbf{r}_1) & \psi_\alpha(\mathbf{r}_2) & \psi_\alpha(\mathbf{r}_3) & \cdots \\ \psi_\beta(\mathbf{r}_1) & \psi_\beta(\mathbf{r}_2) & \psi_\beta(\mathbf{r}_3) & \cdots \\ \psi_\gamma(\mathbf{r}_1) & \psi_\gamma(\mathbf{r}_2) & \psi_\gamma(\mathbf{r}_3) & \cdots \\ \vdots & \vdots & \vdots & \end{vmatrix} \qquad [1.29]$$

The variation of Ψ is again equivalent to varying the set of single-particle wave functions. Each variation in one of the $\psi_\alpha(\mathbf{r})$ gives an equation for a "best-fit" single-particle wave function and an eigenvalue E_α corresponding to it. As in the Hartree case, the set of equations are coupled by their dependence upon the other single-particle wave functions.

The equations are

$$(H_0(\mathbf{r}) + V_{H\alpha}(\mathbf{r}) - E_\alpha)\psi_\alpha(\mathbf{r}) + \int V_{ex}(\mathbf{r},\mathbf{r}')\psi_\alpha(\mathbf{r}')\,d\mathbf{r}' = 0 \qquad [1.30]$$

The Hartree potential $H_{H\alpha}(\mathbf{r})$ is now supplemented by a nonlocal exchange potential derived from the antisymmetric nature of the wave function

$$V_{ex}(\mathbf{r},\mathbf{r}') = \frac{-e^2}{|\mathbf{r} - \mathbf{r}'|} \sum_{\substack{\beta\neq\alpha \\ (\beta \text{ occupied})}} \psi_\beta(\mathbf{r})\psi_\beta^*(\mathbf{r}') \qquad [1.31]$$

where the summation is over the occupied states only (as in the case of the Hartree potential). Each of these single-particle equations has a complete set of solutions in terms of the exchange potential $V_{ex}(\mathbf{r},\mathbf{r}')$ but these are not the excited states of that particle. In exactly the same way as before, the relaxation of the system is just as important in Hartree–Fock as in the Hartree approximation. There is an important new aspect of this potential, however; it is *nonlocal,* i.e., it depends upon two variables and appears within an integral so that the equation to be solved is a complicated integro-differential one.

We can formally define a new local potential for each state $\psi_\alpha(\mathbf{r})$ through the relation

$$V_{ex}^\alpha(\mathbf{r})\psi_\alpha(\mathbf{r}) = \int V_{ex}(\mathbf{r},\mathbf{r}')\psi_\alpha(\mathbf{r}')\,d\mathbf{r}' \qquad [1.32]$$

The potential so formed is sensitively dependent upon the state, even in the limit of the jellium model, whereas the Hartree term could be simplified to a well-behaved, state-independent potential. If the Hartree term simplifies to the averaged electrostatic field due to the other electrons, what is the equivalent physical basis of the exchange term? This is best seen by reference to two particles.

The electrostatic energy is given by [using Eq. (1.28)]

$$\left\langle \Psi(\mathbf{r}_1,\mathbf{r}_2) \left| \frac{1}{|\mathbf{r}_1 - \mathbf{r}_2|} \right| \Psi(\mathbf{r}_1,\mathbf{r}_2) \right\rangle \approx \int \frac{|\psi_\alpha(\mathbf{r}_1)|^2 |\psi_\beta(\mathbf{r}_2)|^2}{|\mathbf{r}_1 - \mathbf{r}_2|}\, d\mathbf{r}_1\, d\mathbf{r}_2 - \int \frac{\psi_\alpha(\mathbf{r}_1)\psi^*{}_\beta(\mathbf{r}_1)\psi^*{}_\alpha(\mathbf{r}_2)\psi_\beta(\mathbf{r}_2)}{|\mathbf{r}_1 - \mathbf{r}_2|}\, d\mathbf{r}_1\, d\mathbf{r}_2 \qquad [1.33]$$

The first term is simply the energy due to the charge densities and appears as the Hartree potential, while the second term allows for the "repulsion" of the single-particle states. Because of the Pauli principle, particles with the same spin are not allowed to be in close proximity, so that in the case of charged particles a local charge deficiency, or hole, is produced leading to a change in the electrostatic energy. The exchange potential is a measure of the effect of that hole on the single-electron states.

Suppose we take our previous example of the Sommerfeld model. The exchange potential is fairly easy to calculate, since the states remain plane waves, and we obtain the relation for the energy of the plane-wave state of momentum **k**.

$$E(\mathbf{k}) = \frac{\hbar^2 k^2}{2m} - \frac{e^2 k_F}{2\pi}\left\{2 + \frac{(k_F^2 - k^2)}{k_F k}\ln\left|\frac{k_F + k}{k_F - k}\right|\right\} \qquad [1.34]$$

The exchange potential, as well as the energy, is a fairly rapidly varying function of the wave vector **k** of the state; this alters many of the properties of the Sommerfeld model drastically. In fact so drastic is the change that one can safely say that the Hartree–Fock solution bears no resemblance to physical reality for the case of solids, in contrast to the atomic situation where it works reasonably well. The most spectacular deviation is in the density of one-electron states. This is given by a term which depends upon the inverse of the derivative of $E(k)$ (i.e., $[\partial E(k)/\partial k]^{-1}$). At the Fermi energy ($k = k_F$) the Hartree–Fock energy has an infinite slope with respect to the wave vector so the density of states goes to zero, thus producing enormous changes in properties which are sensitive to the number of states at the Fermi energy, such as the conductivity (both thermal and electrical) for which there is no experimental evidence. This is shown in Fig. 1.2.

The failure of the Hartree–Fock approximation in solids is nothing if not spectacular. It leaves the problem of where it has gone wrong, for obviously it is necessary that the trial wave function be of the form in Eq. (1.27). The magic word, which crops up, is *correlation.* If we have two electrons, then, quite apart from any Pauli principle working to keep them away from each other, the electrostatic repulsion also does a very good job. Thus, the Slater determinant may not be such a good starting point after all, for the simple reason that it leaves out this electron "cor-

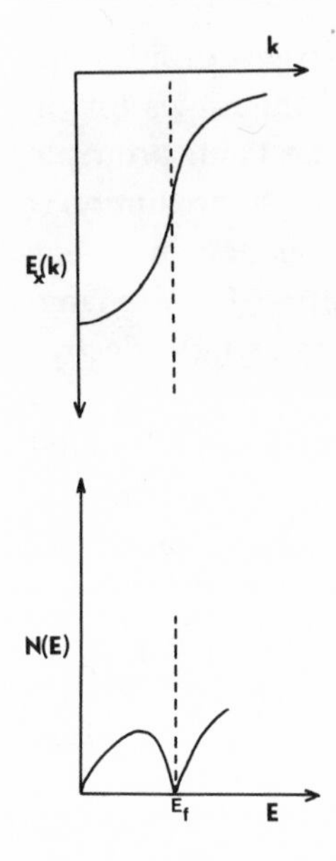

FIGURE 1.2. Hartree–Fock exchange energy $E_x(k)$ and resulting density of states $N(E)$ for an electron gas.

relation." It would carry us too far out of our way in this introductory chapter to discuss the various calculations of the correlation energy.

If we look at the exchange part of the energy of Eq. (1.34) we find that, although the variation is wrong, the absolute magnitude is not far from that required to explain the experimental binding energy of the electron in the solid. (The Hartree potential simply canceled the positive background; the electrons were still unbound.) Slater therefore suggested that one should average the exchange over the electron states to produce a local, i.e., state-independent, exchange, and correlation potential

$$V_{ex}^{\alpha}(\mathbf{r}) \Rightarrow \tilde{V}_{ex}(\mathbf{r}) = \frac{\Sigma_\alpha V_{ex}^{\alpha}(\mathbf{r})\psi_\alpha(\mathbf{r})}{\Sigma_\alpha \psi^*{}_\alpha(\mathbf{r})\psi_\alpha(\mathbf{r})} \qquad [1.35]$$

Since this is also difficult to use, he proposed a further simplification arising out of the density dependence of Eq. (1.35) in the case of plane-wave states. This comes out to be $n^{2/3}$ so that, provided the charge density variation is not too rapid, one could approximate Eq. (1.35) by

$$\tilde{V}_{ex}(\mathbf{r}) = \beta[n(\mathbf{r})]^{2/3} \qquad [1.36]$$

where β is a constant to be determined. This is the potential used extensively as a measure of the exchange and correlation in practical numerical calculations. It has the virtue that, because the β factor has only a tenuous theoretical basis, it can be regarded as an adjustable parameter to be altered to fit the experimental evidence (though it must be admitted that the successes of the method have been spectacular). However, it is not by any means Hartree–Fock and in itself does not give us any basic insight into the real world of interacting particles.

1.5. EXCHANGE AND CORRELATION HOLES

If Hartree–Fock does not lead to a reasonable solution perhaps we should consider further the physical effect of the interaction along the lines of the correlation effect. Consider the jellium solid, pick out one particular electron to study, and consider its effect on all of the other charges (Fig. 1.3). If we neglect the fact that it is a quantum system, the other electrons will move away under classical electrostatic forces until an equal and opposite positive background charge has been exposed so that the electric field at larger distances is removed. In other words the electrons will correlate their motion so as to screen out the electric field. The electron is left surrounded by a hole in the electron density which contains an equal and opposite charge. This is the simplest example of *screening* and the formation of a correlation hole.

The changeover to a quantum system will not alter the situation all that much; the boundary between the hole and the rest of the system

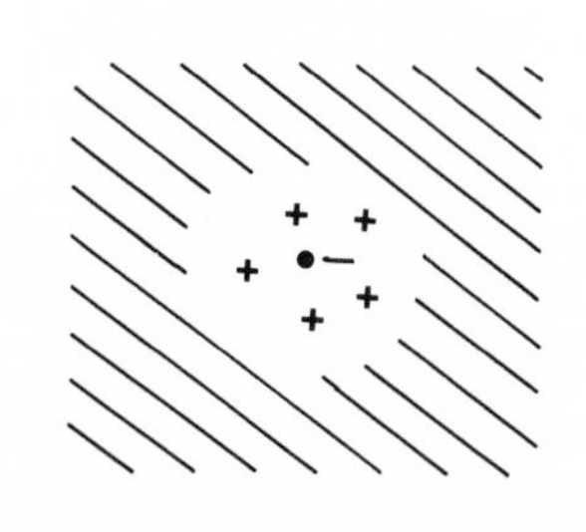

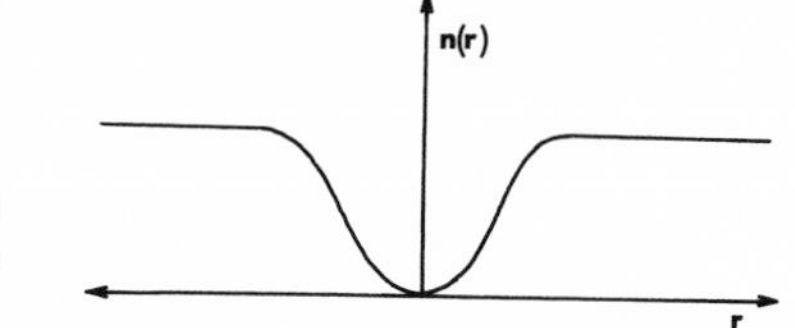

FIGURE 1.3. Schematic representation of the coulomb-induced hole in the electron density [$n(r)$].

will become more diffuse but it will still contain, in total, just enough charge to screen out the electronic charge. If there were no coulomb interaction, only exchange, we would have a similar situation; the test electron would have an exchange repulsion for all other electrons in the same spin state. Thus, the electron would be surrounded by a hole but one containing only one-half an electronic charge since half the electron density is unaltered being of opposite spin. Thus, *exchange* and *coulomb* effects have much the same net result: the production of a hole in the electron density surrounding the test electron.

At the densities encountered in solids, the coulomb effect is as large, if not larger, than the exchange effect. The inclusion, therefore, of exchange through the Pauli principle is not the most sensible approximation if we are to think in terms of an effective single-particle Hamiltonian. Firstly, the electrons simply do not get close enough for the exclusion principle to be effective. Secondly, the interaction between the electrons is no longer simply the coulomb interaction. The appearance of the coulomb and exchange holes introduce screening effects and, unless the electrons are within a very short distance of each other, the interaction is negligible.

The idea of a hole in the electron distribution is the physical basis for the Wigner model for calculating the properties of the alkali metals (i.e., those in which there is only one conduction electron per atom). In this model the electron sees the whole of the ionic potential of the atomic cell in which it is, but outside that cell it sees neutral atomic potentials. Obviously, from Fig. 1.4 it is similar to the coulomb hole (Fig. 1.3) except

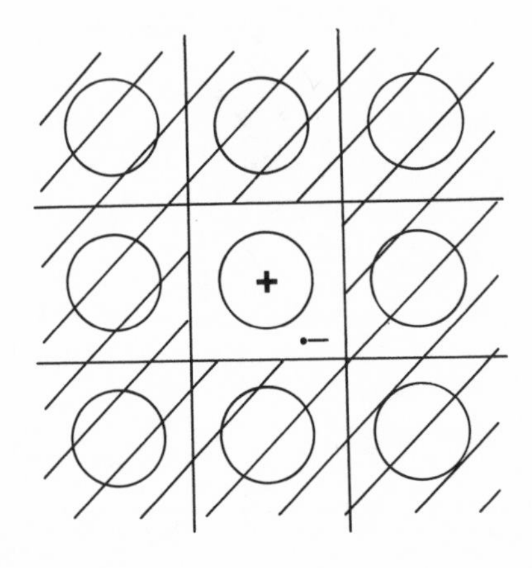

FIGURE 1.4. The Wigner model.

that the neutralizing charge has been concentrated onto the ionic core. It does give very good values for the cohesive energies for these metals, suggesting that the major portion of the interaction energy has gone into the cohesive energy of the electrons in forming the coulomb hole.

If this line of reasoning is to be pursued further, one must have the means of describing the response of the electrons to a given perturbation. This will form a major part of later chapters, so in order to introduce the concepts let us look at the simplest model consistent with quantum ideas, the Thomas–Fermi model.

1.6. CORRELATION EFFECTS AND THE THOMAS–FERMI MODEL

For a solid in equilibrium, the local electron density is determined by the Fermi level, which is a property of the solid as a whole. If we apply a local electrostatic force that disturbs the system (Fig. 1.5), there

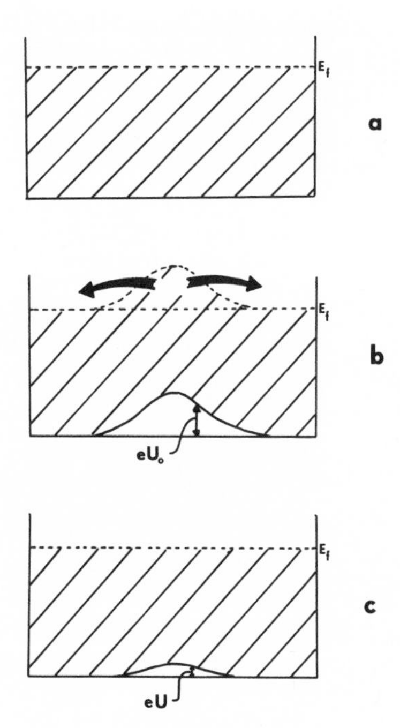

FIGURE 1.5. The Thomas–Fermi model. (a) Equilibrium electron gas. (b) Electrons migrate under influence of applied potential. (c) New equilibrium achieved.

will be a tendency for the electrons to migrate until the density is again consistent with the Fermi level. A new local equilibrium between the applied potential and the electrostatic potential, built up by the charge movement, is produced. Provided the potential varies slowly we can assume that each part of the system acts like the bulk solid but with a different energy zero.

Suppose the applied potential is $U_0(\mathbf{r})$, then the total potential $U(\mathbf{r})$ is given by

$$U(\mathbf{r}) = U_0(\mathbf{r}) + V(\mathbf{r}) \quad [1.37]$$

$V(\mathbf{r})$ is the electrostatic potential due to charge movement and so satisfies Poisson's equation with the change in the charge density $\rho(\mathbf{r})$

$$\nabla^2 V(\mathbf{r}) = 4\pi\rho(\mathbf{r}) \quad [1.38]$$

The simple Sommerfeld model can now be used to obtain the relationship between $\rho(\mathbf{r})$ and the *total* potential. From Fig. 1.6 this is obviously

$$\rho(\mathbf{r}) = eU(\mathbf{r})N(E_F) \quad [1.39]$$

where $N(E_F)$ is the density of states at the Fermi energy. That is, the potential at $\mathbf{r}$ raises the energy of each electron state by $eU(\mathbf{r})$. Those states that rise above the Fermi level are emptied and so the charge den-

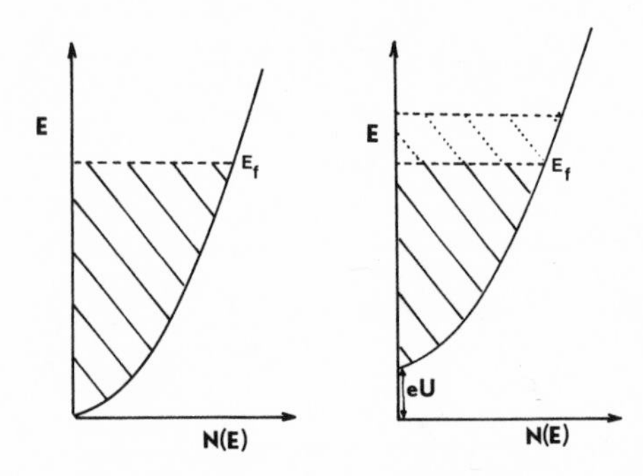

FIGURE 1.6. Calculation of electron density change in Thomas–Fermi model.The local potential raises the energy of the states so that those raised above the Fermi level empty.

sity changes. Thus

$$\nabla^2 V(\mathbf{r}) = 4\pi e U(\mathbf{r}) N(E_F) \qquad [1.40]$$

Therefore, we have

$$\nabla^2 (U(\mathbf{r}) - U_0(\mathbf{r})) = 4\pi e U(\mathbf{r}) N(E_F)$$

or

$$(\nabla^2 - \lambda^2) U(\mathbf{r}) = \nabla^2 U_0(\mathbf{r}) \qquad [1.41]$$

where

$$\lambda^2 = 4\pi e N(E_F) \qquad [1.42]$$

Equation (1.41) is the nonhomogeneous Thomas–Fermi equation, but it is usual to set the right-hand side to zero and consider only the homogeneous equation

$$(\nabla^2 - \lambda^2) U(\mathbf{r}) = 0 \qquad [1.43]$$

The sources of the external potential [described by $\nabla^2 U_0(\mathbf{r})$] are replaced by the appropriate boundary conditions. One very important aspect of this model is carried over into the more exact calculation, self-consistency. The electrons are unable to distinguish between the applied external [$U_0(\mathbf{r})$] and the induced potential created by the correlated movement of the other charges [$V(\mathbf{r})$] so they respond to the total potential $U(\mathbf{r})$.

As an example of the application of the Thomas–Fermi model, consider the response to a point charge so that

$$U_0(\mathbf{r}) = \frac{e}{|\mathbf{r}|} \qquad [1.44]$$

In spherical coordinates, the Thomas–Fermi equation reduces to

$$\left[\frac{1}{r^2}\frac{\partial}{\partial r}\left(r^2 \frac{\partial}{\partial r}\right) - \lambda^2\right] U(r) = 0 \qquad [1.45]$$

with the boundary conditions that

$$U(r) \to \frac{e}{r}, \qquad r \to 0$$
$$U(r) \to 0, \qquad r \to \infty$$

The solution is

$$U(r) = e\,\frac{e^{-\lambda r}}{r} \tag{1.46}$$

We can get something more out of this calculation. Suppose the charge is our test electron, then Eq. (1.46) will give the effective interaction potential between electrons in the Thomas–Fermi model, taking into account the coulomb-induced correlation of the electron motion, i.e., the screening of the coulomb interaction by all of the other electrons. It could then be introduced, instead of the normal interaction potential, into the Schrödinger equation:

$$V_{\mathrm{INT}}(\mathbf{r}_1 \cdots \mathbf{r}_N) \Rightarrow \frac{1}{2}\sum_{i,j}{}' e^2\,\frac{e^{-\lambda|\mathbf{r}_i-\mathbf{r}_j|}}{|\mathbf{r}_i - \mathbf{r}_j|} \tag{1.47}$$

If we now use the Slater determinant as a trial wave function, the resulting "screened" exchange potential is very much better behaved and so forms a reasonable approach to the inclusion of exchange into an effective single-particle picture. The rest of the interaction is tied up in the correlation energy. It corresponds, classically, to the induced potential felt by the electron due to the screening effect of the other electrons. Thus, the correlation energy is given by the induced potential felt by the electron or, more mathematically

$$E_{\mathrm{CORR}} = \lim_{r\to 0}\left(-\frac{e}{2}\,V_{\mathrm{IND}}(r)\right) \tag{1.48}$$

From Eq. 1.46

$$V_{\mathrm{IND}}(\mathbf{r}) = \frac{e}{r}\,(1 - e^{-\lambda r})$$

so

$$E_{\text{CORR}} = -\frac{\lambda e^2}{2} \qquad [1.49]$$

The correlation energy thus appears as an additive constant to the energy of each electron. In order of magnitude terms this is, in fact, quite a good approximation to the correlation energy as evaluated by more exact means (as in Chap. 9) and suggests that the physical picture is a good one.

Another concept we can usefully introduce here is that of a *generalized dielectric function.* In solving Eq. (1.43) we have an expression for the total potential $U(\mathbf{r})$ resulting from the application of a known external potential $U_0(\mathbf{r})$. In classical electrostatics we would express such a relationship by way of a dielectric constant ϵ. Thus

$$U \approx \frac{U_0}{\epsilon} \qquad [1.50]$$

If we wished to continue the analogy it would be equally correct to describe the potential in terms of a dielectric function $\epsilon(\mathbf{r})$

$$U(\mathbf{r}) = \frac{U_0(\mathbf{r})}{\epsilon(\mathbf{r})} \qquad [1.51]$$

In the case of the impurity problem, we would have

$$\epsilon(\mathbf{r}) = e^{-\lambda|r|} \qquad [1.52]$$

This is not a particularly informative way to proceed, however, since for each case we would find a different analytical form for $\epsilon(\mathbf{r})$. It is much better to go back to Eq. (1.41), take Fourier transforms, and use a wave-vector-dependent dielectric function $\epsilon(\mathbf{q})$. Thus, we have

$$(-q^2 - \lambda^2)U(\mathbf{q}) = -U_0(\mathbf{q})q^2 \qquad [1.53]$$

so that

$$U(\mathbf{q}) = \frac{U_0(\mathbf{q})}{(1 + \lambda^2)/q^2} \qquad [1.54]$$

Defining $\epsilon(\mathbf{q})$ by the relationship

$$U(\mathbf{q}) = \frac{U_0(\mathbf{q})}{\epsilon(\mathbf{q})} \qquad [1.55]$$

we have

$$\epsilon(\mathbf{q}) = (1 + \lambda^2)/q^2 \qquad [1.56]$$

Thus, the nearest useful replacement for the expression (1.50) is in reciprocal space. The total potential can now be obtained from

$$U(\mathbf{r}) = \frac{1}{(2\pi)^3} \int d^3q \, e^{i\mathbf{q}\cdot\mathbf{r}} \frac{U_0(\mathbf{q})}{\epsilon(\mathbf{q})} \qquad [1.57]$$

for any $U_0(\mathbf{r})$.

Knowing the external potential *and* the dielectric function, we could have the total potential without recourse to the Thomas–Fermi equation. Alternatively we could say that the basic approximations of the Thomas–Fermi theory result in a dielectric (or *response*) function of the form in Eq. (1.56) and consider the differential equation as secondary. A more sophisticated model of the system would then undoubtedly produce a better dielectric function. The form of Eq. (1.57) would not change, however, only the details of the integration, even though a differential equation might not exist.

The calculation of the correlation energy can be expressed in the form [from (1.48)]

$$E_{\text{CORR}} = \lim_{r\to 0} \frac{e^2}{2(2\pi)^3} \int \frac{4\pi}{q^2} \left(\frac{1}{\epsilon(\mathbf{q})} - 1 \right) e^{i\mathbf{q}\cdot\mathbf{r}} \, d^3q \qquad [1.58]$$

where we have used the Fourier transform of the coulomb potential

$$U_0(\mathbf{q}) = \frac{4\pi e^2}{q^2} \qquad [1.59]$$

This would also not change, so we begin to see that the dielectric function (later to be considered also as a function of energy) serves as both a useful connection to our ideas of electrodynamics and a very important

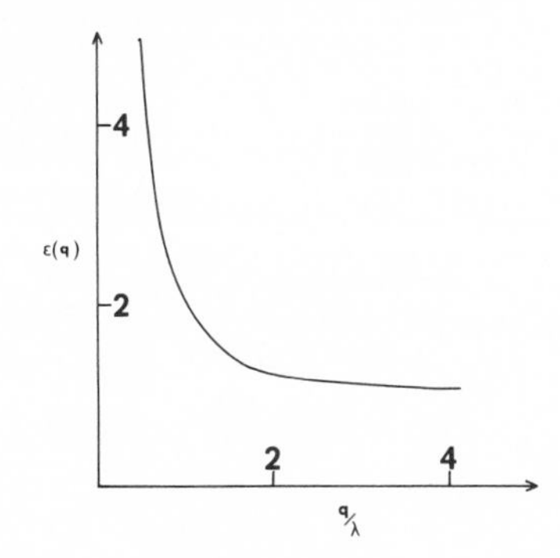

FIGURE 1.7. Thomas–Fermi response function $\epsilon(q)$.

unifying concept for the various approximations encountered in the theory of interacting systems.

The Thomas–Fermi model does treat the response of solids reasonably well in the long-wavelength limit and can be used as a first estimate for many effects. Figure 1.7 shows the Thomas–Fermi response function. Finally, if we take the screening effect (as described in the Thomas–Fermi model) as a reasonable representation, we see that the charge distribution surrounding the test electron appears as in Fig. 1.8. The region of reduced electron density is the correlation hole. If the exchange interaction has a comparable or shorter range, then it must be drastically reduced in strength. This competition between coulomb and exchange effects can be very subtle but both are important aspects of the interaction.

The Thomas–Fermi model, as with the Hartree and Hartree–Fock, unfortunately does not lend itself easily to extensions, so we must go back to basics and develop the immense superstructure of many-body theory in order to provide a rational basis for our theories.

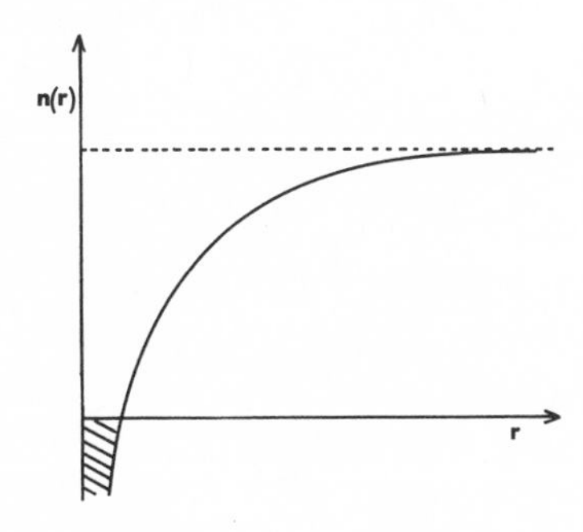

FIGURE 1.8. Electron density $[n(r)]$ surrounding a test charge in Thomas–Fermi theory. The hatched area corresponds to an unphysical overscreening.

BIBLIOGRAPHY

Solid State

ASHCROFT, N. W. and MERMIN, N. D., *Solid State Physics,* Holt, Rinehart and Winston, New York, 1976.

KITTEL, C., *Introduction to Solid State Physics,* Wiley, New York, 1971.

ZIMAN, J. M., *Principles of the Theory of Solids,* Cambridge University Press, New York, 1969.

Quantum Mechanics

LANDAU, L. D. and LIFSHITZ, E. M., *Quantum Mechanics,* Pergamon, Elmsford: NY, 1977.

MESSIAH, A., *Quantum Mechanics,* North Holland, Amsterdam, 1966.

SCHIFF, L. I., *Quantum Mechanics,* McGraw-Hill, New York, 1968.

PROBLEMS

1. Use the variational principle to derive the Hartree and Hartree–Fock equations for the Hamiltonian

$$H = \sum_k -\frac{\hbar^2}{2m}\nabla_k^2 + \frac{1}{2}\sum_{k,l}\frac{e^2}{|\mathbf{r}_k - \mathbf{r}_l|}$$

2. A useful approximation for the dielectric function for a semiconductor is

$$\epsilon(q) = 1 + \frac{\epsilon(0) - 1}{(1 + q^2\epsilon(0)/\gamma^2)}$$

where

$$\gamma^2 = \lambda_{\mathrm{T,F}}^2 \frac{\epsilon(0)}{\epsilon(0) - 1}$$

and $\epsilon(0)$ is the dielectric constant.

(i) Show that the corresponding Thomas–Fermi equation to this is

$$(\nabla^2 - \gamma^2)V = \left(\nabla^2 - \frac{\gamma^2}{\epsilon(0)}\right)V_{\mathrm{ext}}$$

(ii) Show that the screened potential of a point charge Q is given by

$$V = \frac{Q}{\epsilon(0)r}(1 + (\epsilon(0) - 1)e^{-\gamma r})$$

and check this by direct application of Eq. (1.57).

3. Prove that the Hartree–Fock contribution to the energy of the state of momentum k in a uniform electron gas is given by Eq. (1.34).

4. Use the screened interaction of Eq. (1.47) to calculate the screened exchange energy of the lowest ($k = 0$) state of a uniform electron gas and compare this with the Thomas–Fermi correlation hole contribution in the limits of low and high electron density.

CHAPTER TWO

GREEN'S FUNCTIONS OF THE SINGLE-PARTICLE SCHRÖDINGER EQUATION

In the theory of interacting systems the *Green's function*, or propagator, plays a crucial role. In its basic definition it is a much more complex function than the "simple" Green's function, familiar from the theory of partial differential equations, but many of its properties do bear a very close relationship to the simple function. It is worthwhile, therefore, to review our theory of Green's functions within the framework of the Schrödinger equation and perturbation theory. In this way we can introduce more of the concepts that appear in the full theory of interacting systems within a well-understood framework.

2.1. GREEN'S FUNCTIONS OF THE SCHRÖDINGER EQUATION

Suppose we have a partial differential equation of the general form

$$(H(\mathbf{r}) - E)\psi(\mathbf{r}) = 0 \qquad [2.1]$$

where $H(\mathbf{r})$ is a general Hermitian operator. The Green's function is defined as the solution of the equation

$$(H(\mathbf{r}) - E)G(\mathbf{r},\mathbf{r}',E) = -\delta(\mathbf{r} - \mathbf{r}') \qquad [2.2]$$

which also satisfies the same boundary conditions imposed on the original problem. In other words it satisfies the same equation and boundary conditions as the wave function $\psi(\mathbf{r})$ but with an additional "source" at an arbitrary position $\mathbf{r}'$. It is this extra degree of freedom which makes the Green's function so very useful.

Consider for instance the Helmholtz equation

$$(\nabla^2 + k^2)\psi(\mathbf{r}) = 0 \qquad [2.3]$$

The equation for the Green's function is

$$(\nabla^2 + k^2)G(\mathbf{r},\mathbf{r}',k) = -\delta(\mathbf{r} - \mathbf{r}') \qquad [2.4]$$

There are in general no unique solutions—we must introduce the boundary conditions imposed by the particular problem. Suppose we consider spherical symmetry ($\mathbf{r} - \mathbf{r}' = \mathbf{R}$); then we have

$$\left(\frac{1}{R^2}\frac{d}{dR}R^2\frac{d}{dR} + k^2\right)G(R,k) = \frac{-1}{4\pi R^2}\delta(R) \qquad [2.5]$$

for which the general solutions are

$$G(R,k) = \pm\frac{e^{\pm ikR}}{4\pi R}\,\frac{1}{4\pi}\,\frac{\sin kR}{R}\,\frac{1}{4\pi}\,\frac{\cos kR}{R} \qquad [2.6]$$

Which of these solutions is used depends upon the problem treated—i.e., are incoming, outgoing, or standing wave solutions required?

A better known expression for the Green's function is in terms of the eigenvalues E_n and eigenfunctions $\psi_n(\mathbf{r})$ of the defining operator:

$$(H(\mathbf{r}) - E_n)\psi_n(\mathbf{r}) = 0 \qquad [2.7]$$

where the $\psi_n(\mathbf{r})$ satisfy all of the boundary conditions of the system. The $\psi_n(\mathbf{r})$ will then form a complete set of orthonormal functions, in terms of which we can write, quite generally.

$$G(\mathbf{r},\mathbf{r}',E) = \sum_{n,n'} G_{nn'}\psi_n(\mathbf{r})\psi^*_{n'}(\mathbf{r}') \qquad [2.8]$$

then

$$(H(\mathbf{r}) - E)G(\mathbf{r},\mathbf{r}',E) = \sum_{n,n'} G_{nn'}(E_n - E)\psi_n(\mathbf{r})\psi^*_{n'}(\mathbf{r}')$$

i.e.,

$$\delta(\mathbf{r} - \mathbf{r}') = \sum_{n,n'} G_{nn'}(E_n - E)\psi_n(\mathbf{r})\psi^*_{n'}(\mathbf{r}')$$

which, since

$$\sum_n \psi_n(\mathbf{r})\psi^*_n(\mathbf{r}') = \delta(\mathbf{r} - \mathbf{r}')$$

requires that

$$G_{nn'} = \frac{\delta_{nn'}}{E - E_n}$$

and

$$G(\mathbf{r},\mathbf{r}',E) = \sum_n \frac{\psi_n(\mathbf{r})\psi^*_n(\mathbf{r}')}{E - E_n} \qquad [2.9]$$

Having obtained the Green's function, what are its uses? The most common one in the theory of differential equations is that it tells us about the solution to the inhomogeneous problem. If we have an equation of the form

$$(H(\mathbf{r}) - E)\psi(\mathbf{r}) = f(\mathbf{r}) \qquad [2.10]$$

where $f(\mathbf{r})$ is a known function, $\psi(\mathbf{r})$ is given by

$$\psi(\mathbf{r}) = -\int f(\mathbf{r}')G(\mathbf{r},\mathbf{r}',E)\, d\mathbf{r}' \qquad [2.11]$$

This is easily verified by operating on both sides with $(H(\mathbf{r}) - E)$:

$$\begin{aligned}(H(\mathbf{r}) - E)\psi(\mathbf{r}) &= -\int(H(\mathbf{r}) - E)G(\mathbf{r},\mathbf{r}',E)f(\mathbf{r}')\, d\mathbf{r}' \\ &= \int\delta(\mathbf{r} - \mathbf{r}')f(\mathbf{r}')\, d\mathbf{r}' \\ &= f(\mathbf{r})\end{aligned}$$

Thus, knowing the Green's function for the homogeneous equation one can calculate, with the minimum of extra work, the properties of the solution under a whole range of different conditions.

2.2. GREEN'S FUNCTIONS AND PERTURBATION THEORY

The most common form of inhomogeneity in quantum mechanics is the presence of a perturbation. In this case we have, in addition to the Hamiltonian $H(\mathbf{r})$, which is assumed soluble, an extra, usually small potential $V(\mathbf{r})$, so that

$$(H(\mathbf{r}) - E + V(\mathbf{r}))\psi(\mathbf{r}) = 0 \qquad [2.12]$$

In terms of Eq. (2.10) this means that

$$f(\mathbf{r}) = -V(\mathbf{r})\psi(\mathbf{r}) \qquad [2.13]$$

and we have an integral equation to solve for the wave function rather than the simple integral solution of Eq. (2.11).

$$\psi(\mathbf{r}) = \psi_0(\mathbf{r}) + \int G(\mathbf{r},\mathbf{r}',E)V(\mathbf{r}')\psi(\mathbf{r}')\, d\mathbf{r}' \qquad [2.14]$$

where $\psi_0(\mathbf{r})$ is a solution to the unperturbed equation. There are a number of ways of solving this integral equation. Supposing we know the solution to the unperturbed equation

$$(H(r) - E)\psi_0(\mathbf{r}) = 0 \qquad [2.15]$$

then, provided the effect of $V(\mathbf{r})$ is small, we might solve Eq. (2.14) by successive approximations. This gives

$$\text{(i)} \quad \psi(\mathbf{r}) = \psi_0(\mathbf{r}) \qquad [2.16a]$$

$$\text{(ii)} \quad \psi(\mathbf{r}) = \psi_0(\mathbf{r}) + \int G(\mathbf{r},\mathbf{r}',E)V(\mathbf{r}')\psi_0(\mathbf{r}')\, d\mathbf{r}' \qquad [2.16b]$$

$$\text{(iii)} \quad \psi(\mathbf{r}) = \psi_0(\mathbf{r}) + \int G(\mathbf{r},\mathbf{r}',E)V(\mathbf{r}')\psi_0(\mathbf{r}')\, d\mathbf{r}' + \int G(\mathbf{r},\mathbf{r}',E)V(\mathbf{r}')G(\mathbf{r}',\mathbf{r}'',E)V(\mathbf{r}'')\psi_0(\mathbf{r}'')\, d\mathbf{r}'\, d\mathbf{r}'' \qquad [2.16c]$$

and so on.

In the case of a scattering problem, this series is in fact identical to the Born approximation series, but in general would be more familiar as the *Wigner–Brillouin perturbation series* for $\psi(\mathbf{r})$, since substituting for the Green's functions from Eq. (2.9) gives

$$\psi(\mathbf{r}) = \psi_0(\mathbf{r}) + \sum_n \frac{\psi_n(\mathbf{r})[\int \psi_n^*(\mathbf{r}')V(\mathbf{r}')\psi_0(\mathbf{r}')\, dr']}{E - E_n} + \cdots \qquad [2.17]$$

In treating series like these it is very easy to be sidetracked by the presence of various multiple integrals. There is a simpler way of writing these equations, however. Suppose we consider $V(\mathbf{r})$ a continuous function of the variable $\mathbf{r}$. Let us now instead define $V(\mathbf{r})$ by its value on an array defined by an infinite set of points $\mathbf{r}_n$, differing only by an infinitesimal ($\mathbf{r}_{n+1} - \mathbf{r}_n = \delta$):

$$V(\mathbf{r}) \Rightarrow \{V(\mathbf{r}_1), V(\mathbf{r}_2), V(\mathbf{r}_3), \ldots\} \qquad [2.18a]$$

Having defined it as an infinite array of numbers, the next step would be to write it as a vector in the infinite dimensional (Hilbert) space defined by the array:

$$V(\mathbf{r}) \Rightarrow \mathbf{V} \qquad [2.18b]$$

whose components are the $V(\mathbf{r}_n)$. The Green's functions are defined in terms of two numbers (i.e., $\mathbf{r},\mathbf{r}'$) so $G(\mathbf{r},\mathbf{r}',E)$ will become a matrix function in Hilbert space:

$$G(\mathbf{r},\mathbf{r}',E) \Rightarrow \mathbf{G}$$

Integrals like

$$\int G(\mathbf{r},\mathbf{r}',E)V(\mathbf{r}')\, d\mathbf{r}'$$

then revert to being matrix products so that we have

$$\int G(\mathbf{r},\mathbf{r}',E)V(\mathbf{r}')\, d\mathbf{r}' \Rightarrow \mathbf{GV}$$

where now, of course, the order is important. If we now rewrite the perturbation series as $[\mathbf{V} \equiv V(\mathbf{r})\delta(\mathbf{r} - \mathbf{r}')]$

$$\psi = \psi_0 + \mathbf{GV}\psi_0 + \mathbf{GVGV}\psi_0 + \mathbf{GVGVGV}\psi_0 + \cdots \qquad [2.19]$$

This enables us to see clearly that we can, for instance, write

$$\psi = \psi_0 + \mathsf{GV}[\psi_0 + \mathsf{GV}\psi_0 + \mathsf{GVGV}\psi_0 + \cdots] \qquad [2.20]$$

or

$$\psi = \psi_0 + \mathsf{GV}\psi \qquad [2.21]$$

which is equivalent to Eq. (2.14). This integral equation is the simplest form of what is generally termed a *Dyson equation.*

In many cases we are not interested in the wave function itself. The Green's function contains so much of interest that it is usually far better to work with it alone. Supposing we consider the same problem as before, but in terms of Green's functions. Suppose we know the solution to the problem

$$(E - H(\mathbf{r}))G_0(\mathbf{r},\mathbf{r}',E) = \delta(\mathbf{r} - \mathbf{r}') \qquad [2.22]$$

and wish to solve for the Green's function of the equation.

$$(E - H(\mathbf{r}) - V(\mathbf{r}))G(\mathbf{r},\mathbf{r}',E) = \delta(\mathbf{r} - \mathbf{r}') \qquad [2.23]$$

which is equivalent to solving for the eigenfunctions. Then again, in perturbation theory, by successive approximations, we have

$$G(\mathbf{r},\mathbf{r}',E) = G_0(\mathbf{r},\mathbf{r}',E) + \int G_0(\mathbf{r},\mathbf{r}'',E)V(\mathbf{r}'')G_0(\mathbf{r}'',\mathbf{r}',E)\,d\mathbf{r}''$$
$$+ \int G_0(\mathbf{r},\mathbf{r}'',E)V(\mathbf{r}'')G_0(\mathbf{r}'',\mathbf{r}''',E)V(\mathbf{r}''')G_0(\mathbf{r}''',\mathbf{r}',E)\,d\mathbf{r}''\,d\mathbf{r}''' + \cdots \qquad [2.24]$$

In matrix terms we would write (2.22) and (2.23), respectively, as

$$(E\mathbf{1} - \mathsf{H})\mathsf{G}_0 = 1 \qquad [2.22a]$$
$$(E1 - \mathsf{H} - \mathsf{V})\mathsf{G} = 1 \qquad [2.23a]$$

so that (2.24) becomes

$$\mathsf{G} = \mathsf{G}_0 + \mathsf{G}_0\mathsf{V}\mathsf{G}_0 + \mathsf{G}_0\mathsf{V}\mathsf{G}_0\mathsf{V}\mathsf{G}_0 + \cdots$$

or

$$\mathsf{G} = \mathsf{G}_0 + \mathsf{G}_0\mathsf{V}\mathsf{G} \qquad [2.25]$$

by the same reasoning as before. This is again a "Dyson equation," this time for the Green's function. Really, however, it is nothing more than a reaffirmation of the original property of the Green's function represented in Eq. (2.10) and (2.11) for the case where the perturbation is

$$V(\mathbf{r})G(\mathbf{r},\mathbf{r}',E)$$

and the solution is a deviation from $G_0(\mathbf{r},\mathbf{r}',E)$.

2.3. TIME-DEPENDENT GREEN'S FUNCTIONS

One must consider, in many cases, the time-dependent Schrödinger equation. This is written as

$$\left(i\hbar\frac{\partial}{\partial t} - H(\mathbf{r})\right)\Psi(\mathbf{r},t) = 0 \qquad [2.26]$$

and has the formal solution

$$\Psi(\mathbf{r},t) = \psi(\mathbf{r})e^{(-iEt)/\hbar} \qquad [2.27]$$

The defining equation for this Green's function is

$$\left(i\hbar\frac{\partial}{\partial t} - H(\mathbf{r})\right)G\,(\mathbf{r},\mathbf{r}',t,t') = \hbar\,\delta(\mathbf{r}-\mathbf{r}')\,\delta(t-t') \qquad [2.28]$$

Notice, that in this case, where H is not a function of time, the Green's function will only depend upon $t - t'$. The Fourier transform of the energy-dependent Green's function defined through

$$G(\mathbf{r},\mathbf{r}',t-t') = \frac{1}{2\pi}\int G(\mathbf{r},\mathbf{r}',E)e^{[-iE(t-t')]/\hbar}\,dE \qquad [2.29]$$

can be seen to be a solution to Eq. (2.28) and so is the function we require. If we substitute the eigenfunction expansion form of the Green's function into Eq. (2.29), we have

$$G(\mathbf{r},\mathbf{r}',t-t') = \frac{1}{2\pi}\int\left\{\sum_n\frac{\psi_n(\mathbf{r})\psi_n^*(\mathbf{r}')}{E-E_n}\right\}e^{[-iE(t-t')]/\hbar}\,dE \qquad [2.30]$$

As it stands (for real energies) this integral is undefined; what are missing are the boundary conditions.

If we evaluate Eq. (2.30) by a contour integral, then from Fig. 2.1 we have a number of possibilities, according to how we move the poles given by the zeros of the denominator. We define two new Green's functions given by

(i)
$$G^{R}(\mathbf{r},\mathbf{r}',E) = \sum_n \frac{\psi_n(\mathbf{r})\psi_n^*(\mathbf{r}')}{E - E_n + i\epsilon} \qquad [2.31]$$

which on Fourier transforming gives

$$\begin{aligned} G^{R}(\mathbf{r},\mathbf{r}',E) &= \sum_n \psi_n(\mathbf{r})\psi_n^*(\mathbf{r}')e^{[-iE_n(t-t')]/\hbar} \qquad (t > t') \\ &= 0 \qquad (t < t') \end{aligned} \qquad [2.32]$$

and

(ii)
$$G^{A}(\mathbf{r},\mathbf{r}',E) = \sum_n \frac{\psi_n(\mathbf{r})\psi_n^*(r')}{E - E_n - i\epsilon} \qquad [2.33]$$

which leads to

$$\begin{aligned} G^{A}(r,r',t - t') &= -\sum_n \psi_n(\mathbf{r})\psi_n(\mathbf{r}')e^{[iE_n(t-t')]/\hbar} \qquad (t < t') \\ &= 0 \qquad (t > t') \end{aligned} \qquad [2.34]$$

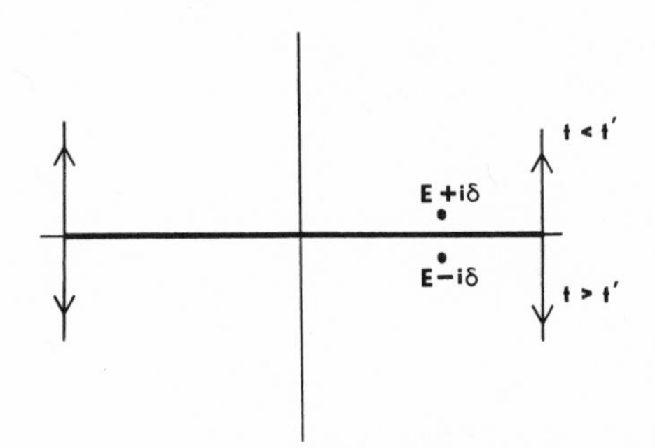

FIGURE 2.1. Contours in the complex energy plane for evaluating time-dependent Green's functions.

These two functions are called the *retarded* and *advanced Green's functions*, respectively. There is a further Green's function, which we will come to later, called the *time-ordered* Green's function, but this need not concern us yet. The two new Green's functions are used to fit the appropriate time boundary conditions in the problem, just as in the Helmholtz equation we had the choice of a number of forms for the Green's function to fit the appropriate spatial boundary conditions posed by the system.

The time-dependent Green's function has one very interesting property, which gives rise to the term propagator. Suppose we have a problem in which we know the eigenfunction at some particular space–time point $\mathbf{r}',t'$, then we can obtain the wave function at a later time t from the integral

$$\Psi(\mathbf{r},t) = \int G^{R}(\mathbf{r},\mathbf{r}',t - t')\Psi(\mathbf{r}',t')\, d\mathbf{r}' \qquad [2.35]$$

This is easy to show. From Eq. (2.32)

$$\Psi(\mathbf{r},t) = \int \sum_{n} \psi_n(\mathbf{r})\psi_n^*(\mathbf{r}')e^{[-iE_n(t-t')]/\hbar}\Psi(\mathbf{r}',t')\, d\mathbf{r}' \qquad [2.36]$$

Expanding $\Psi(\mathbf{r}',t')$ in terms of the eigenfunctions gives

$$\Psi(\mathbf{r}',t') = \sum_{m} \alpha_m\psi_m(\mathbf{r}')e^{(-iE_mt')/\hbar} \qquad [2.37]$$

where α_m are the expansion coefficients. Then substituting

$$\Psi(\mathbf{r},t) = \int \sum_{m,n} \psi_n(\mathbf{r})\psi_n^*(\mathbf{r}')e^{[-iE_n(t-t')]/\hbar}\alpha_m\psi_m(\mathbf{r}')e^{(-iE_mt')/\hbar}\, d\mathbf{r}'$$

$$= \sum_{n,m} \alpha_m\psi_n(\mathbf{r})e^{(-iE_nt)/\hbar}\left[\int d\mathbf{r}'\, \psi_n^*(\mathbf{r}')\psi_m(r')\right] e^{[i(E_n-E_m)t']/\hbar}$$

using

$$\int \psi_n(\mathbf{r}')\psi_m(\mathbf{r}')\, d\mathbf{r}' = \delta_{nm}$$

we have

$$\Psi(\mathbf{r},t) = \Sigma\alpha_n\psi_n(\mathbf{r})e^{(-iE_nt)/\hbar} = \Psi(\mathbf{r},t)$$

Thus, in a very real sense, the Green's function tells us how the wave function develops in time or how the particle described by the wave function propagates. One thing to notice is that the propagation is entirely causal and single valued in the sense that given $\Psi(\mathbf{r}',t')$ we always get $\Psi(\mathbf{r},t)$ at a later time, even though both wave functions have to be considered in terms of probabilities.

2.4. GREEN'S FUNCTION DIAGRAMS

The use of Green's functions as propagators enables us to develop a simple but powerful visual aid to our understanding of processes in many-body theory, which we can simply illustrate with respect to the perturbation series of the previous section. Restricting ourselves to the advanced Green's function we can Fourier transform the whole of Eq. (2.24) to obtain the series

$$\begin{aligned} G(\mathbf{r},\mathbf{r}',t-t') = {} & G_0(\mathbf{r},\mathbf{r}',t-t') \\ & + \iint d\mathbf{r}''\,dt''\,G_0(\mathbf{r},\mathbf{r}'',t-t'')V(\mathbf{r}'')G_0(\mathbf{r}'',\mathbf{r}',t''-t') \\ & + \iiiint d\mathbf{r}''\,d\mathbf{r}'''\,dt''\,dt'''\,G_0(\mathbf{r},\mathbf{r}'',t-t'')V(\mathbf{r}'')G_0 \\ & \times (\mathbf{r}'',\mathbf{r}''',t''-t''')V(\mathbf{r}''')G(\mathbf{r}''',\mathbf{r}',t'''-t') \end{aligned} \qquad [2.38]$$

If we express the motion of the particle in the absence of the perturbation, described by $G_0(\mathbf{r},\mathbf{r}',t-t')$, as a line (Fig. 2.2a) then we can express the effect of the perturbation on the motion in the visual form shown in Fig. 2.2b. Thus, we have a picture of the motion of the particle as being a series of scattering events caused by the potential $V(\mathbf{r})$, separated by periods in which the motion is free. The total propagation between any two space–time points is then the sum of all possible scattering processes. Suppose we take, as an example, the case of a delta function potential

$$V(\mathbf{r}) = A\,\delta(\mathbf{r}) \qquad [2.39]$$

then Eq. (2.38) can be simplified to

$$\begin{aligned} G(\mathbf{r},\mathbf{r}',t-t') = {} & G_0(\mathbf{r},\mathbf{r}',t-t') + A\int G_0(\mathbf{r},0,t-t'')G_0(0,\mathbf{r}',t''-t')\,dt'' \\ & + A^2\iint G_0(\mathbf{r},0,t-t'')G_0(0,0,t''-t''')G_0(0,\mathbf{r}',t'''-t')\,dt''\,dt''' + \cdots \end{aligned} \qquad [2.40]$$

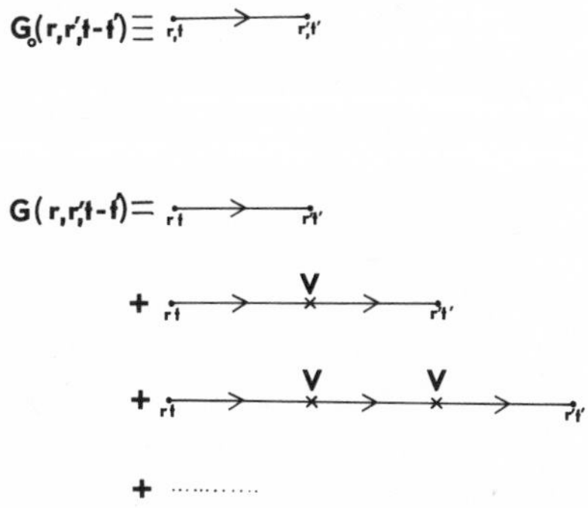

$G_0(r,r',t-t') \equiv$ r,t ⟶ r',t'

$G(r,r',t-t') =$ r,t ⟶ r',t'

\+ r,t ⟶ V ⟶ r',t'

\+ r,t ⟶ V ⟶ V ⟶ r',t'

\+ ⋯⋯

FIGURE 2.2. (a) Pictorial representation of a Green's function. (b) Representation of the perturbation expansion for the Green's function.

In pictorial form we have the series of Fig. 2.3. If the factor A is *not* small, then each of the successive terms, where the propagation takes place around the potential, may be important. This is obviously reminiscent of a bound particle, since the propagation is being dominated by a small region of space. Let us look at the energy-dependent Green's function:

$$\begin{aligned} G(\mathbf{r},\mathbf{r}',E) = {} & G_0(\mathbf{r},\mathbf{r}',E) + AG_0(\mathbf{r},0,E)G_0(0,\mathbf{r}',E) \\ & + A^2G_0(\mathbf{r},0,E)G_0(0,0,E)G_0(0,\mathbf{r}',E) + \cdots \\ & + A^3G_0(\mathbf{r},0,E)G_0(0,0,E)G_0(0,0,E)G_0(0,\mathbf{r}',E) + \cdots \end{aligned} \quad [2.41]$$

which can be summed to give

$$G(\mathbf{r},\mathbf{r}',E) = G_0(\mathbf{r},\mathbf{r}',E) + \frac{AG_0(\mathbf{r},0,E)G_0(0,\mathbf{r}',E)}{1 - AG_0(0,0,E)} \quad [2.42]$$

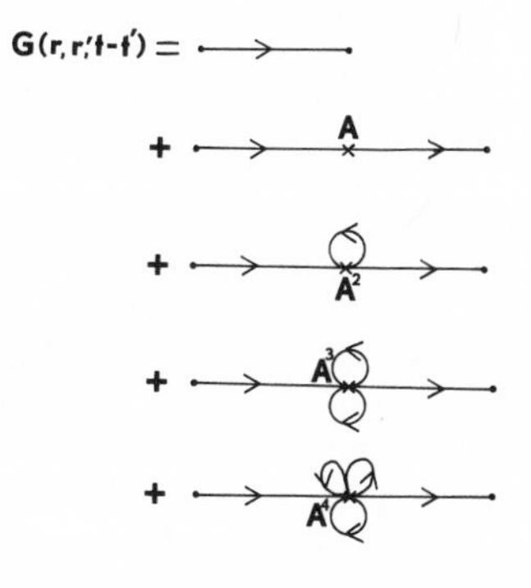

FIGURE 2.3. The perturbation series for a localized (delta function) potential.

We can now investigate the properties of this new Green's function. In particular, from Sec. 2.1 we know that the poles of $G(\mathbf{r},\mathbf{r}',E)$ will give the new eigenvalues for the perturbed system. The first term (and the numerator of the second) adds no new poles but if

$$1 - AG_0(0,0,E) = 0 \qquad [2.43]$$

then there will be another new pole in the Green's function in addition to those present in $G_0(\mathbf{r},\mathbf{r}',E)$. This new pole corresponds to a new eigenvalue of the system, which, if the potential is attractive, we would expect to correspond to a bound state (see Prob. 3 and Sec. 13.1). Thus, again there is the possibility of a bound state for the case of A not too small. But notice also that the presence of a bound state is associated with the failure of the series expansion (2.41). We will see that in the general case, where one cannot so easily see the physics of the processes involved, the failure of the series expansion is the first indication of new types of states brought about by the interaction. In any case, however, the ability to pictorially represent the Green's function, as a propagation from one point to another, is often an important aid in the understanding of complicated physical systems.

2.5. GREEN'S FUNCTIONS OR WAVE FUNCTIONS?

Simple quantum mechanics depends to a large extent on the evaluation of the eigenfunction and eigenvalue solutions of Schrödinger's equation. From these we evaluate observables by calculating expectation values of various operators. The Green's function is, of course, closely related to the wave functions and also contains information about the eigenvalues [cf. Eq. (2.9)] and the time development of the system [cf. Eq. (2.35)]. It might be useful, therefore, to catalogue exactly what sort of information we can usefully obtain from the Green's function.

2.5.1. Eigenvalues. The Green's function has poles at the eigenvalues of the system, i.e.,

$$G(\mathbf{r},\mathbf{r}',E) = \sum_n \frac{\psi_n(\mathbf{r})\psi_n^*(\mathbf{r}')}{E - E_n}$$

has poles at

$$E = E_n$$

2.5.2. The Density Matrix. Considering the advanced Green's function, we have

$$\text{Im } G^A(\mathbf{r},\mathbf{r}',E) = \text{Im}\left[\sum_n \frac{\psi_n(\mathbf{r})\psi_n(\mathbf{r}')}{E - E_n - i\epsilon}\right]$$

We use the standard identity

$$\frac{1}{\omega \pm i\delta} = P\left(\frac{1}{\omega}\right) \pm i\pi\delta(\omega) \qquad [2.44]$$

(which strictly is true only under an integral sign) to give

$$\text{Im } G^A(\mathbf{r},\mathbf{r}',E) = \pi\psi_n(\mathbf{r})\psi_n^*\mathbf{r}')\delta(E - E_n) \qquad [2.45]$$

The density matrix is defined as

$$\rho(\mathbf{r},\mathbf{r}',E) = \sum_n \psi_n(\mathbf{r})\psi_n^*(\mathbf{r}')\delta(E - E_n)$$

and so we can write

$$\rho(\mathbf{r},\mathbf{r}',E) = \frac{1}{\pi}\text{Im } G^A(\mathbf{r},\mathbf{r}',E) \qquad [2.46a]$$

$$= \frac{1}{2\pi}\text{Im } (G^A(\mathbf{r},\mathbf{r}',E) - G^R(\mathbf{r},\mathbf{r}',E)) \qquad [2.46b]$$

2.5.3. Propagation. The Green's function tells us how a particle propagates in the system, as we saw

$$\Psi(\mathbf{r},t) = \int G^R(\mathbf{r},\mathbf{r}',t - t')\Psi(\mathbf{r}',t')\, d\mathbf{r}' \qquad [2.35]$$

2.5.4. Expectation Values. The single-particle Green's function does not easily give the expectation value of an operator in a general state.

Supposing, however, that we have an operator $\hat{O}(\mathbf{r})$, then we can write the following related function

$$\lim_{\substack{t' \to t+\delta \\ \mathbf{r} \to \mathbf{r}'}} \mathrm{Tr}[\hat{O}(\mathbf{r})G(\mathbf{r},\mathbf{r}',t-t')] = \lim_{\substack{t' \to t+\delta \\ \mathbf{r} \to \mathbf{r}'}} \mathrm{Tr}\left[\hat{O}(\mathbf{r}) \sum_n \psi_n(\mathbf{r})\psi_n^*(\mathbf{r}')e^{[-iE_n(t-t')]/\hbar}\right]$$

$$= \lim_{\mathbf{r} \to \mathbf{r}'} \mathrm{Tr}\left[\sum_n \psi_n^*(\mathbf{r}')\hat{O}(\mathbf{r})\psi_n(\mathbf{r})\right] \qquad [2.47]$$

$$= \sum_n \int \psi_n^*(\mathbf{r})\hat{O}(\mathbf{r})\psi_n(\mathbf{r})\, dr$$

$$= \sum_n \langle n|\hat{O}|n\rangle$$

(notice the use of the trace Tr as a generalization of the sum of the diagonal elements of a matrix), i.e., for the simple single-particle Green's function such an expression gives the sum of the expectation values between the eigenfunctions. This is because all of the properties of the physical system have not yet been included into the Green's function. In particular, the expectation value requires a knowledge of which states are occupied. In later sections we will see that an expression similar to the above gives the expectation value.

Thus, we see that in many respects the Green's function can replace the wave functions in a single-particle system. This is not surprising, of course, in view of the similarity of the wave equations satisfied by the two functions. The reason for introducing Green's function is, however, that even when the concept of single particle states have lost their meaning the Green's function can be defined and calculated in such a way that many of the properties seen in this chapter are retained.

BIBLIOGRAPHY

Generally any good mathematical physics text book but especially

MORSE, P. M., and FESHBACK, H., *Methods of Theoretical Physics,* McGraw-Hill, New York, 1962.

ARFKEN, G., *Mathematical Methods for Physicists,* Academic, New York, 1970.

PROBLEMS

1. Show that the Green's functions of the Helmholtz equation given in Eq. (2.6) are compatible with the eigenfunction expansion by using the complete set of eigenstates

$$\phi_k(\mathbf{r}) = \frac{1}{\sqrt{\Omega}} e^{i\mathbf{k}\cdot\mathbf{r}}$$

2. Investigate the one-dimensional free-particle system subject to a perturbation

$$V(x) = V_0 \cos Gx$$

and show that the poles of the Green's function develop a gap of magnitude $2V_0$.

3. Investigate the eigenvalue spectrum in the presence of the one-dimensional perturbation $A\delta(x)$ using Eq. (2.42) when

(i) $A > 0$

(ii) $A < 0$

If $H_0 = (-\hbar^2/2m)\, d^2/dx^2$, show that the bound state corresponds to the direct solution of the Schrödinger equation.

CHAPTER THREE

QUANTIZATION OF WAVES (SECOND QUANTIZATION)

3.1. WAVES AND PARTICLES

In the introduction to elementary quantum mechanics, much of the experimental evidence presented concerns the quantum (or particle) nature of light (Compton effect, photoelectric effect, infrared catastrophe, etc.), yet the wave nature of particles and the Schrödinger equation dominates most considerations from then on. In this chapter we will look at the particle nature of waves and begin the development of a unified approach to fields and particles suitable for most applications but especially to condensed matter, where many of the interesting effects concern oscillatory phenomena.

The particle nature of wave motion is a very ubiquitous phenomenon, and it is important to realize the huge range of wave motion that must be treated in a quantum form. Table 3.1 shows a range of examples of waves where the quantum of energy and momentum (or elementary excitation) is known. We see that they range from waves excited within a matrix composed of particles, which are themselves governed by quantum mechanics, through the electromagnetic field, to situations in elementary particle physics, where the particle, or elementary excitation is better known than the field which supports it. The very wide variety of examples that occur in solids and liquids makes this a very important area for study.

In all cases where we have a good classical understanding of the field and the wave produced, the energy and momentum of the elementary excitations are given by $\hbar\omega_k$ and $\hbar\mathbf{k}$, where ω_k and $\mathbf{k}$ are the classical frequency and wave vector of the wave. This suggests that the process of

TABLE 3.1

Background matrix	Type of wave	Associated quantized object (elementary excitation)
1. *Waves in media*		
Atoms in solids	Vibrations (sound waves)	Phonon
Electrons in solids	Plasma waves	Plasmon
Magnetic dipoles in a solid	Spin waves	Magnon
Atoms in liquids	Gravity waves (^{4}He)	Riplon
Atoms in liquids	Eddies (^{4}He)	Roton
2. *"Free" space*		
Electromagnetic field	Electromagnetic wave	Photon
Strong nuclear force	?	Meson
Dirac field	?	Electron and positron
Gravitation field	?	Graviton (?)

quantization is not dependent upon the microscopic details of the wave motion but only upon the gross properties of the wave. Therefore, it seems sensible first to consider the simplest possible example to see if we can understand the process of quantization.

3.2. THE LINEAR CHAIN OF ATOMS

Figure 3.1 shows schematically a one-dimensional regular linear chain of atoms of mass M and separation a interacting with nearest neighbors through a harmonic potential

$$V(R_{i+1} - R_i) = \tfrac{1}{2}K(q_{i+1} - q_i)^2 \qquad [3.1]$$

The treatment of this system is a standard exercise in normal-mode theory. The Hamiltonian is given by

$$H = \sum_i \frac{p_i^2}{2M} + \frac{1}{2} K(q_{i+1} - q_i)^2 \qquad [3.2]$$

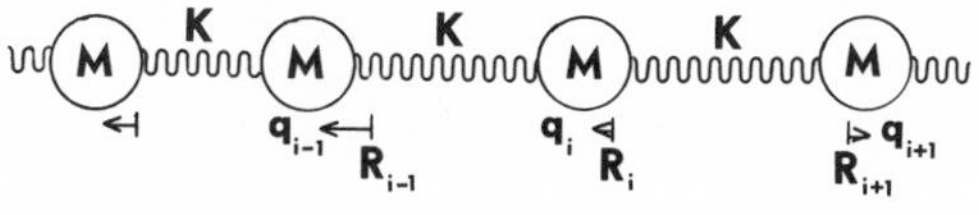

FIGURE 3.1. The linear chain of atoms. The symbols are as defined in the text.

We use the expected wave nature of the motion to write the momentum p_i and position q_i of each atom as the sum of a set of vibrations of the chain

$$p_i = \frac{1}{\sqrt{N}} \sum_k p_k e^{+ik \cdot R_i} \tag{3.3a}$$

$$q_i = \frac{1}{\sqrt{N}} \sum_k q_k e^{+ik \cdot R_i} \tag{3.3b}$$

These have the inverses

$$p_k = \frac{1}{\sqrt{N}} \sum_i p_i e^{-ik \cdot R_i} \tag{3.4a}$$

$$q_k = \frac{1}{\sqrt{N}} \sum_i q_i e^{-ik \cdot R_i} \tag{3.4b}$$

since

$$\frac{1}{N} \sum_i e^{i(k-k')R_i} = \delta(k - k') \tag{3.5}$$

If we substitute for p_i and q_i in the Hamiltonian we have

$$\begin{aligned} H &= \sum_k \frac{p_k p_{-k}}{2M} + q_k q_{-k} \frac{K}{2} (1 - e^{-ika})(1 - e^{+ika}) \\ &= \sum_k \frac{p_k p_{-k}}{2M} + q_k q_{-k} \left(\frac{M\omega_k^2}{2} \right) \end{aligned} \tag{3.6}$$

where ω_k is the oscillation frequency and is given by

$$\omega_k^2 = \frac{4K}{M} \sin^2 \left(\frac{ka}{2} \right) \tag{3.7}$$

H is now in the standard classical harmonic oscillator form for each of the normal modes of wave vector k.

Let us consider now the consequences of quantizing, as we must, the motion of the atoms. The coordinates q_i, p_i become operators

$$q_i \rightarrow \hat{q}_i; \qquad p_i \rightarrow \hat{p}_i = \frac{\hbar}{i}\frac{\partial}{\partial q_i} \tag{3.8}$$

Alternatively we can use the much more useful commutator relation

$$[\hat{p}_i,\hat{q}_j] = \frac{\hbar}{i}\delta_{ij} \tag{3.9}$$

If we now apply the commutator to the transformation equations [(3.3) and (3.4)], we can define operators $\hat{q}_k$, $\hat{p}_k$, and we have

$$\begin{aligned}[\hat{p}_k,\hat{q}_{k'}] &= \frac{1}{N}\sum_{R_i,R_j}[\hat{p}_i,\hat{q}_j]e^{-i(k\cdot R_i+k'\cdot R_j)} \\ &= \frac{\hbar}{iN}\sum_{R_i,R_j}\delta_{ij}e^{-i(k\cdot R_i+k'\cdot R_j)} \\ &= \frac{\hbar}{i}\delta_{k,-k'}\end{aligned} \tag{3.10}$$

It is as well to note that the transformation gives us non-Hermitian operators:

$$\left.\begin{aligned}\hat{p}_i^+ &= \hat{p}_i \\ \hat{q}_i^+ &= \hat{q}_i\end{aligned}\right\} \Rightarrow \left\{\begin{aligned}\hat{p}_k^+ &= \hat{p}_{-k} \\ \hat{q}_k^+ &= \hat{q}_{-k}\end{aligned}\right. \tag{3.11}$$

The resulting Hamiltonian looks very promising, however

$$\hat{H} = \sum_k \frac{\hat{p}_k\hat{p}_k^+}{2M} + \frac{M\omega_k^2}{2}\hat{q}_k\hat{q}_k^+ \tag{3.12}$$

We now introduce two new operators

$$\hat{a}_k^+ = \frac{1}{(2M\hbar\omega_k)^{1/2}}(\hat{p}_k^+ + iM\omega_k\hat{q}_k^+) \tag{3.13a}$$

$$\hat{a}_k = \frac{1}{(2m\hbar\omega_k)^{1/2}}(\hat{p}_k - iM\omega_k\hat{q}_k) \qquad [3.13b]$$

These have the property that

$$[\hat{a}_k, \hat{a}_{k'}^+] = \hat{a}_k\hat{a}_{k'}^+ - \hat{a}_{k'}^+\hat{a}_k = \delta_{k,k'} \qquad [3.14]$$
$$[\hat{a}_k^+, \hat{a}_{k'}^+] = [\hat{a}_k, \hat{a}_{k'}] = 0$$

which can be seen by substituting in the definitions and using Eq. (3.9). More importantly, we can rewrite (3.12) as

$$\hat{H} = \sum_k \hbar\omega_k\left(\hat{a}_k^+\hat{a}_k + \frac{1}{2}\right) \qquad [3.15]$$

This is the operator form of the Hamiltonian for a set of *quantized* independent harmonic oscillators, one for each k value.

To see that this is so, we need two further relationships, which follow from Eqs. (3.14) and (3.15).

$$[\hat{H}, \hat{a}_j^+] = \sum_k \hbar\omega_k(\hat{a}_k^+\hat{a}_k\hat{a}_k^+ - \hat{a}_j^+\hat{a}_k^+\hat{a}_k)$$
$$= \sum_k \hbar\omega_k\hat{a}_k^+[\hat{a}_k, \hat{a}_j^+]$$

Therefore

$$[\hat{H}, \hat{a}_j^+] = \hbar\omega_j\hat{a}_j^+ \qquad [3.16]$$

Similarly

$$[\hat{H}, \hat{a}_j] = -\hbar\omega_j\hat{a}_j \qquad [3.17]$$

Suppose we describe the linear chain by an eigenequation

$$\hat{H}|\zeta\rangle = E_\zeta|\zeta\rangle \qquad [3.18]$$

then from Eq. (3.16)

$$\hat{H}\hat{a}_k^+|\zeta\rangle = (\hbar\omega_k\hat{a}_k^+ + \hat{a}_k^+\hat{H})|\zeta\rangle \qquad [3.19]$$
$$= (\hbar\omega_k + E_\zeta)\hat{a}_k^+|\zeta\rangle$$

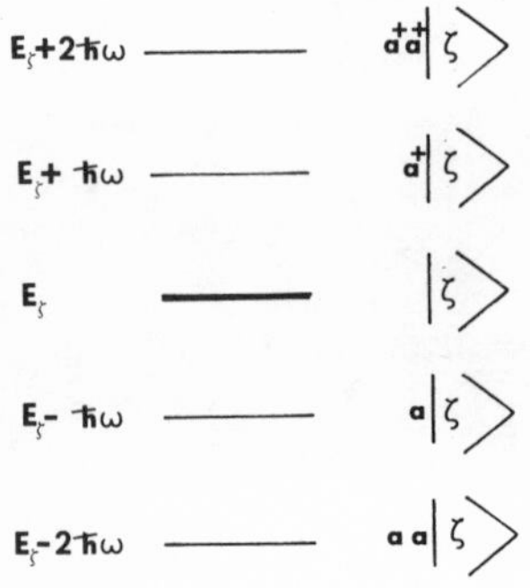

$E_\zeta + 2\hbar\omega$ — $a^+a^+|\zeta\rangle$

$E_\zeta + \hbar\omega$ — $a^+|\zeta\rangle$

E_ζ — $|\zeta\rangle$

$E_\zeta - \hbar\omega$ — $a|\zeta\rangle$

$E_\zeta - 2\hbar\omega$ — $aa|\zeta\rangle$

FIGURE 3.2. The eigenvalue–eigenfunction ladder.

and similarly

$$\hat{H}\hat{a}_k|\zeta\rangle = (E_\zeta - \hbar\omega_k)\hat{a}_k|\zeta\rangle \qquad [3.20]$$

Thus, if $|\zeta\rangle$ is the eigenfunction for energy E_ζ, $\hat{a}_k^+|\zeta\rangle$ is the eigenfunction for energy $E_\zeta + \hbar\omega_k$, and $\hat{a}_k|\zeta\rangle$ is the eigenfunction for energy $E_\zeta - \hbar\omega_k$. Thus, we can build up a ladder of eigenvalues *for each normal mode* k from the energy E_ζ, obtaining eigenvalues and eigenfunctions simply by repeatedly operating on our wave function $|\zeta\rangle$ with the set of operators $\hat{a}_k^+$, $\hat{a}_k$ (Fig. 3.2). For this reason $\hat{a}_k^+$, $\hat{a}_k$ are often referred to as *ladder operators* for the normal mode of wave vector k.

Suppose we define the ground state for the normal mode k by the relation

$$a_k|\text{ground state } k\text{; everything else}\rangle = 0 \qquad [3.21]$$

Since there can be no eigenstate below the ground state, we write this as

$$|0_k, \text{everything else}\rangle$$

A total ground state would then require

$$\hat{a}_k|0\rangle = 0 \qquad [3.22]$$

for *every* normal-mode wave vector k. Applying the Hamiltonian we have

$$\begin{aligned}\hat{H}|0\rangle &= \sum_k \hbar\omega_k\left(\hat{a}_k^+\hat{a}_k + \frac{1}{2}\right)|0\rangle \\ &= \sum_k \frac{1}{2}\hbar\omega_k|0\rangle\end{aligned} \qquad [3.23]$$

This naturally brings in the concept of a ground-state energy equal to the sum of the zero-point energies ($\frac{1}{2}\hbar\omega_k$) of each of the normal modes k as being the minimum possible energy for the linear chain. Looking at our ladder again, the energy of each normal mode is now $\hbar\omega_k(n_k + \frac{1}{2})$, where n_k is the number identifying the step of the ladder we are on. Equally, since n_k is the number of times the operator $\hat{a}_k^+$ has to be applied to reach the eigenfunction appropriate to the energy labeled by n_k, it is a natural choice of quantum number for the eigenfunction. Thus, it is convenient to write the eigenfunction and eigenvalue for the linear chain as

$$|\zeta\rangle = |n_1, n_2, \ldots, n_k, \ldots\rangle \qquad [3.24a]$$

$$E_\zeta = \sum_k \hbar\omega_k\left(n_k + \frac{1}{2}\right) \qquad [3.24b]$$

where

$$\hat{H}|\zeta\rangle = E|\zeta\rangle \qquad [3.18]$$

We can define another useful operator and also ensure normalization of the eigenfunctions, in the following way. Consider

$$\hat{a}_k^+\hat{a}_k|\cdots n_k \cdots\rangle = \alpha\hat{a}_k^+\hat{a}_k(\hat{a}_k^+)^{nk}|\cdots 0_k \cdots\rangle \qquad [3.25]$$

where α is a numerical factor to allow for the normalization; then

$$\begin{aligned}\hat{a}_k^+\hat{a}_k|\cdots n_k \cdots\rangle &= \alpha\hat{a}_k^+\underbrace{\hat{a}_k\hat{a}_k^+}(\hat{a}_k^+)^{nk-1}|\cdots 0_k \cdots\rangle \\ &= \alpha(\hat{a}_k^+)^2\hat{a}_k(\hat{a}_k^+)^{nk-1}|\cdots 0_k \cdots\rangle + |\cdots n_k \cdots\rangle\end{aligned} \qquad [3.26]$$

on using the commutator relationship on the coupled pair of operators. This procedure can be repeated until we have

$$\hat{a}_k^+ \hat{a}_k | \cdots n_k \cdots \rangle = n_k | \cdots n_k \cdots \rangle + \alpha(\hat{a}_k^+)^{nk+1} \hat{a}_k | \cdots 0_k \cdots \rangle \qquad [3.27]$$

the second term being zero because of Eq. (3.21). Thus, $\hat{a}_k^+ \hat{a}_k$ is the operator with eigenvalue n_k and is termed the *number operator* $\hat{n}_k$. From this and the normalization criterion

$$\begin{aligned} \langle \cdots n_k + 1 \cdots | \cdots n_k + 1 \cdots \rangle &= \langle \cdots n_k \cdots | \cdots n_k \cdots \rangle \\ &= \langle \cdots n_k - 1 \cdots | \cdots n_k - 1 \cdots \rangle \\ &= 1 \end{aligned} \qquad [3.28]$$

we have

$$\hat{a}_k^+ | \cdots n_k \cdots \rangle = (n_k + 1)^{1/2} | \cdots n_k + 1 \cdots \rangle \qquad [3.29a]$$

$$\hat{a}_k | \cdots n_k \cdots \rangle = (n_k)^{1/2} | \cdots n_k - 1 \cdots \rangle \qquad [3.29b]$$

The general state can now be written in terms of the ground state as

$$| n_1, n_2, \ldots, n_k, \ldots \rangle = \frac{(\hat{a}_1^+)^{n1}}{(n_1!)^{1/2}} \cdots \frac{(\hat{a}_k^+)^{nk}}{(n_k!)^{1/2}} \cdots | 0_1 \cdots 0_k \cdots \rangle \qquad [3.30]$$

Thus, applying quantum mechanics to the motion of the atoms in the chain gives us a quantum mechanical description of the system in terms of a set of independent harmonic oscillator systems labeled by the wave vector of the classical normal mode. Within this the eigenfunction and eigenvalue can be labeled completely by the quantum numbers n_k. Furthermore *any* state can be constructed by the use of the set of ladder operators $\hat{a}_k^+$ and $\hat{a}_k$.

This description contains the details of the motion of the atoms only in the frequency of the normal modes so we would expect that the general features of ladder operators, ground-state energies, and the description of the normal modes would have a much wider application.

3.3. THE GENERAL QUANTIZATION OF A WAVE SYSTEM

In order to understand how such different systems as the electromagnetic field and the linear chain of atoms may produce mathematically similar systems on quantization, it is best to go back to our concept of how quantization is performed.

The "normal" Schrödinger equation corresponds to the replacement within the Hamiltonian of the position and momentum variables **r, p** by the operators $\hat{\mathbf{r}}$, $\hat{\mathbf{p}}$ in which either

$$\left.\begin{aligned} \hat{\mathbf{r}} &= \mathbf{r} \\ \hat{\mathbf{p}} &= \frac{\hbar}{i}\frac{\partial}{\partial \mathbf{r}} \end{aligned}\right\} \quad \text{or} \quad \left\{\begin{aligned} \hat{\mathbf{r}} &= -\frac{\hbar}{i}\frac{\partial}{\partial \mathbf{p}} \\ \hat{\mathbf{p}} &= \mathbf{p} \end{aligned}\right. \qquad [3.31]$$

The Hamiltonian operator then acts upon the wave function, which is a function of either the position or momentum (but not both). Either way we lose half the classical variables. The familiar examples of this operator replacement are

(i) position and linear momentum

$$\hat{p}_x = \frac{\hbar}{i}\frac{\partial}{\partial x} \qquad [3.32]$$

(ii) angle and angular momentum

$$\hat{l}_z = \frac{\hbar}{i}\frac{\partial}{\partial \phi} \qquad [3.33]$$

where $\hat{l}_z$ is the angular momentum around the z axis and ϕ is the spherical polar coordinate.

It is obviously more satisfactory to have a general technique. To do this we borrow from classical mechanics. Hamilton's equations of motion are

$$\dot{p} = -\frac{\partial H}{\partial q} \quad \text{and} \quad \dot{q} = \frac{\partial H}{\partial p} \qquad [3.34]$$

where p and q are conjugate variables defined in terms of the Lagrangian:

$$p = \frac{\partial L}{\partial \dot{q}}(q,\dot{q},t) \qquad [3.35]$$

We quantize on the basis that if p and q are *classically conjugate variables* then we make the operator substitutions

$$\hat{p} = \frac{\hbar}{i}\frac{\partial}{\partial q}; \qquad \hat{q} = q \qquad [3.36]$$

Alternatively we can express the same information in the commutator

$$[\hat{p},\hat{q}] = \frac{\hbar}{i} \qquad [3.37]$$

As long as we have a system of discrete particles this prescription is sufficient, but many wave systems are, of course, continuous. If we consider the Hamiltonian and Lagrangian for a dense system of particles

$$H = \sum_i \left[\frac{p_i^2}{2m_i} + V(q_i)\right] \qquad [3.38]$$

$$L = \sum_i \left[\frac{m_i(\dot{q}_i)^2}{2} - V(q_i)\right] \qquad [3.39]$$

where $V(q_i)$ is the potential energy. Then this suggests we go over to the continuous system by the replacement of the sum over particles by a spatial integration

$$L = \int \mathcal{L}\, d^3r \qquad [3.40]$$

$$H = \int \mathcal{H}\, d^3r \qquad [3.41]$$

where $\mathcal{L}$ and $\mathcal{H}$ are Lagrangian and Hamiltonian densities. The discrete coordinates q_i, $\dot{q}_i$ become functions of space and time

$$q_i,\dot{q}_i \Rightarrow \phi(\mathbf{r},t),\dot{\phi}(\mathbf{r},t) \qquad [3.42]$$

and

$$\mathcal{L} = \mathcal{L}(\phi(\mathbf{r},t),\dot{\phi}(r,t)) \qquad [3.43]$$

We then define a conjugate momentum $\pi(\mathbf{r},t)$ by

$$\pi(\mathbf{r},t) = \frac{\partial \mathcal{L}}{\partial \dot{\phi}(\mathbf{r},t)} \qquad [3.44]$$

The Hamiltonian density can now be written as

$$\mathcal{H} = \pi(r,t)\dot{\phi}(r,t) - \mathcal{L} \qquad [3.45]$$

equivalent to

$$H = \sum_i p_i \dot{q}_i - L \qquad [3.46]$$

and the field equations are given by Lagrange's equation

$$\frac{\partial \mathcal{L}}{\partial \phi} - \sum_{j=1}^{4} \frac{\partial}{\partial x_j} \left[\frac{\partial \mathcal{L}}{\partial(\partial \phi / \partial x_j)} \right] = 0 \qquad [3.47]$$

$$[(x_1,x_2,x_3,x_4) = (x,y,z,t)]$$

The momentum of the system P is now given by

$$P = -\int d^3r \, \pi(\mathbf{r},t)\nabla\phi(\mathbf{r},t) \qquad [3.48]$$

We *assume* that the system may be quantized by the basic relation [comparable to Eqs. (3.36) and (3.37)]

$$\hat{\pi}(\mathbf{r},t) = \frac{\hbar}{i} \frac{\partial}{\partial \phi(\mathbf{r},t)} \qquad [3.49]$$

or in commutator form

$$[\hat{\pi}(\mathbf{r},t),\hat{\phi}(\mathbf{r}',t)] = \frac{\hbar}{i} \delta(\mathbf{r} - \mathbf{r}') \qquad [3.50]$$

$$[\hat{\pi}(\mathbf{r},t),\hat{\pi}(\mathbf{r}',t)] = [\hat{\phi}(\mathbf{r},t),\hat{\phi}(\mathbf{r}',t)] = 0 \qquad [3.51]$$

(NOTE: Time values must always be the same.) Thus, the procedure for quantization of a general field is as follows:

(i) Write the Lagrangian of the system either in terms of the generalized coordinates q_i, $\dot{q}_i$ of the discrete components or, if continuous, the field variables $\phi(\mathbf{r},t)$, $\dot{\phi}(\mathbf{r},t)$.

(ii) Identify the conjugate momentum variable p_i [or $\pi(\mathbf{r},t)$] through the relationships

$$p_i = \frac{\partial L}{\partial \dot{q}_i}(q_i,\dot{q}_i) \quad \text{or} \quad \pi(r,t) = \frac{\partial \mathcal{L}}{\partial \dot{\phi}(r,t)}$$

(iii) Write the Hamiltonian

$$H = \sum_i q_i p_i - L$$

or

$$H = \int d^3r\, (\pi(\mathbf{r},t)\phi(\mathbf{r},t) - \mathcal{L})$$

(iv) Make the transformation to normal coordinates to produce a Hamiltonian for a set of independent oscillator systems in the form of Eq. (3.6).

(v) Quantize by the operator substitutions

$$\hat{p}_i = \frac{\hbar}{i}\frac{\partial}{\partial q_i}, \qquad \hat{q}_i = q_i$$

or

$$\hat{\pi}(\mathbf{r},t) = \frac{\hbar}{i}\frac{\partial}{\partial \phi(\mathbf{r},t)}, \qquad \hat{\phi}(\mathbf{r},t) = \phi(\mathbf{r},t)$$

(or use the commutator relationships) and hence define the operator equivalents of the normal-mode coordinates and the Hamiltonian operator.

(vi) Define the ladder operators so as to produce a Hamiltonian in the form

$$H = \sum_k \hbar\omega_k \left(\hat{a}_k^+ \hat{a}_k + \frac{1}{2} \right)$$

where k is a general normal-mode label.

Everything can then proceed as in the case of the linear chain.

Before going on to give an example we *must* emphasize that step (v) is a basic assumption. Its justification lies (just as for the whole of quantum mechanics) in its accuracy as confirmed by experiment.

3.4. QUANTIZATION OF THE ELECTROMAGNETIC FIELD

Suppose we take as the archetype of a continuous system the electromagnetic field and follow the prescription outlined above. (It also serves to illustrate that we are not confined to scalar fields.)

1. The Lagrangian for the electromagnetic field is given by

$$L = \tfrac{1}{2}\int (\epsilon_0 |\mathbf{E}(\mathbf{r},t)|^2 - \mu_0 |\mathbf{H}(\mathbf{r},t)|^2)\, d^3r \qquad [3.52]$$

That is

$$\mathcal{L}(\mathbf{r},t) = \tfrac{1}{2}(\epsilon_0 |\mathbf{E}(\mathbf{r},t)|^2 - \mu_0 |\mathbf{H}(\mathbf{r},t)|^2) \qquad [3.53]$$

In terms of the vector potential $\mathbf{A}(\mathbf{r},t)$ this can be written in the form

$$L = \frac{1}{2\mu_0} \int \left(\frac{|\dot{\mathbf{A}}(\mathbf{r},t)|^2}{c^2} + |\nabla \mathbf{A}(\mathbf{r},t)|^2 \right) d^3r \qquad [3.54]$$

which suggests $\mathbf{A}(\mathbf{r},t)$ as the natural generalized coordinate. With vector quantities it is helpful to define a polarization vector so that

$$\mathbf{A}(\mathbf{r},t) = \mathbf{e} A_e(\mathbf{r},t) \qquad [3.55]$$

and take $A_e(\mathbf{r},t)$ as the scalar generalized coordinate.

2. The generalized momentum $\boldsymbol{\pi}(\mathbf{r},t)$ becomes

$$\boldsymbol{\pi}(\mathbf{r},t) = \frac{\partial \mathcal{L}}{\partial \dot{\mathbf{A}}(\mathbf{r},t)} \qquad [3.56]$$

or

$$\begin{aligned}\pi_e(\mathbf{r},t) &= \frac{\partial \mathcal{L}}{\partial \dot{A}_e(\mathbf{r},t)} \\ &= \frac{\dot{A}_e(\mathbf{r},t)}{\mu_0 c^2}\end{aligned} \qquad [3.57]$$

3. The Hamiltonian can now be expressed as

$$H = \frac{1}{2\mu_0} \int (|\boldsymbol{\pi}(\mathbf{r},t)|^2 \mu_0^2 c^2 + |\nabla \mathbf{A}(\mathbf{r},t)|^2)\, d^3 r \qquad [3.58]$$

4. The obvious normal modes are

$$\mathbf{A}(\mathbf{r},t) = \sum_\lambda A_\lambda(t)\mathbf{e}_\lambda e^{i\mathbf{k}_\lambda \cdot \mathbf{r}} \qquad [3.59a]$$

$$\boldsymbol{\pi}(\mathbf{r},t) = \sum_\lambda \pi_\lambda(t)\mathbf{e}_\lambda e^{i\mathbf{k}_\lambda \cdot \mathbf{r}} \qquad [3.59b]$$

(the λ subscript is necessary because of the polarization) and their inverses

$$A_\lambda(t) = \frac{1}{\sqrt{V}} \int A_{e\lambda}(\mathbf{r},t) e^{-i\mathbf{k}_\lambda \cdot \mathbf{r}}\, d\mathbf{r} \qquad [3.60a]$$

$$\pi_\lambda(t) = \frac{1}{\sqrt{V}} \int \frac{A_{e\lambda}(\mathbf{r},t)}{\mu_0 c^2} e^{-i\mathbf{k}_\lambda \cdot \mathbf{r}}\, d\mathbf{r} \qquad [3.60b]$$

where V is the volume of the system.

Substituting into the Hamiltonian, remembering that

$$\nabla \times \mathbf{A}(\mathbf{r},t) = \sum_\lambda (\mathbf{k}_\lambda \times \mathbf{e}_\lambda) e^{i\mathbf{k}_\lambda \cdot \mathbf{r}} A_\lambda(t) \qquad [3.61]$$

gives

$$H = \frac{1}{2}\sum_{\lambda}\left[\pi_{\lambda}(t)\pi_{-\lambda}(t)c^2\mu_0 + \frac{k_{\lambda}^2}{\mu_0}A_{\lambda}(t)A_{-\lambda}(t)\right] \qquad [3.62]$$

where we have defined $-\lambda$ such that

$$\begin{aligned} \mathbf{k}_{\lambda} &= -\mathbf{k}_{-\lambda} \\ \mathbf{e}_{\lambda} &= \mathbf{e}_{-\lambda} \\ A_{\lambda}(\mathbf{r},t) &= A^{*}_{-\lambda}(\mathbf{r},t) \end{aligned} \qquad [3.63]$$

which has the advantage of ensuring that $\mathbf{A}(\mathbf{r},t)$ is explicitly real, as it must be.

5. Quantization is obtained from

$$[\hat{\pi}(\mathbf{r},t),(\hat{A}\mathbf{r}',t)] = \frac{\hbar}{i}\,\delta(r - r') \qquad [3.64]$$

which gives

$$[\hat{\pi}_{\lambda}(t),\hat{A}_{\lambda'}(t)] = \frac{\hbar}{i}\,\delta_{\lambda,-\lambda'} \qquad [3.65]$$

Substituting we have

$$\hat{H} = \frac{1}{2}\sum_{\lambda}\left[c^2\mu_0\pi_{\lambda}(t)\pi_{\lambda}^{+}(t) + \frac{\omega_{\lambda}^2}{\mu_0 c^2}\hat{A}_{\lambda}(t)\hat{A}_{\lambda}^{+}(t)\right] \qquad [3.66]$$

where the frequency ω_{λ} is given by

$$\omega_{\lambda} = \frac{|\mathbf{k}_{\lambda}|}{c} \qquad [3.67]$$

and is the frequency of the light.

6. The ladder operators $\hat{a}_{\lambda}^{+}$, $\hat{a}_{\lambda}$ can now be defined as

$$\hat{a}_{\lambda}^{+} = \left(\frac{\mu_0 c^2}{2\hbar\omega_{\lambda}}\right)^{1/2}\left[\hat{\pi}_{\lambda}^{+}(t) + \frac{i\omega_{\lambda}}{\mu_0 c^2}\hat{A}_{\lambda}^{+}(t)\right] \qquad [3.68a]$$

$$\hat{a}_{\lambda} = \left(\frac{\mu_0 c^2}{2\hbar\omega_{\lambda}}\right)^{1/2}\left[\hat{\pi}_{\lambda}(t) - \frac{i\omega_{\lambda}}{\mu_0 c^2}\hat{A}_{\lambda}(t)\right] \qquad [3.68b]$$

Comparing these relationships with Eqs. (3.12) and (3.13) we see that

$$[\hat{a}_\lambda,\hat{a}_\lambda^+] = 1 \qquad [3.69]$$

and

$$\hat{H} = \sum_\lambda \hbar\omega_\lambda \left[\hat{a}_\lambda^+\hat{a}_\lambda + \frac{1}{2}\right] \qquad [3.70]$$

From this point on we see that the result of our quantization is an energy spectrum that has discrete levels for each normal mode λ separated by the constant energy $\hbar\omega_\lambda$. We can also label the total energy and wave function for the system in terms of the quantum numbers n_λ, which gives the position on the "ladder" for each of the normal modes:

$$E_\zeta = \sum_\lambda \hbar\omega_\lambda \left(n_\lambda + \frac{1}{2}\right) \qquad [3.71]$$

$$|\zeta\rangle = |n_1,n_2,\ldots,n_\lambda,\ldots\rangle \qquad [3.72]$$

The electromagnetic field has a zero-point oscillation energy given by

$$E_0 = \sum_\lambda \frac{1}{2}\hbar\omega_\lambda \qquad [3.73]$$

Since the number of modes is infinite (in the linear chain the number of modes is equal to the number of degrees of freedom) the zero-point energy is also infinite. This is one of many unmeasurable infinities one comes across in this subject (since it is unmeasurable it is of course ignored).

The momentum operator can be derived from Eq. (3.48) in the same way, to give

$$\hat{p} = \sum_\lambda \hbar\mathbf{k}_\lambda \left(\hat{a}_\lambda^+\hat{a}_\lambda + \frac{1}{2}\right) \qquad [3.74]$$

Comparing this with the Hamiltonian we see that the momentum will also have equally spaced eigenvalues for each normal mode separated by a value of $\hbar\mathbf{k}_\lambda$. We also have the concept of a zero-point momentum of $\frac{1}{2}\hbar\mathbf{k}_\lambda$ associated with each normal mode.

3.5. ELEMENTARY EXCITATIONS AND "PARTICLES"

It should be obvious now that using the extended idea of quantization any system described by a wave motion will be quantized so that the energy and momentum in each normal mode is characterized by discrete levels equally spaced. Furthermore, the state of the system is described by a set of quantum numbers, which give the "rung of the ladder" on which each normal mode is situated.

At this stage it is very beneficial to make a *conceptual* change in our description of the system. Consider one normal mode only; then

$$H = \hbar\omega(\hat{a}^{+}\hat{a} + \tfrac{1}{2}) \quad [3.75]$$

$$E = \hbar\omega(n + \tfrac{1}{2}) \quad [3.76]$$

If we try to think classically what the eigenvalue ladder corresponds to for changing n, it is very clear that increasing n corresponds to larger and larger amplitudes of the same oscillating system (Fig. 3.3). That is, n labels the energy and amplitude of the oscillation. Thus, if we change the system from state n to m, the energy changes by $(m - n)\hbar\omega$ and the amplitude adjusts accordingly.

The above description obviously connects with the classical picture, but because the energy and momentum always change by multiples of some fixed amount, we can conceive of a different description. Ignoring the zero-point energy and momentum (which are constants anyway) we can consider our oscillator at level n as being n oscillators at level 1 (Fig. 3.4). The total energy and momentum is the same. Changing our system from level n to level m ($n < m$) corresponds, in this scheme, to adding $m - n$ oscillators at level 1. Thus, changes in the system are thought of as adding or subtracting packets of $\hbar\omega$ and $\hbar k$ rather than changing the amplitude and energy of the normal mode. These packets are then referred to as *elementary excitations* of the normal mode of the system and

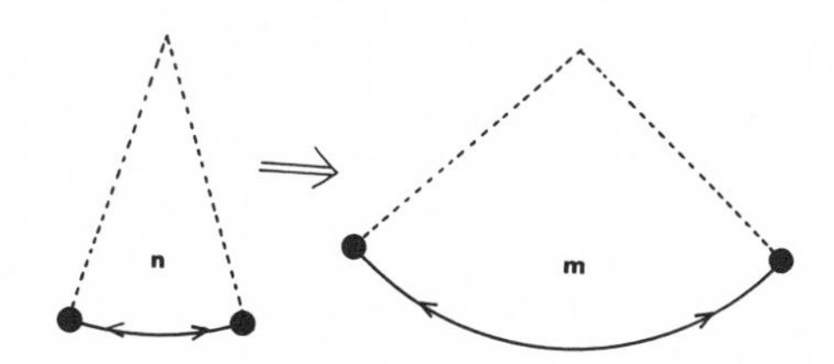

FIGURE 3.3. The classical interpretation of the eigenvalue ladder as a change in amplitude of oscillation.

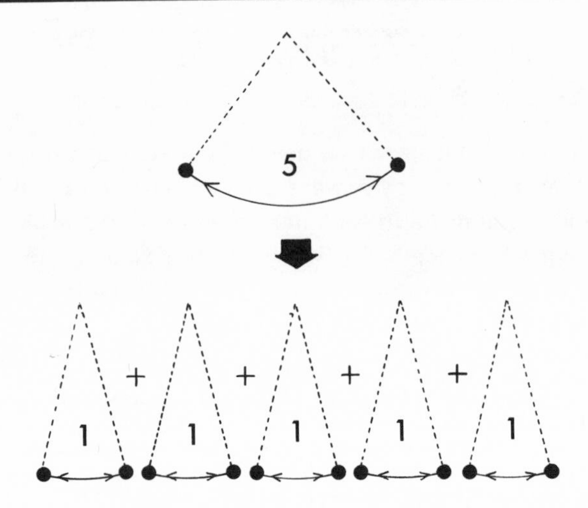

FIGURE 3.4. The interpretation of an $n = 5$ oscillation as 5 elementary ($n = 1$) excitations.

have a definite energy and momentum associated with them. Changes in the system, which we would classically describe as quantized changes in the energy and momentum (or amplitude) of the normal modes, become the creation and destruction (or annihilation) of these elementary excitations.

Thus, a state described by the quantum number n has to be thought of as n elementary excitations each of energy $\hbar\omega$ and momentum $\hbar k$ (plus the zero-point energy and momentum). This concept of separate elementary excitations is very close to our idea of a particle, in the sense of being discrete packets carrying energy and momentum; but these particles may be created and destroyed.

The conceptual change in our description of the mathematics of quantization suggests a change in our terminology. Since the operators $\hat{a}_k^+$, $\hat{a}_k$ increase and decrease by one, respectively, the quantum number for their normal mode, in our new terminology they create and annihilate an elementary excitation of that normal mode. Hence $\hat{a}_k^+$ is termed a *creation operator* and adds to the system an elementary excitation of the normal mode along with its associated packets of energy and momentum, while $\hat{a}_k$ is termed an *annihilation operator.* The total ground state $|0_1,0_2,\ldots,0_k,\ldots\rangle$ can be considered as the state with no elementary excitations in it, i.e., it is the vacuum (empty) state. We do not actually use the vacuum state so that its internal structure can be ignored. The whole conceptual picture depends upon the fact that we require *changes* between states, and these can be calculated in terms of the creation and annihilation operators and hence thought of in terms of the elementary excitations.

In the two examples we have studied the elementary excitations are termed *photons* for the electromagnetic field and *phonons* for the linear chain (when it is extended to be three-dimensional so as to describe a solid).

3.6. PERTURBATIONS AND THE ELEMENTARY EXCITATIONS

In reality the Hamiltonian is really harmonic only in very special cases. It is necessary, therefore, to consider how the concept of particles, or elementary excitations, will survive in a situation where there are nonharmonic effects and interactions with other systems. We must leave the second possibility until later, but consider the linear chain Hamiltonian again:

$$\begin{aligned} H &= \sum_i \frac{p_i^2}{2M} + \frac{1}{2} K(q_{i+1} - q_i)^2 + \epsilon(q_i - q_{i+1})^3 \\ &= H_0 + H_1 \end{aligned} \qquad [3.77]$$

H_0 is the harmonic oscillator Hamiltonian while H_1 is the first anharmonic term. If ϵ is a small quantity we can consider solving H_0 in terms of the $\hat{a}_k^+$, $\hat{a}_k$ and *then* think about H_1 as a perturbation. Changing to the normal coordinates we have

$$H_1 = \frac{\epsilon}{N^{1/2}} \sum_{k,l,m} q_k q_l q_m (1 - e^{ika})(1 - e^{ila})(1 - e^{ima})\delta(l + k + m) \qquad [3.78]$$

Making the transformation to quantized operators $q_k \rightarrow \hat{q}_k$ leaves H_1 formally unaltered, but we can write $\hat{H}_1$ also in terms of the creation and annihilation operators. From Eq. (3.13)

$$\hat{q}_k = \frac{1}{i}\left(\frac{\hbar}{2M\omega_k}\right)^{1/2} (\hat{a}_{-k}^+ - \hat{a}_k) \qquad [3.79]$$

So, putting all of the numerical constants into a factor γ, we have

$$\hat{H}_1 = \sum_{k,l,m} \gamma(k,l,m)\delta(l + k + m)(\hat{a}^+_{-k} - \hat{a}_k)(\hat{a}^+_{-l} - \hat{a}_j)(\hat{a}^+_{-m} - \hat{a}_m) \quad [3.80]$$

This consists of three parts:

(i) $\gamma(k,l,m)$: this is a matrix element and gives the strength of the particular process.

(ii) $\delta(l + k + m)$: this is obviously equivalent to the equation

$$(l + k + m) = 0 \quad [3.81a]$$

or

$$\hbar l + \hbar k + \hbar m = 0 \quad [3.81b]$$

In terms of our elementary excitations the interpretation is obviously that each process described in $\hat{H}_1$ conserves the momentum of the elementary excitations involved.

(iii) $(\hat{a}^+_{-k} - \hat{a}_k)(\hat{a}^+_{-l} - \hat{a}_l)(\hat{a}^+_{-m} - \hat{a}_m)$: the first thing to note is that within each of these parentheses the momentum change suggested by each term is the same. The creation of momentum $\hbar k$ is the same as the removal of momentum $\hbar(-k)$, etc. Expanding the terms out we have four typical terms

(a) $\hat{a}^+_{-k}\hat{a}^+_{-l}\hat{a}^+_{-m}$—creation of three phonons,

(b) $\hat{a}_k\hat{a}_l\hat{a}_m$—destruction of three phonons,

(c) $\hat{a}^+_{-k}\hat{a}^+_{-l}\hat{a}_m$—creation of two phonons from the annihilation of one, and

(d) $\hat{a}_k\hat{a}_l\hat{a}^+_{-m}$—two phonons coalesce to form one new one.

So we see that the introduction of the anharmonic term causes the elementary excitations to interact with each other. As long as the interaction is not too big, the excitations do not lose their meaning. Too much interaction, of course, and the system becomes essentially anharmonic anyway. Figure 3.5 shows a pictorial way of describing the above excitation processes akin to the Green's function diagrams at the end of the last chapter. It is interesting to consider the classical interpretation of the terms in (a), (b), (c), and (d). The terms in (c) and (d) correspond to the mixing of frequencies and wave vectors by the anharmonic part of the

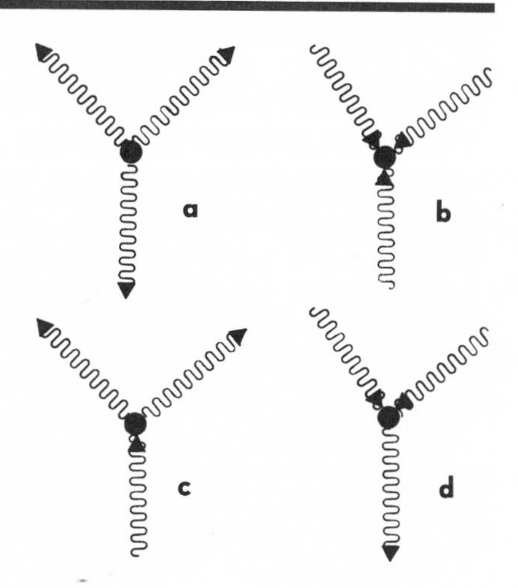

FIGURE 3.5. Pictorial description of the interaction between the elementary excitations caused by the perturbation.

potential, while those in (a) and (b) correspond to a change in frequency of the normal modes in the presence of an anharmonic restoring force. This is not obvious from the form of the perturbation, but we shall solve a similar but simpler example later which illustrates this aspect.

In general, where perturbations are concerned, we will be able to write

$$H = H_0(q_i, p_i) + H_1(q_i, p_i)$$

Once one has transformed the oscillator Hamiltonian to the creation and annihilation operator form, it is then possible to express the operator $\hat{H}_1$ in terms of those operators. This enables one to discuss the perturbation in terms of the scattering, creation, and annihilation of the elementary excitations, be they phonons, photons, or whatever.

3.7. SUMMARY

We have seen that the quantization of fields leads directly to the concept of elementary excitations as a way of interpreting the mathematics. This is an immensely powerful concept and serves to unify the treatment of many different types of field. It is well to remember, however, that this may break down in particular cases. Once nonharmonic effects become large in a system (near a phase change, for instance), the elementary excitations may lose their meaning.

Now that fields may be treated within quantum mechanics, it is possible to consider waves and particles on the same footing. The formalism we have is not, however, similar to the standard quantum mechanics. The creation and annihilation operators are some way from Schrödinger wave functions. There are creation and annihilation operators applicable to electrons and positrons—they can be considered as the elementary excitations of a four-component field—but because of the rest mass, the energies involved are very high, well beyond our interest in solids. It is necessary to develop the concepts of creation and annihilation operators, however, if we are to treat together particles and waves and their interaction.

BIBLIOGRAPHY

Elementary

MESSIAH, A., *Quantum Mechanics,* North Holland, Amsterdam, 1966.

FEYNMAN, R. P., LEIGHTON, R. B., and SANDS, M., *Feynman Lectures on Physics,* vol. 3, Addison-Wesley, Reading: MA, 1963.

More Advanced

HEITLER, W., *Quantum Theory of Radiation,* Oxford University Press, New York, 1964.

HENLEY, E. M., and THIRRING, W., *Elementary Quantum Field Theory,* McGraw-Hill, New York, 1962.

Classical Theory of Fields

GOLDSTEIN, H., *Classical Mechanics,* Addison-Wesley, Reading: MA, 1972.

PROBLEMS

1. Repeat the steps of the quantization procedure for the two-component chain where the masses of the atoms are alternatively M_1 and M_2 but the force constants remain the same. Show that one has now *two* sets of creation and annihilation operators for each k value and

$$\hat{H} = \sum_k \hbar\omega_k^{(O)}\left(\hat{a}_k^+\hat{a}_k + \frac{1}{2}\right) + \sum_k \hbar\omega_k^{(A)}\left(\hat{b}_k^+\hat{b}_k + \frac{1}{2}\right)$$

where $\omega_k^{(A)}$ $\omega_k^{(O)}$ are the acoustic and optical phonon frequencies.

2. Use the quantization procedure on the Klein–Gordon equation

$$\left(\nabla^2 - \frac{1}{c^2}\frac{\partial^2}{\partial t^2} + \frac{m_0^2 c^4}{\hbar^2}\right)\phi(\mathbf{r},t) = 0$$

which may be derived from the Lagrangian density

$$\mathcal{L} = \frac{1}{2c^2}(\dot{\phi}(\mathbf{r},t))^2 - c^2(\nabla\phi(\mathbf{r},t))^2 - \left(\frac{m_0^2 c^2}{\hbar}\right)(\phi(\mathbf{r},t))^2$$

and show that

$$\hat{H} = \sum_{\mathbf{k}} \hbar\omega_{\mathbf{k}}\left(\hat{a}_{\mathbf{k}}^{+}\hat{a}_{\mathbf{k}} + \frac{1}{2}\right)$$

where

$$\hbar\omega_k = [\hbar^2 k^2 c^2 + m_0^2 c^4]^{1/2}$$

Interpret your result in terms of the expected properties of the "elementary excitation" of the Klein–Gordon field.

3. For a single one-dimensional anharmonic oscillator

$$H(x) = \frac{p^2}{2m} + \alpha x^2 + \beta x^3 + \gamma x^4$$

where β and γ are small

(i) write the nonharmonic part of the Hamiltonian in terms of the creation and annihilation operators of the harmonic part

$$H = \frac{p^2}{2m} + \alpha x^2$$

(ii) show that the first-order shift in the ground-state energy of the system, due to the nonharmonic terms, is given by

$$\Delta E = \frac{3}{4}\beta\left(\frac{\hbar}{m\omega}\right)^2$$

4. Show that the momentum operator for the electromagnetic field is indeed given by

$$\hat{P} = \sum_{\lambda} \hbar \mathbf{k}_{\lambda} \left(\hat{a}_{\lambda}^{+} \hat{a}_{\lambda} + \frac{1}{2} \right)$$

CHAPTER FOUR

REPRESENTATIONS OF QUANTUM MECHANICS

In ordinary quantum mechanics, the description of a particle is in terms of its wave function whose time development is determined by the Schrödinger equation. This is sometimes neither the best nor the most convenient way. In order to change the description we start from the basic concept that reality must be independent of its mathematical description. Reality corresponds to the measurable quantities, which in turn correspond to the expectation values. Consider a measurable quantity O; corresponding to this there is an operator $\hat{O}$ and the measured quantity, the expectation value $\langle O \rangle$, is given by

$$\langle O(t) \rangle = \int \Psi^{+}(q_i,t)\hat{O}(q_i)\Psi(q_i,t)\, dq_i \qquad [4.1]$$

$\Psi(q_i,t)$ is the wave function, q_i represent the set of commuting variables, and dq_i represents the integration over all such variables.

As long as we keep $\langle O(t) \rangle$ invariant, we are free to

(i) partition the information about the system between operator and wave function in any way we like and
(ii) describe $\hat{O}$ and Ψ in different ways.

A particular way of describing the system through its wave function and the operators is termed a *representation*. Obviously there are an infinite number of possible representations but we will be primarily concerned with four: the *Schrödinger, Heisenberg, interaction,* and *occupation number representations.* The identifications are not exclusive, the occupation number representation can exist in both Heisenberg and Schrödinger forms but this will become clearer as we go on.

4.1. SCHRÖDINGER REPRESENTATION

The Schrödinger representation is familiar from elementary quantum mechanics. The wave function $\Psi_S(q_i,t)$ describing the system is time dependent and the development of the system is determined by Schrödinger's equation

$$\hat{H}(q_i)\Psi_S(q_i,t) = i\hbar \frac{\partial}{\partial t} \Psi_S(q_i,t) \qquad [4.2]$$

The operator $\hat{O}_S(q_i)$ is a function of the independent variables but *not* of time. The expectation value of the property O is given by

$$\langle O(t)\rangle = \int \Psi_S^+(q_i,t)\hat{O}_S(q_i)\Psi_S(q_i,t)\, dq_i \qquad [4.3]$$

and is a function of time through the wave function.

4.2. HEISENBERG REPRESENTATION

Since wave functions are less immediately connected with observable quantities than the operators themselves, the Heisenberg representation seeks to transfer the time dependences from the wave functions to the operators, leaving the wave function as a time-independent framework against which the expectation values are calculated.

An operator is defined such that

$$\Psi_S(q_i,t) = \hat{A}\Psi_H(q_i) \qquad [4.4]$$

where $\Psi_H(q_i)$ is the new Heisenberg time-independent wave function. Thus, we require

$$\frac{\partial}{\partial t} \Psi_H(q_i) = 0 \qquad [4.5]$$

From which it immediately follows that

$$\frac{\partial}{\partial t} \Psi_S(q_i,t) = \frac{\partial \hat{A}}{\partial t} \Psi_H(q_i) \qquad [4.6]$$

From Eqs. (4.2) and (4.4)

$$-\frac{i}{\hbar}\hat{H}\hat{A}\Psi_H(q_i,t) = \frac{\partial}{\partial t}\hat{A}\Psi_H(q_i) \qquad [4.7]$$

or

$$\left(i\frac{\hat{H}\hat{A}}{\hbar} + \frac{\partial \hat{A}}{\partial t}\right)\Psi_H(q_i) = 0 \qquad [4.8]$$

This gives an operator equation for $\hat{A}$ in terms of the Hamiltonian operator $\hat{H}$. For many systems $\hat{H}$ is independent of time so that

$$\hat{A} = e^{(-i\hat{H}t)/\hbar} \qquad [4.9]$$

but this should never be assumed. The Heisenberg form for the operator can now be obtained, since from Eqs. (4.3) and (4.4)

$$\langle O(t)\rangle = \int \Psi_H^+(q_i)\hat{A}^+\hat{O}_S(q_i)\hat{A}\Psi_H(q_i)\, dq_i \qquad [4.10]$$

The Heisenberg operator becomes

$$\hat{O}_H(q_i,t) = \hat{A}^+\hat{O}_S(q_i)\hat{A} \qquad [4.11]$$

The time development of the observable is now contained in the operator so, taking the time derivative

$$\begin{aligned}\frac{\partial}{\partial t}\hat{O}_H(q_i,t) &= \frac{\partial}{\partial t}(\hat{A}^+\hat{O}_S(q_i)\hat{A}) \\ &= \left(\frac{\partial A^+}{\partial t}\right)\hat{O}_S(q_i)\hat{A} + \hat{A}^+\hat{O}_S(q_i)\left(\frac{\partial \hat{A}}{\partial t}\right)\end{aligned}$$

Substituting from Eq. (4.8) we have (since $\hat{A}$ and $\hat{H}$ must commute)

$$\frac{\partial}{\partial t}\hat{O}_H(q_i,t) = \frac{i}{\hbar}(\hat{H}\hat{A}^+\hat{O}_S(q_i)\hat{A} - \hat{A}^+\hat{O}_S(q_i)\hat{A}\hat{H})$$

Therefore

$$\frac{\hbar}{i}\frac{\partial}{\partial t}\hat{O}_H(q_i,t) = [\hat{H},\hat{O}_H(q_i,t)] \qquad [4.12]$$

This is an operator equation for $\hat{O}_H$; it replaces the Schrödinger equation and is the Heisenberg equation of motion. Thus, the time dependence is now in the operator, and once the wave function is known at some time, the system is completely determined.

4.3. INTERACTION REPRESENTATION

This representation is designed primarily for the situation where the Hamiltonian is most conveniently written in the form

$$\hat{H} = \hat{H}_0 + \hat{H}_1 \tag{4.13}$$

$\hat{H}_0$ is considered soluble and time independent. We can write, in the Schrödinger representation

$$\hat{H}(q_i)\Psi_S(q_i,t) = i\hbar \frac{\partial}{\partial t} \Psi_S(q_i,t) \tag{4.14}$$

If we define [cf. Eq. (4.9)]

$$\Psi_I(q_i,t) = e^{(i\hat{H}_0 t)/\hbar}\Psi_S(q_i,t) \tag{4.15}$$

this, on substitution, gives

$$i\hbar \frac{\partial}{\partial t} \Psi_I(q_i,t) = e^{(i\hat{H}_0 t)/\hbar}\hat{H}_1 e^{(-i\hat{H}_0 t)/\hbar}\Psi_I(q_i,t) \tag{4.16}$$

Thus, using a Heisenberg type of transformation of the operator, we can write

$$\hat{H}_1(t) = e^{(i\hat{H}_0 t)/\hbar}\hat{H}_1 e^{(-i\hat{H}_0 t)/\hbar} \tag{4.17}$$

and

$$i\hbar \frac{\partial}{\partial t} \Psi_I(q_i,t) = \hat{H}_1(t)\Psi_I(q_i,t) \tag{4.18}$$

Returning to the expectation value, we have, using (4.15)

$$\langle O(t)\rangle = \int \Psi_I^+(q_i,t)e^{(i\hat{H}_0t/\hbar}\hat{O}_S(q_i)e^{(-i\hat{H}_0t)/\hbar}\Psi_I(q_i,t)\, dq_i \qquad [4.19]$$

from which we deduce an interaction representation operator as

$$\hat{O}_I(q_i,t) = e^{(i\hat{H}_0t)/\hbar}\hat{O}_S(q_i)e^{(-i\hat{H}_0t)/\hbar} \qquad [4.20]$$

The equation of motion now follows, as in the Heisenberg representation

$$i\hbar \frac{\partial}{\partial t} \hat{O}_I(q_i,t) = [\hat{O}_I(q_i,t),\hat{H}_0] \qquad [4.21]$$

Because the $H_I(t)$ will not normally commute at different times, the *time ordering operator* T is necessary to define the expansion of the exponential. It orders the terms by their time values (cf. Section 6.1).

The interaction picture is most useful in treating perturbations. Consider Eq. (4.18); a formal solution is

$$\Psi_I(q_i,t) = \left[T \exp\left(-\frac{i}{\hbar}\int_{t_0}^{t} \hat{H}_1 dt'\right)\right]\Psi_I(q_i,t_0) \qquad [4.22]$$

Without $\hat{H}_1$ acting, we have the unperturbed wave functions

$$\Psi_S^0(q_i,t) = e^{[i\hat{H}_0(t-t_0)]/\hbar}\Psi_S^0(q_i,t_0) \qquad [4.23]$$

Suppose t_0 is taken far enough back in time so that $\hat{H}_1$ had not acted; then

$$\begin{aligned}\Psi_I(q_i,t_0) &= e^{(i\hat{H}_0t_0)/\hbar}\Psi_S(q_i,t_0) \\ &= e^{(i\hat{H}_0t_0)/\hbar}\Psi_S^0(q_i,t_0)\end{aligned} \qquad [4.24]$$

Therefore, from Eqs. (4.9) and (4.4), since $\hat{H}_0$ is not a function of time

$$\Psi_I(q_i,t_0) = \Psi_H^0(q_i)$$

where $\Psi_H(q_i)$ is the Heisenberg wave function for the *unperturbed* system.

In terms of this wave function we have

$$\Psi_I(q_i,t) = \left[T \exp\left(-\frac{i}{\hbar} \int_{-\infty}^{t} \hat{H}_1 dt' \right) \right] \Psi_H(q_i) \qquad [4.25]$$

where we have extended t_0 to $-\infty$ to ensure that H_1 had not commenced to act. It is equally true that

$$\Psi_I(q_i,t) = \left[T \exp\left(\frac{i}{\hbar} \int_{t}^{\infty} \hat{H}_1 \, dt' \right) \right] \Psi_H^0(q_i) \qquad [4.26]$$

for many of the perturbations considered because Eq. (4.26) is simply a statement that after the removal of the perturbation, the system will revert to its original state (and implies an adiabatic, i.e., slow, introduction and removal of the perturbation).

The relationship between the Heisenberg and interaction operators also needs to be considered. From Eq. (4.8), we have formally

$$\hat{A}(t) = T \exp\left(-\frac{i}{\hbar} \int^{t} \hat{H} \, dt' \right) \qquad [4.27]$$

This gives

$$\hat{O}_H(t) = T \exp\left(\frac{i}{\hbar} \int^{t} \hat{H} \, dt' \right) \hat{O}_S(q_i) \, T \exp\left(-\frac{i}{\hbar} \int^{t} \hat{H} \, dt' \right) \qquad [4.28]$$

or

$$\hat{O}_H(t) = T \exp\left(\frac{i}{\hbar} \int^{t} \hat{H}_1 \, dt' \right) \hat{O}_I(t) \, T \exp\left(-\frac{i}{\hbar} \int^{t} \hat{H}_1 dt' \right) \qquad [4.29]$$

The interaction representation is essentially a mixed representation, but the separation of the action of the Hamiltonian into two parts, one of which acts upon the wave function while the other acts upon the operator, makes it ideal, as indicated, for the treatment of perturbations.

4.4. OCCUPATION NUMBER REPRESENTATION

Having removed the time dependence from the wave function in the Heisenberg representation, it is useful to ask whether it is possible to remove any more information. Since the wave function $\Psi_H(q_i)$ describes a system with possibly many particles, it will be a very complicated object, as we saw in the first chapter. Suppose we construct a complete set of one-electron states for the system governed by a Hamiltonian $\hat{H}_0$ and some set of boundary conditions. In a metal, for instance, it may be the set of plane-wave states but the details do not really matter as long as it is a complete set. Thus, we have a set of eigenfunctions ϕ_m and eigenvalues E_m defined through the equation

$$\hat{H}_0\phi_m = E_m\phi_m \qquad [4.30]$$

We can now describe the system in terms of the *"occupancy"* of these levels.

With this set we may write for the wave function in a one-electron system

$$\Psi(q_i) = \sum_m \alpha_m\phi_m(q_i) \qquad [4.31]$$

where $|\alpha_m|^2$ could be considered as the probability that the state ϕ_m is occupied. For many-electron systems, Ψ would be made up of appropriately symmetrized products (for instance a Slater determinant)

$$\Psi = \sum_S \alpha_S\Gamma_S(\phi_a(q_1)\phi_b(q_2)\phi_c(q_3)\cdots) \qquad [4.32]$$

where each set Γ_S would contain a subset of the complete set of states. This could be interpreted as a probability $|\alpha_S|^2$ that the *set* $\phi_a, \phi_b, \phi_c, \ldots,$ were occupied by particles. Considering the eigenvectors as a basis, we could describe the wave functions in terms of the set of numbers α. Thus, the wave function of Eq. (4.31) could be written as

$$\Psi \Rightarrow |\alpha_1, \alpha_2, \ldots, \alpha_m, \ldots\rangle$$

In order to avoid fractional occupancy we would then write Eq. (4.31) as

$$\Psi = \sum_{m} \alpha_m |0_1, 0_2, 0_3, \ldots, 1_m, \ldots\rangle \qquad [4.33]$$

The extension to more than one particle would follow in the same way. The appropriately symmetrized sets of wave functions $\Gamma_S(\phi_a(q_1)\phi_b(q_2) \cdots)$ could be rewritten as $|0_1, 0_2, \ldots, 1_a, \ldots, 1_b, 1_c \ldots\rangle$, the symmetrization being implicit, and Eq. (4.32) would become

$$\Psi = \sum_{S} \alpha_S |0_1, 0_2, \ldots, 1_a, \ldots, 1_b, \ldots, 1_c, \ldots\rangle \qquad [4.34]$$

i.e., in general we build up a wave function by a combination of basis set wave functions, each identified by the occupation of the states described by the chosen complete set $\{\phi_m\}$. A general basis state can now be written as

$$\Phi = |n_1, n_2, \ldots, n_a, \ldots\rangle \qquad [4.35]$$

In order to develop a quantum mechanics in this represenation, we must have a way of changing from one state to another. By analogy with the previous chapter, we can *define* creation and annihilation operators to add and subtract particles from eigenstates; that is

$$\hat{b}_m^+ |n_1, n_2, \ldots, n_m, \ldots\rangle = \eta |n_1, n_2, \ldots, n_m + 1, \ldots\rangle \qquad [4.36a]$$

$$\hat{b}_m |n_1, n_2, \ldots, n_m, \ldots\rangle = \xi |n_1, n_2, \ldots, n_m - 1, \ldots\rangle \qquad [4.36b]$$

where η, ξ are normalization constants. With these two types of operators we can transfer from one state to another as in Fig. 4.1. We can define a vacuum state for each eigenfunction

$$\hat{b}_m |n_1, n_2, \ldots, 0_m, \ldots\rangle = 0 \qquad [4.37]$$

and a complete vacuum state if

$$\hat{b}_m |0\rangle = 0 \qquad [4.38]$$

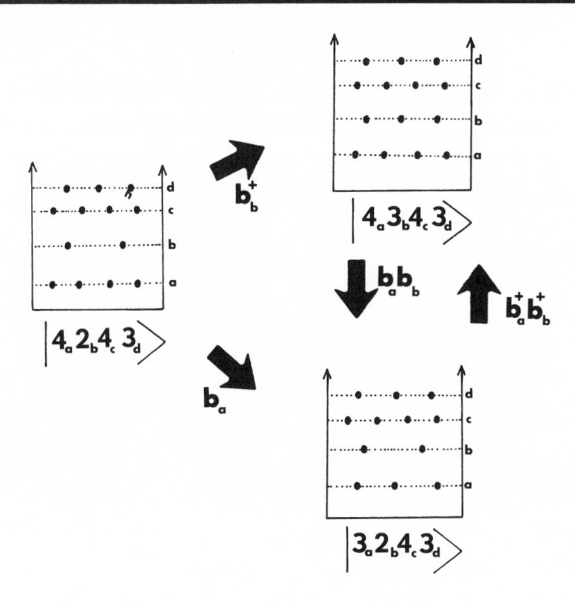

FIGURE 4.1. Schematic illustration of the action of the creation and annihilation operators.

for all m. The definition of the vacuum state now enables us to define any state by the action of the operators on the vacuum state; thus

$$\begin{aligned}\Phi &= |n_1, n_2, \ldots n_m, \ldots\rangle \\ &= \chi \prod_m (\hat{b}_m^+)^{n_m} |0\rangle\end{aligned} \quad [4.39]$$

where χ is a normalization constant.

The state we produce *must* be uniquely defined by the set of quantum numbers $\{n_m\}$ otherwise the transformation will be useless. Consider a two-particle system, for instance, with a vacuum state $|0_1, 0_2\rangle$ then, using the notation $\phi_i(j)$ to correspond to the ith state, jth particle

$$\hat{b}_1^+ |0_1, 0_2\rangle = |1_1, 0_2\rangle = \phi_1(1) \quad [4.40]$$

Introducing the second particle, we have

$$\hat{b}_2^+\hat{b}_1^+|0_1,0_2\rangle = |1_1,1_2\rangle \equiv \frac{1}{\sqrt{2}}(\phi_1(1)\phi_2(2) \pm \phi_2(1)\phi_1(2)) \quad [4.41]$$

where we have $\pm$ according to whether the particles are bosons or fermions. Thus, at this stage, we are introducing the properties of the particles into the state function automatically. To be consistent, if we put the "first" particle in state 2 and the "second" particle in state 1, since the particles are indistinguishable, we *must* get the same state, i.e.,

$$\hat{b}_1^+\hat{b}_2^+|0_1,0_2\rangle \equiv \frac{1}{\sqrt{2}}(\phi_1(2)\phi_2(1) \pm \phi_2(2)\phi_1(1)) \quad [4.42]$$

should also be $|1_1,1_2\rangle$. Comparing Eqs. (4.41) and (4.42), we must have

$$(\hat{b}_1^+\hat{b}_2^+ \mp \hat{b}_2^+\hat{b}_1^+)|0_1,0_2\rangle = 0 \quad [4.43]$$

Similarly

$$(\hat{b}_1\hat{b}_2 \mp \hat{b}_2\hat{b}_1)|1_1,1_2\rangle = 0 \quad [4.44]$$

Thus, creation and annihilation operators for different states commute if the particles are bosons and anticommute if the particles are fermions in order that the description of the state be unique. This extends to any number of states. Since we have already considered the case of commuting operators, the elementary excitation operators of the previous chapter, we will consider only fermions now.

For fermions (using $\hat{c}^+$, $\hat{c}$ for the operators, since b^+, b referred to either type), from (4.43) and (4.44)

$$\hat{c}_m^+\hat{c}_{m'}^+ + \hat{c}_{m'}^+\hat{c}_m^+ = 0 \quad [4.45]$$

and

$$\hat{c}_m\hat{c}_{m'} + \hat{c}_{m'}\hat{c}_m = 0 \quad [4.46]$$

If the two states are the same

$$(\hat{c}_m^+)^2 = 0 \quad \text{and} \quad (\hat{c}_m)^2 = 0 \tag{4.47}$$

Thus, we are not allowed to put or take more than one particle from any fermion state, as expected from the statistics we have built in. That is, there are only 0 or 1 particles in each state (m of course includes a spin index if it defines a complete state). From the identities

$$\hat{c}_m^+|\cdots 1_m \cdots\rangle = \hat{c}_m|\cdots 0_m \cdots\rangle = 0 \tag{4.48a}$$
$$\hat{c}_m\hat{c}_m^+|\cdots 0_m \cdots\rangle = |\cdots 0_m \cdots\rangle \tag{4.48b}$$
$$\hat{c}_m^+\hat{c}_m|\cdots 1_m \cdots\rangle = |\cdots 1_m \cdots\rangle \tag{4.48c}$$

we see that

$$\hat{c}_m^+\hat{c}_m + \hat{c}_m\hat{c}_m^+ = 1 \tag{4.49}$$

or

$$\{\hat{c}_m^+, \hat{c}_{m'}\} = \delta_{m,m'} \tag{4.50}$$

where { } is the anticommutator bracket and Eq. (4.50) corresponds to Eq. (3.14) for bosons. The number operator $\hat{n}_m$ ($= \hat{c}_m^+\hat{c}_m$) has, trivially, only two eigenvalues, 0 and 1. Now that there is a prescription to define the wave function, it remains only to define the operator within this representation.

Consider how the operator $\hat{O}$ acts on the single-particle state $\phi_m(q_i)$

$$\hat{O}\phi_m(q_i) = \psi(q_i) \tag{4.51}$$

where $\psi(q_i)$ is a new state which we expand in the complete set of wave functions

$$\hat{O}\phi_m(q_i) = \sum_l \hat{O}_{lm}\phi_l(q_i) \tag{4.52}$$

where

$$\hat{O}_{lm} = \int \phi_l^+(q_i)\hat{O}\phi_m(q_i)\, dq_i \tag{4.53}$$

The action of the operator $\hat{O}$ on the state ϕ_m may now be represented by the operator expression

$$\sum_{l} \hat{O}_{lm}\hat{b}_l^+\hat{b}_m \qquad [4.54]$$

where $\hat{b}_m$ destroys state $\phi_m(q_i)$ and $\hat{b}_l^+$ creates the state $\phi_l(q_i)$, O_{lm} gives the matrix element for the process, and the summation covers all the possible states. That is, in terms of Eq. (4.33), (4.51) becomes

$$\psi \equiv \sum_{l} O_{lm}\hat{b}_l^+\hat{b}_m|0_1,0_2,\ldots,1_m,\ldots\rangle \qquad [4.55]$$

Constructing the general state $|n_1,n_2,\ldots,n_m,\ldots\rangle$ from the single-particle states gives us, immediately, that the operator $\hat{O}$, acting upon the general state Ψ, is equivalent to the operator

$$\hat{O} = \sum_{l,m} O_{lm}\hat{b}_l^+\hat{b}_m \qquad [4.56]$$

We can see this more easily, perhaps, by considering the expectation value

$$\begin{aligned}\langle O\rangle &= \langle n_1,n_2,\ldots,n_m,\ldots|\sum_{l,m} O_{lm}\hat{b}_l^+\hat{b}_m|n_1,n_2,\ldots,n_m,\ldots\rangle \\ &= \sum_{l,m} O_{l,m}\langle\cdots n_l \cdots n_m|\hat{b}_l^+\hat{b}_m|\cdots n_l \cdots n_m \cdots\rangle\end{aligned} \qquad [4.57]$$

If $l \neq m$, the two eigenstates will be orthogonal, having different occupations. With $l = m$, $\hat{b}_m^+\hat{b}_m$ is the number operator and then

$$\begin{aligned}\langle O\rangle &= \sum_{l,m} O_{lm}n_m\delta_{lm} \\ &= \sum_{m} O_{mm}n_m\end{aligned} \qquad [4.58]$$

This is the sum over the expectation values of the occupied states, exactly what one expects. As an example, take the Hamiltonian itself

$$\begin{aligned}H_{lm} &= \int\phi_l^+(q_i)\hat{H}\phi_m(q_i)\,dq_i \\ &= E_m\delta_{lm}\end{aligned} \qquad [4.59]$$

then

$$\hat{H} = \sum_m E_m \hat{b}_m^+ \hat{b}_m \qquad [4.60]$$

Before we go on let us clear up a possible source of confusion. The $\hat{b}_m^+$, $\hat{b}_m$ refer to an eigenstate, and the expressions for the operator transformation (4.56) apply equally to Bose or fermion systems. The eigenstates for the quantized field are not, however, created by the creation operators in the above sense. The creation operator $\hat{a}_k^+$ of Chap. 3 changes one eigenstate of wave vector k into another. It is because of the constant energy and momentum packets that we *prefer* to consider $\hat{a}_k^+$, $\hat{a}_k$ as creating and destroying elementary excitations. Since both approaches serve to define a set of quantum numbers specifying the state, there is no real conflict, but it is well to remember that if we wished to use Eq. (4.60) for the bosons created by the quantization of the field, we would have to change the definition of the operators. In practice the only effect of the change is to replace

$$(m\hbar\omega_k + \tfrac{1}{2})\hat{b}_{m,k}^+ \hat{b}_{m,k} \quad \text{by} \quad \hbar\omega_k(\hat{a}_k^+ \hat{a}_k + \tfrac{1}{2})$$

where we need two labels m and k to specify the eigenstate and the normal mode but only one to specify the elementary excitation.

Thus, the use of creation and annihilation operators can be extended to particles as well as fields. For historical reasons this formalism is often termed the *second quantization formulation.* What we have done is really replace the wave function by a background framework (the vacuum state) and creation and annihilation operators. Now, in evaluating matrix elements between states, we see very clearly that we have not gained anything in the amount of *algebra* required; we still need to evaluate the set O_{lm}, but we do gain conceptually.

Going back to our definition of the representation, we see that it does not matter whether we connect the creation operator b_m^+ with the time-dependent state $\phi_m(q_i)\exp[(-iE_m t)/\hbar]$ or the time-independent state as long as we keep all of the commutators as equal time relationships (as in the electromagnetic field). Thus, the occupation number representation can be used in Heisenberg or Schrödinger representation, though following the policy of removing as much information as possible from the wave function, it is more natural to use the Heisenberg representation. The equation of motion for the operators is then

$$\frac{\hbar}{i}\frac{\partial}{\partial t}\hat{b}_m = [\hat{H},\hat{b}_m] \tag{4.61}$$

Substituting from (4.60) we have

$$\begin{aligned}\frac{\hbar}{i}\frac{\partial}{\partial t}\hat{b}_m &= \sum_n E_n(\hat{b}_n^+\hat{b}_n\hat{b}_m - \hat{b}_m\hat{b}_n^+\hat{b}_n)\\ &= \sum_n E_n(\pm\hat{b}_n^+\hat{b}_m\hat{b}_n - \hat{b}_m\hat{b}_n^+\hat{b}_n)\\ &= -\sum_n E_n\hat{b}_n\delta_{nm}\\ &= -E_m\hat{b}_m =\end{aligned} \tag{4.62}$$

That is

$$\hat{b}_m(t) = \hat{b}_m(0)e^{(-iE_mt)/\hbar} \tag{4.63a}$$

and

$$\hat{b}_m^+(t) = \hat{b}_m^+(0)e^{(iE_mt)/\hbar} \tag{4.63b}$$

for both boson and fermion systems. Thus, we also have our time-dependent operator $\hat{O}_H(t)$ in the form

$$\hat{O}_H(t) = \sum_{l,m} O_{lm}\hat{b}_l^+(t)\hat{b}_m(t) \tag{4.64}$$

or

$$\hat{O}_H(t) = \sum_{l,m} O_{lm}\hat{b}_l^+\hat{b}_m e^{[i(E_l-E_m)t]/\hbar} \tag{4.65}$$

and so on.

4.5. INTERACTION BETWEEN WAVES AND PARTICLES

Now that we have a consistent formulation for particles and waves, let us see how we can treat the interaction between them. It is best to consider a specific example first. Consider a free particle (i.e., electron) moving in an electromagnetic field. The Hamiltonian consists of two parts:

(i) The Hamiltonian of the electromagnetic field, which we write as

$$\hat{H}_{em} = \sum_{\lambda} \hbar \omega_{\lambda} \left(\hat{a}_{\lambda}^{+} \hat{a}_{\lambda} + \frac{1}{2} \right) \tag{4.66}$$

(ii) The particle Hamiltonian is

$$H_{\text{PART}} = \frac{1}{2m} \left(\mathbf{p} - \frac{e\mathbf{A}}{c} \right)^2 \tag{4.67}$$

This in turn can be expanded to give

$$\begin{aligned} H_{\text{PART}} &= \frac{|\mathbf{p}|^2}{2m} + \frac{e^2 |\mathbf{A}|^2}{2mc^2} - \frac{e}{2mc} (\mathbf{p} \cdot \mathbf{A} + \mathbf{A} \cdot \mathbf{p}) \\ &= H_{\text{FREE PART}} + H_{\text{INT}} \end{aligned} \tag{4.68}$$

Thus, the total Hamiltonian consists of three parts, the Hamiltonian of the individual components and the interaction Hamiltonian H_{INT}. Now from Eqs. (3.59a) and (3.68)

$$\mathbf{A}(\mathbf{r},t) = \sum_{\lambda} A_{\lambda}(t) \mathbf{e}_{\lambda} e^{i\mathbf{k}_{\lambda} \cdot \mathbf{r}} \tag{3.59a}$$

and so

$$\hat{\mathbf{A}}(\mathbf{r},t) = \sum_{\lambda} e^{i\mathbf{k}_{\lambda} \cdot \mathbf{r}} \mathbf{e}_{\lambda} \frac{1}{i} \left(\frac{\hbar \mu_0 c^2}{2\omega_{\lambda}} \right)^{1/2} (\hat{a}_{-\lambda}^{+} - \hat{a}_{\lambda}) \tag{4.69}$$

We can thus write the interaction Hamiltonian, remembering that in free space the polarization $\mathbf{e}_\lambda$ is perpendicular to the direction of propagation (in the $\mathbf{k}_\lambda$ direction).

$$\hat{H}_{\text{INT}} = -\sum_{\lambda,j} e^{i(\mathbf{k}_\lambda\cdot\mathbf{r}+\mathbf{k}_j\cdot\mathbf{r})}\mathbf{e}_\lambda\cdot\mathbf{e}_j\left(\frac{\hbar\mu_0}{4m}\right)\frac{1}{(\omega_\lambda\omega_j)^{1/2}}(\hat{a}_{-\lambda}^{+}-\hat{a}_\lambda)(\hat{a}_{-j}^{+}-\hat{a}_j)$$
$$+\sum_{\lambda}\left(e^{i\mathbf{k}_\lambda\cdot\mathbf{r}}\mathbf{e}_\lambda\cdot\frac{\hbar}{i}\frac{\partial}{\partial\mathbf{r}}\right)\frac{1}{j}\left[\frac{\hbar\mu_0c^2}{2\omega_\lambda}\right]^{1/2}(\hat{a}_{-\lambda}^{+}-\hat{a}_\lambda) \quad [4.70]$$

Both of these terms consist of two parts: a spatial part, which acts on the electrons, and an operator part which acts on the field. Consider the electron part and define two operators $\hat{R}^\lambda$, $\hat{S}^{\lambda,j}$

$$\hat{R}^\lambda = \frac{\hbar}{i}e^{i\mathbf{k}_\lambda\cdot\mathbf{r}}\mathbf{e}_\lambda\cdot\frac{\partial}{\partial\mathbf{r}} \quad [4.71]$$

$$\hat{S}^{\lambda,j} = e^{i(\mathbf{k}_\lambda\cdot\mathbf{r}+\mathbf{k}_j\cdot\mathbf{r})} \quad [4.72]$$

We can take as a suitable basis set for the free particle the plane-wave states

$$\phi_l(\mathbf{r}) = \frac{1}{\sqrt{V}}e^{i\mathbf{l}\cdot\mathbf{r}} \quad [4.73]$$

The free-particle Hamiltonian becomes

$$\hat{H}_{\text{FREE PART}} = \sum_{l}\frac{\hbar^2l^2}{2m}\hat{c}_{\mathbf{l}}^{+}\hat{c}_{\mathbf{l}} \quad [4.74]$$

and then

$$\hat{R}^\lambda = \sum_{l,m} R_{\mathbf{lm}}^{\lambda}\hat{c}_{\mathbf{l}}^{+}\hat{c}_{\mathbf{m}} \quad [4.75]$$

$$\hat{S}^{\lambda j} = \sum_{l,m} S_{lm}^{\lambda j}\hat{c}_{\mathbf{l}}^{+}\hat{c}_{\mathbf{m}} \quad [4.76]$$

where

$$R^{\lambda}_{\mathbf{l},\mathbf{m}} = \frac{\hbar}{Vi} \int e^{-i\mathbf{l}\cdot\mathbf{r}} e^{i\mathbf{k}_\lambda\cdot\mathbf{r}} \mathbf{e}_\lambda \cdot \frac{\partial}{\partial \mathbf{r}} e^{i\mathbf{m}\cdot\mathbf{r}}\, d\mathbf{r}$$
$$= \frac{\hbar}{V} \mathbf{e}_\lambda \cdot \mathbf{m}\, \delta(\mathbf{k}_\lambda + \mathbf{m} - \mathbf{l}) \quad [4.77]$$

and

$$S^{\lambda j}_{\mathbf{lm}} = \frac{1}{V} \delta(\mathbf{k}_\lambda + \mathbf{k}_j + \mathbf{m} - \mathbf{l}) \quad [4.78]$$

Subsuming all of the constants into two functions $\alpha^{\lambda j}_{\mathbf{lm}}$, $\beta^{\lambda j}_{\mathbf{lm}}$, the interaction Hamiltonian can now be written as

$$\hat{H}_{\mathrm{INT}} = \sum_{\substack{\lambda,j\\l,m}} \alpha^{\lambda j}_{\mathbf{lm}}\, \delta(\mathbf{k}_\lambda + \mathbf{k}_j + \mathbf{m} - \mathbf{l}) \hat{c}^{+}_{\mathbf{l}} \hat{c}_{\mathbf{m}} (\hat{a}^{+}_{-\lambda} - \hat{a}_\lambda)(\hat{a}^{+}_{-j} - \hat{a}_j)$$
$$+ \sum_{\substack{\lambda,j\\l,m}} \beta^{\lambda j}_{\mathbf{lm}}\, \delta(\mathbf{k}_\lambda + \mathbf{m} - \mathbf{l}) \hat{c}^{+}_{\mathbf{l}} \hat{c}_{\mathbf{m}} (\hat{a}^{+}_{-\lambda} - \hat{a}_\lambda) \quad [4.79]$$

In both terms the momentum conservation is explicit [cf. (3.81)], and the "active" parts consist of products of electron and photon operators along with a matrix element for the process. We will consider the effect of the interaction on the total eigenstate, which has the form

$$|\text{electron states; photon states}\rangle$$

(a) $\mathbf{A} \cdot \mathbf{p}$ interaction
   (i) $\hat{c}^{+}_{\mathbf{l}} \hat{c}_{\mathbf{m}} \hat{a}^{+}_{-\lambda}$: the creation of a photon (i.e., emission by the electron) with a subsequent scattering of the electron from state **m** to state **l**.
   (ii) $\hat{c}^{+}_{\mathbf{l}} \hat{c}_{\mathbf{m}} \hat{a}_{\lambda}$: the absorption of a photon with subsequent scattering of the electron from state **m** to state **l**.

(b) A^2 interaction
   Typical terms are
   (i) $\hat{c}^{+}_{\mathbf{l}} \hat{c}_{\mathbf{m}} \hat{a}^{+}_{-\lambda} \hat{a}^{+}_{-j}$: the emission of two photons by the electron with a scattering of the electron from state **m** to **l**.

(ii) $\hat{c}_l^+ \hat{c}_m \hat{a}_{-\lambda}^+ \hat{a}_j$: The scattering of both the electron and the photon.

(iii) $\hat{c}_l^+ \hat{c}_m \hat{a}_\lambda \hat{a}_{-j}^+$: the scattering of the electron and photon as in (ii).

(iv) $\hat{c}_l^+ \hat{c}_m \hat{a}_\lambda \hat{a}_j$: the absorption of two photons by the electron with a resulting scattering.

These are all illustrated schematically in Fig. 4.2. We could, of course, have considered the interaction as a perturbation on either the electron or photon system, but the description in terms of a series of scattering events is much more useful.

In a general system we will usually be presented with a Hamiltonian in the form

$$H = H_{\text{PART}}(q_i, p_i) + H_{\text{FIELD}}(\phi, \pi) + H_{\text{INT}}(q_i, p_i, \phi, \pi) \quad [4.80]$$

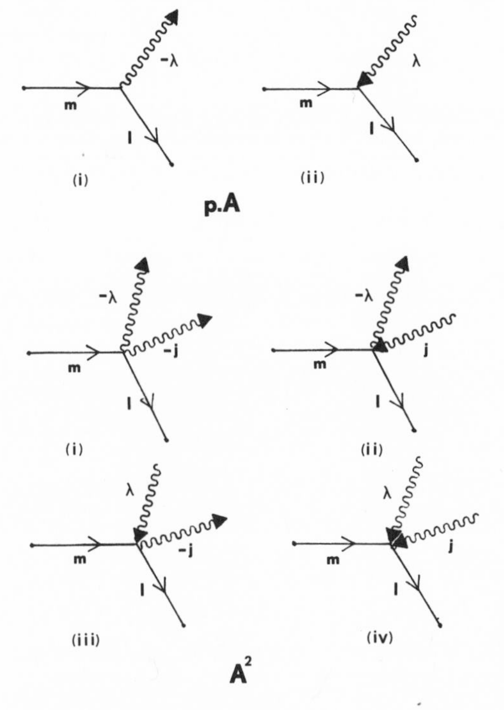

FIGURE 4.2. Possible interactions between electrons and photons as described in the text.

where q_i, p_i are particle variables ϕ, π are those of the field. On quantizing the system we will be left with an interaction term

$$\hat{H}_{\mathrm{INT}} = \sum_{\substack{l,m,\ldots \\ p,q,\ldots}} M(l,m,\ldots)f(\hat{a}_l^+,\hat{a}_m,\ldots,\hat{c}_l^+,\hat{c}_m,\ldots) \qquad [4.81]$$

consisting of a matrix element (including conservation laws) and a set of creation and annihilation operators for the particles and field. Each term in the interaction Hamiltonian can then be considered as a scattering event. Just as for the anharmonic terms, if the scattering becomes very large the description in terms of an identifiable particle or field excitation becomes meaningless; in some cases it is then possible to define a new elementary excitation of the total system, but not always.

4.6. FIELD OPERATORS

Although creation and annihilation operators are, in some ways, excellent reformulations of quantum mechanics, they do suffer from the drawback of being defined in terms of a set of eigenstates and so do not have a spatial interpretation. The diagrams we drew, for instance, in Fig. 4.2 were meant to describe scattering events, but they were scattering from one eigenstate to another. It is much easier to visualize, in an interacting system, the process of scattering in quasiballistic terms. What is required is a form of creation and annihilation operator which will conserve the features of the occupation number representation yet allow the spatial development of the system to be shown explicitly.

The solution to this problem is the definition of a new operator, the *field operator*, in the form

$$\hat{\psi}(q_i,t) = \sum_m \phi_m(q_i)\hat{b}_m(t) \qquad [4.82a]$$

$$\hat{\psi}^+(q_i,t) = \sum_m \phi_m^+(q_i)\hat{b}_m^+(t) \qquad [4.82b]$$

These operators act upon the *occupation number* wave function; the eigenfunctions ϕ_m (which correspond of course to the operators $\hat{b}_m^+$, $\hat{b}_m$) must be thought of as amplitude and phase factors only, which weight the various creation and annihilation operators.

We can derive the properties of the field operators from those of the creation and annihilation operators. We will write out only the boson operator case and leave the fermion case as an exercise.

$$[\hat{\Psi}(q_i,t),\hat{\Psi}^+(q_i',t)] = \sum_{m,n} \phi_m(q_i)\phi_n^+(q_i')[\hat{b}_m(t)\hat{b}_n^+(t) - \hat{b}_n^+(t)\hat{b}_m(t)]$$
$$= \sum_{m,n} \phi_m(q_i)\phi_n^+(q_i')\,\delta_{mn}$$
$$= \sum_{m} \phi_m(q_i)\phi_m^+(q_i')$$

Therefore

$$[\hat{\Psi}(q_i,t),\hat{\Psi}^+(q_i',t)] = \delta(q_i - q_i') \qquad [4.83]$$

Similarly

$$[\hat{\Psi}(q_i,t),\hat{\Psi}(q_i',t)] = [\hat{\Psi}^+(q_i,t)\hat{\Psi}^+(q_i',t)] = 0 \qquad [4.84]$$

The next task is to write the other operators in terms of the field operators. The general single-particle operator relation was

$$\hat{O}(t) = \sum_{l,m} O_{lm}\hat{b}_l^+(t)\hat{b}_m(t) \qquad [4.85]$$

Consider the expression

$$\int dq_i\, \hat{\Psi}^+(q_i,t)\hat{O}_S(q_i)\hat{\Psi}(q_i,t)$$

Substituting for the field operators

$$\int dq_i\, \hat{\Psi}^+(q_i,t)\hat{O}_S(q_i)\hat{\Psi}(q,t) = \sum_{l,m}\left(\int dq_i\, \phi_l^+(q_i)\hat{O}_S(q_i)\phi_m(q_i)\right)\hat{b}_l^+(t)\hat{b}_m(t)$$
$$= \sum_{l,m} O_{lm}\hat{b}_l^+(t)\hat{b}_m(t)$$

So in terms of the field operators

$$\hat{O}(t) = \int dq_i\, \hat{\Psi}^+(q_i,t)\hat{O}(q_i)\hat{\Psi}(q_i,t) \qquad [4.86]$$

in particular

$$\hat{H}(t) = \int dq_i\, \hat{\psi}^+(q_i,t)\hat{H}(q_i)\hat{\psi}(q_i,t) \qquad [4.87]$$

Like any other operator, the field operators must satisfy the Heisenberg equation of motion.

$$\frac{\hbar}{i}\frac{\partial\hat{\psi}}{\partial t}(q,t) = [\hat{H},\hat{\psi}(q,t)] \qquad [4.88]$$

From Eq. (4.87)

$$\frac{\hbar}{i}\frac{\partial}{\partial t}\hat{\psi}\,(q,t) = \int dq'\,[\hat{\psi}^+(q',t)\hat{H}(q')\hat{\psi}(q',t)(q,t) - \hat{\psi}(q,t)\hat{\psi}^+(q',t)\hat{H}(q')\hat{\psi}(q',t)]$$

Using the commutator relationships (4.83) and (4,84)

$$\begin{aligned}\frac{\hbar}{i}\frac{\partial}{\partial t}\hat{\psi}(q,t) &= \int dq'\,[\hat{\psi}^+(q',t),\hat{\psi}(q,t)]\hat{H}(q')\hat{\psi}(q',t)\\ &= -\hat{H}(q)\hat{\psi}(q,t)\end{aligned}$$

or

$$\hat{H}(q)\hat{\psi}(q,t) = i\hbar\,\frac{\partial}{\partial t}\hat{\psi}(q,t) \qquad [4.89]$$

Thus, the field operator satisfies Schrödinger's equation but acts upon the occupation number representation. This is obviously an important development. Let us consider the implications further. Since the vacuum state is an arbitrary framework and is not used explicitly, it means that all of the spatial and time dependence of the system now resides in the field operator whose equation of motion is the Schrödinger equation. In fact, we have come full circle, in a sense, because the field operator in many respects appears to be the Schrödinger wave function interpreted as an operator. The fact that it is an operator which acts in the occupation number representation is, however, of vital importance. The Schrödinger wave functions do not, for instance, include the statistics of the system inherent in the commutator of Eq. (4.83) and its fermion anticommutator equivalent. Furthermore, from the form of the field operator, its action upon a system is to add or subtract a particle to it. This follows from the fact that in

$$\hat{\psi}^{+}(\mathbf{r},t) = \sum_{m} \phi_m^*(\mathbf{r})\hat{b}^{+}(t) \qquad [4.90]$$

$\hat{b}_m^+$ acts to create the particle, and the total probability over all states is given by

$$\sum_{m} |\phi_m(\mathbf{r})|^2 = 1 \qquad [4.91]$$

In the case of interacting systems this is important because by studying how the system reacts to that extra particle a great deal may be learned about it.

BIBLIOGRAPHY

MESSIAH, A., *Quantum Mechanics,* North Holland, Amsterdam, 1965.
FEYNMAN, R. P., *Quantum Electrodynamics,* Benjamin, Menlo Park: CA, 1961.

PROBLEMS

1. Show that the two-particle interaction $v(\mathbf{r}_1,\mathbf{r}_2)$ is given in terms of the creation and annihilation operators as

$$\hat{v} = \sum_{m,n,p,q} v_{mnpq}\hat{a}_m^+\hat{a}_n^+\hat{a}_q\hat{a}_p$$

where

$$v_{mnpq} = \int \phi_m^*(\mathbf{r}_1)\phi_n^*(\mathbf{r}_2)v(\mathbf{r}_1,\mathbf{r}_2)\phi_p(\mathbf{r}_2)\phi_q(\mathbf{r}_1)\, d\mathbf{r}_1\, d\mathbf{r}_2$$

and in terms of the field operators by

$$\hat{v} = \int \hat{\psi}^+(\mathbf{r}_1)\hat{\psi}^+(\mathbf{r}_2)v(\mathbf{r}_1,\mathbf{r}_2)\hat{\psi}(\mathbf{r}_2)\hat{\psi}(\mathbf{r}_1)\, d\mathbf{r}_1\, d\mathbf{r}_2$$

2. Show that if an operator commutes with the Hamiltonian, the quantity corresponding to it is a constant of the motion.

3. Prove the identity

$$e^{-\lambda\hat{H}}\hat{b}_n^+ e^{+\lambda\hat{H}} = e^{-\lambda E_n}\hat{b}_n^+$$

for the Hamiltonian

$$\hat{H} = \sum_n E_n \hat{b}_n^+ \hat{b}_n$$

4. Confirm that the anticommutator relationships equivalent to Eqs. (4.83) and (4.84) do hold for fermion field operators. Show also that the equation of motion for the fermion field operators is still given by Eq. (4.89).

5. Show that the two relationships for bosons

$$\hat{H} = \sum_k \hbar\omega_k \left(\hat{b}_k^+ \hat{b}_k + \frac{1}{2} \right)$$

and

$$\hat{H} = \sum_{k,n} E_{k,n} \hat{d}_{k,n}^+ \hat{d}_{k,n}$$

(where $E_{k,n}$ is the energy of the nth eigenstate of the kth normal mode and $\hat{d}_{k,n}^+ \hat{d}_{k,n}$ are the creation and annihilation operators corresponding to that eigenstate) are compatible.

CHAPTER FIVE

INTERACTING SYSTEMS AND QUASIPARTICLES

5.1. SINGLE-PARTICLE STATES

The "classical" picture we have of a solid consists of the ion cores fixed in a matrix around which the interacting electrons are essentially free to move throughout the whole crystal. In quantum terms we would like to replace this, as far as possible, by a set of single-particle states (in the sense of the Hartree or Hartree–Fock approximation), so that we could calculate the various properties of the solid. The true eigenstates of the system are too difficult to work with, however, and, even if we knew them, they would certainly not be single-particle states. The desire to use a single-particle approach, coupled with the real success of the single-particle calculations of the properties of solids, requires us, as far as possible, to think in terms that approximate the ideal single-particle system. Suppose we derived a set of single-particle states, then a useful criterion as to how good they were would be how long the electron was able to remain in that state.

Consider starting from the independent-particle system, then as the interaction was slowly turned on, the particles would start to be scattered from their original states and gradually the true many-particle eigenfunctions would be formed. It is not trivial, however, to ask in the many-electron system whether there are any states which are single-particle in nature that approximate the true states of the system. This would require that the electron, on being introduced into those states, remain there for a time comparable to some process time (e.g., the application and removal of an external probe). Supposing, for instance, we take the related case of a noninteracting electron system in which electron–

phonon scattering forms a means of both energy loss and momentum change. The initial and final energy and momentum $E(\mathbf{k}_1)$, $\hbar\mathbf{k}_1$, $E(\mathbf{k}_2)$, $\hbar\mathbf{k}_2$, respectively, will be related by

$$E(\mathbf{k}_1) - E(\mathbf{k}_2) = \hbar\omega(\mathbf{q}) \qquad [5.1]$$

$$\hbar\mathbf{k}_1 - \hbar\mathbf{k}_2 = \hbar\mathbf{q} \qquad [5.2]$$

where $\hbar\mathbf{q}$ and $\omega(\mathbf{q})$ are the momentum and frequency of the phonon. Then, on introducing the electron at an energy well above the Fermi surface, the electron will gradually lose both energy and momentum by repeated phonon emission until the electron cannot lose any further energy, simply by virtue of being within a phonon energy of the Fermi surface. Once this situation is reached, even though the electron–phonon interaction remains, the electron is forbidden by conservation laws from scattering out of the state. Thus, the states close to the Fermi surface will be stable against the phonon interaction and essentially the same as if the phonon emission process did not exist.

Now this is quite different from the situation considered in simple quantum mechanics. If there are two states which interact, one forms combinations of those states which are noninteracting; i.e., if we have two states ψ_0, ψ_1 such that an interaction term gives

$$\langle\psi_0|H_{\mathrm{INT}}|\psi_1\rangle = \alpha \qquad [5.3]$$

then combinations can normally be formed

$$\phi_0 = a\psi_0 + b\psi_1 \qquad [5.4a]$$

$$\phi_1 = c\psi_0 + d\psi_1 \qquad [5.4b]$$

such that

$$\langle\phi_0|\phi_1\rangle = 0; \qquad \langle\phi_0|\phi_0\rangle = \langle\phi_1|\phi_1\rangle = 1 \qquad [5.5]$$

and

$$\langle\phi_0|H_{\mathrm{INT}}|\phi_1\rangle = 0 \qquad [5.6]$$

Thus, the eigenstates ψ_0, ψ_1 are replaced by new eigenstates ϕ_0, ϕ_1 on the introduction of the interaction, and everything conceptually is unchanged apart from some relabeling. If we try to do this in the electron–phonon case, however, we have the following difficulties:

1. The electron–phonon interaction does not lead us to form new eigenstates because there is no requirement that the electron will remain sufficiently long in the scattered state to interact again (i.e., absorb) with the same phonon; i.e., cascade processes are the order of the day.
2. The phonons are bosons and their equilibrium number is a function of the temperature; i.e., the boson emitted by the electron will not remain as it is but will eventually become part of an equilibrium phonon distribution.

The result is that if an electron has energy to lose, it will lose it by emission of phonons; the absorption of phonons will, however, be a rare event unless their density is comparable to that of the electron—a very unlikely situation at normal temperatures. The phonons thus act as a sponge for the energy, which is only inoperative when conservation laws prevent the electron–phonon interaction process from taking place. The second consequence is that there will be no easy determination of the electron's final state from the initial conditions, since there will be many possible combinations in the events needed to bring it down to the Fermi level. It is thus not useful to try to consider the electron after it has been scattered from its initial state. To try to do so is comparable with trying to investigate an essentially statistical system by way of a causal approach. Such attempts are never successful. It would thus seem that the lifetime in the original state is probably the only useful thing we can say about it. If it is long, in the region of the Fermi surface, then it forms a good single-particle state; the shorter it gets though the less use it will be (unless the process we wish to consider has an even shorter time scale, that is).

As an introduction to the idea of quasiparticles (or effective single-particle states) we should then consider the consequences of a situation in which the states have a finite lifetime. We will do this first by reference to a simple phenomenological model system.

5.2. ABSORBING MEDIA

Suppose we take a simple one-particle system in which we have an electron moving in a one-dimensional potential well. We have

$$V(x) = V, \quad 0 < x < a$$
$$= \infty, \quad x < 0, x > a$$

The solution of the Schrödinger equation

$$\left[\frac{\hbar^2}{2m}\frac{d^2}{dx^2} + (E - V(x))\right]\psi(x) = 0 \qquad [5.7]$$

is simply given by

$$\psi(x) = \left(\frac{2}{a}\right)^{1/2} \sin kx, \quad 0 < x < a \qquad [5.8]$$

with

$$k^2 = \frac{2m}{\hbar^2}(E - V) \qquad [5.9]$$

and

$$k = \frac{n\pi}{a} \qquad [5.10]$$

provided V is real. If V is complex how do we interpret the solutions? If we go through the motions, the general solution of Eq. (5.7) is in the form

$$\phi = Ae^{ikx} \qquad [5.11]$$

with k given by

$$k = \pm\frac{2m}{\hbar^2}(E - V_R - iV_I)^{1/2} \qquad [5.12a]$$

i.e., k is of the form

$$k = \pm(l + im) \qquad [5.12b]$$

Now no combination of solutions of the form in Eq. (5.11), for a given energy, will satisfy the boundary condition. We are forced, therefore, to maintain the reality of k and allow the energy to become complex. The solution is then

$$E = E_R + iV_I \qquad [5.13]$$

$$= E_R - i|E_i| \qquad [5.14]$$

$$k = \pm \frac{2m}{\hbar^2}(E_R - V_R)^{1/2}$$

with E_R determined by the boundary condition through Eq. (5.10). Thus, an absorbing medium, characterized by a complex potential, requires a complex energy to describe an eigenstate. (This is entirely different from the case of say an evanescent (surface-state) wave in a solid where we require a real eigenvalue to satisfy the boundary condition, which causes k to be complex.) The eigenfunction now becomes

$$\begin{aligned} \psi(x,t) &= \left(\frac{2}{a}\right)^{1/2} (\sin kx)e^{(-iEt)/\hbar} \\ &= \left(\frac{2}{a}\right)^{1/2} [(\sin kx)e^{(-iE_Rt)/\hbar}]e^{(-Eit)/\hbar} \end{aligned} \qquad [5.15]$$

which satisfies our intuitive feeling for an oscillation decreasing in intensity with time due to the absorption. It is worth stressing the difference between the evanescent case and the absorptive medium. If we took, for instance, the case of an infinite medium so that we were "untroubled" by boundary conditions, then the k value determining the eigen solutions would still not be of the form of Eq. (5.12b), but rather the real, running waveform. This is important when we come to consider wave packets and their propagation. The confusion usually arises because if we have a wave function of the form

$$\psi(x,t) = e^{ikx}e^{(-iEt)/\hbar} \qquad [5.16]$$

then the substitutions

$$E = E_R - i|E_i| \qquad [5.17]$$
$$t = x/v$$

gives

$$\psi(x,t) \approx e^{[i(k+iE_i)/\hbar v]x} e^{(-iE_R t)/\hbar} \qquad [5.18]$$
$$\approx e^{i(k+i\gamma)x} e^{(-iE_R t)/\hbar}$$

which is the same form as the wave functions we would have had if we had demanded that the energy be real. By keeping the energy complex, however, it is much clearer that the amplitude of the wave always decays in the direction of the propagation of the particle.

5.3. EXACT AND APPROXIMATE EIGENSTATES

We introduced an absorbing medium into consideration as an artifact, so perhaps we should consider what it means and where it could come about. The complex potential V had the effect of removing the particle from the state in which it had been put (i.e., decreasing $|\psi|^2$). There are two connected ways in which this can come about. The example we have already mentioned was the case of a state in which the particle (electron) is scattered out of the state due to its interactions with some other particle. Provided there is some other energy loss mechanism associated with that other particle then there is no requirement that it be scattered back in with an equal probability. The Hamiltonian for the electron then becomes, on introducing the scattering

$$\hat{H} = \sum_k E(k)\hat{a}_k^+ \hat{a}_k + \sum_{l,k} \alpha_l \hat{a}_l^+ \hat{a}_k + \beta_l \hat{a}_k^+ \hat{a}_l \qquad [5.19]$$

with no requirement that the α, β be the same. The Hamiltonian becomes non-Hermitian and there is no necessity that the eigenvalues be real—the energies may be and probably will be complex. A similar situation occurs in the excited states of atoms. The decay process is essentially one way, with no equilibrium between the electron and the electromagnetic field. The complex energy, in that case, shows up as a "spread" in energy of possible emitted photons giving a finite linewidth dependent upon the lifetime of the state.

Another case, much simpler, can be illustrated by considering the real one-dimensional potential. We have a set of eigenfunctions and eigenvalues, as before, which are well known. But supposing we had no way of knowing these functions and instead were under the impression that the eigenfunction was, say, the function

$$\Psi_n(x,i) = \frac{1}{\sqrt{a}} A_n(x) \sin nkxe^{(-iE_nt)/\hbar}, \qquad 0 < x < a \qquad [5.20]$$

[where $A_n(x)$ is some known function of x and measures the deviation of the supposed eigenfunction from the "unknown" exact one]. With an exact eigenfunction we would expect the product $\Psi_n(x,t)\Psi_n^*(x,t)$ to be a constant with respect to time. In the case of a "false" eigenfunction, however, we would have a time variation reflecting the change in the relative phase relationships of the component (true) eigenfunctions. We have, expanding the supposed eigenfunction

$$\Psi_n\,(x) = \sum_n \alpha_{nj} \sin jkx \qquad [5.21]$$

with

$$\alpha_{nj} = \int_0^a A_n(x) \sin jkx \sin nkx \; dx \qquad [5.22]$$

This gives a time variation in the product

$$\Psi_n(x,t)\Psi_n^*(x,t) = \sum_{l,j} \alpha_{nj}\alpha_{nl}^* \sin jkx \sin lkxe^{-[i(E_j-E_l)t]/\hbar} \qquad [5.23]$$

This can be rewritten in the form

$$\Psi_n(x,t)\Psi_n^*(x,t) = \Psi_n(x,0)\Psi_n^*(x,0)F(x,t) \qquad [5.24]$$

Where the function of time $F(x,t)$ will depend upon the precise form of the eigenfunction assumed. The important point, however, is that in the absence of any knowledge of the exact eigenstates, the presence of a time dependence in the wave function product would be interpreted as a decay process and would, in a linearized form, be written in terms of an absorptive part of the potential. In the example we have taken $F(x,t)$ must of necessity be periodic, but in the limit of a large system, as in the transition from Fourier series to transform, this periodicity disappears and all the observer would be able to deduce would be the average lifetime in the state. The particle would not return to its original state in the time scale of the observation.

The analogy to the system we have is obvious—in a real solid we can never know the exact eigenstates. When we talk of putting, or considering, the electron in an eigenstate what we really mean is that it is the nearest thing we can consider to a real eigenstate. One-electron states do not really exist so we would expect the system to develop and change as the wave packet of eigenstates, which made up the inexact state, develop at their own rate. If the rate of development is slow compared to the process we are considering, however, there is no reason why we should not consider that inexact eigenstate as a perfectly useful entity.

The two cases we have considered are, of course, different aspects of the same problem. If we took the one-electron eigenstates as our basis and then turned on the electron–electron interaction, we could consider the lifetime of the electron state as being either due to the probability of being scattered out of it by the interactions with the other electrons, or as a consequence of the continual mixing in of other eigenfunctions (again due to the interaction) leading to an N-particle eigenfunction which cannot be treated. One query which must arise is, why should we not put the electron in the exact eigenstate or at least consider the exact eigenstates? There are a number of reasons for this, the principal one is, of course, that they are just too complicated to evaluate useful things like matrix elements. The second, which we have already mentioned, is that in many cases most of the information entailed in the true eigenfunctions is not relevant for the sort of properties measured in experiments. A third reason is that such eigenstates of the solid may not exist at all in the strict sense. If we have a solid, we can only define eigenstates if we have exact boundary conditions around the edges. These do not exist if we have the solid in thermal equilibrium with its surroundings; random mixing of the eigenstates must be the order of the day. The existence of a finite lifetime for the eigenstate then becomes comparable to the thermalization encountered in simple kinetic theory.

If we introduce a particle into a container of gas, how long is it before that particle becomes thermalized; that is, with an average energy characteristic of the system parameters rather than its initial conditions? The answer is a few scattering times at most. A further question might be, what is the overall effect on the system of the particle? Very little again. Within a short time the energy of the particle will have been dissipated, firstly among the other particles and then through the walls to the immediate environment. The correspondence between statistical mechanics and quantum mechanics is a strong one here, as it must be, both dealing with large systems with incomplete information. The idea of finite lifetime states and corresponding mixing, of even the many-particle eigenfunctions, has its roots in the *ergodic theorem* in statistical mechanics. This theorem states that any system will, given time, be found in all of the available configurations subject to the constraints of constant energy (or temperature).

The idea then of an *absorptive potential* is really a simple way of describing a number of possible causes for the existence of a finite lifetime for any single-particle state considered. What we have not done, however, is argue that such states should exist in general. Such an argument is not possible exactly, but we can indicate that it is reasonable. Supposing we start with a uniform system of N noninteracting fermion particles. Then they will have a Fermi level and a density distribution which is of the form

$$\begin{aligned} n(k) &= 1, \qquad k < k_F \\ &= 0, \qquad k > k_F \end{aligned}$$

Supposing now we turn on the interaction slowly. There are a number of effects which can be classified into two categories.

1. Interaction effects which tend to move the energy of the state without scattering and can be expressed in the form of a real effective one-electron potential.
2. Interaction effects which tend to scatter the electron out of the state and result in a finite lifetime for the state.

From our previous discussion it would seem that we should place the Hartree and (less obviously) the Hartree–Fock potentials in category 1 while phonon emission might go into the second. In general, any interaction might produce both. As we turn on the interaction we could argue, however, that there is one category of states which would be least affected by the interaction, those close to the Fermi level. Here the scattering possibility is highly restricted by the simple fact that most of the states which the electron would be scattered into are already filled. If the scattering is restricted, it follows that the lifetime must be high and will remain so as the interaction increases in strength—the exclusion principle will always dominate in this region. Thus, it seems plausible to argue that if one-electronlike (termed quasiparticle) states are going to exist, they will be close to the Fermi surface of the system. Fortunately, this is just the region which dominates many of the interesting properties, so we might expect the resulting quasiparticle state to be a useful entity.

Consider now a more practical aspect of a quasiparticle state; if we take the Hartree–Fock approximation to a uniform system we have, replacing all the states by plane-wave states and evaluating the integrals, that the energy of the particle is given by

$$E(k) = \frac{\hbar^2 k^2}{2m} + V_{\text{ex}}(k) + V_H \qquad [5.25]$$

where V_H is the constant Hartree potential, $V_{\text{ex}}(k)$ is the exchange energy given by

$$V_{\text{ex}}(k) = -\frac{e^2 k_F}{2\pi}\left[2 + \frac{(k_F^2 - k^2)}{k_F k} \ln\left|\frac{k_F + k}{k_F - k}\right|\right] \qquad [5.26]$$

The effective mass of a particle m^* in such a system, given by

$$\frac{1}{m^*} = \frac{1}{\hbar^2}\frac{\partial^2 E(k)}{\partial k^2} \qquad [5.27]$$

would thus be

$$\frac{1}{m^*} = \frac{1}{m} + \frac{1}{\hbar^2}\frac{\partial^2 V_{\text{ex}}(k)}{\partial k^2} \qquad [5.28]$$

This is now momentum dependent.

$$\frac{1}{m^*} = \frac{1}{m} + \frac{e^2}{\hbar^2 3\pi k_F} \qquad \text{for } k \ll k_F \qquad [5.29a]$$

$$= \frac{1}{m} + \frac{e^2}{\hbar^2 \pi \delta} \qquad \text{for } k = k_F - \delta \qquad (\delta \text{ small}) \qquad [5.29b]$$

Thus, the mass can vary from being less than the free electron value to zero as the state considered tends towards the Fermi energy. Thus, even though in this approximation the single-particle states are of infinite lifetime, the "mass" of the particle has changed as a result of the interaction. One can think of it physically as being the mass of the particle plus the inertial effects of the surrounding exchange hole created by the interaction, but one must be careful in such reasoning since one's physical intuition allows an increase, but not a decrease, in the mass. We obviously need a more consistent approach to the idea of quasiparticles and the development of their properties.

5.4. LANDAU QUASIPARTICLES

Suppose we have a set of single-particle states of energy $E(\mathbf{k})$, occupied according to some distribution function $n(\mathbf{k})$, where $\mathbf{k}$ is some convenient label, in this case the momentum. Then the total energy is given by

$$E = \sum_k E(\mathbf{k}) n(\mathbf{k}) \qquad [5.30]$$

If we measure deviations from the ground state, characterized by an energy E_0 and distribution function $n_0(k)$, the excitation energy can be written as

$$\Delta E = E - E_0 \qquad [5.31]$$
$$= \sum_k E(\mathbf{k})\, \Delta n(\mathbf{k})$$

where

$$\Delta n(\mathbf{k}) = n(\mathbf{k}) - n_0(\mathbf{k}) \qquad [5.32]$$

The essence of the *Landau theory* is that, for small deviations from the equilibrium, one can have a one-to-one correspondence with a *model* system of single-particle energy states. Thus, it is assumed that the new elementary excitations of the interacting system are close enough to those of a noninteracting system that one retains the idea of single-particlelike states, the quasiparticles. These quasiparticles are allowed to interact with each other, however. Thus, instead of Eq. (5.31) we put

$$\Delta E = \sum_k E_0(\mathbf{k})\, \Delta n(\mathbf{k}) + \frac{1}{2} \sum_{\mathbf{k},\mathbf{k}'} f(\mathbf{k},\mathbf{k}')\, \Delta n(\mathbf{k})\, \Delta n(\mathbf{k}') \qquad [5.33]$$

The excitation energy consists of two terms:

1. A single-particle term consisting of the sum of the additional quasiparticle energies.
2. A term consisting of the contribution from the quasiparticle interactions. Notice we do not consider the interaction with the ground-state "quasiparticles." This is subsumed in the $E_0(\mathbf{k})$. It is only deviations from the ground state which are important. The $f(\mathbf{k},\mathbf{k}')$ is considered as the interaction but one must be careful—it contains much more than the bare coulomb interaction, as we will see when we analyze the system by way of Green's functions and the array of many-body techniques.

Obviously this can only work if the quasiparticle concept is a useful one for the process we consider and if the number of quasiparticles remain small—there is no purpose in reducing one many-body problem to another one. It must also be clear to what extent we are sweeping the interactions under the carpet. As we have said before, it is deviations from the ground state which are important. By concentrating on those deviations and expressing them in terms of a quasiparticle distribution, the Landau theory succeeds in giving an empirical model of many properties.

An alternative way of expressing Eq. (5.33) is to consider the energy of the quasiparticle state in the presence of some distribution of quasiparticles $E(\mathbf{k})$; this is given by

$$E(\mathbf{k}) = \frac{\delta(\Delta E)}{\delta\, \Delta n(\mathbf{k})} = E_0(\mathbf{k}) + \sum_{k'} f(\mathbf{k},\mathbf{k}')\, \Delta n(\mathbf{k}') \qquad [5.34]$$

and says that the energy of the quasiparticle state is affected by the presence of the other quasiparticles. Thus, suppose we start with the ground state and add a quasiparticle in state **k** (Fig. 5.1), which then goes in at the energy $E_0(\mathbf{k})$. On introducing another particle in the state $\mathbf{k}'$ we have the contribution $E_0(\mathbf{k})$, which it would have on its own, plus the change due to the mutual interaction. Similarly, the energy that the second quasiparticle goes into is altered from the energy $E_0(\mathbf{k}')$ by the presence of the first.

In thermal equilibrium we would expect the quasiparticles to be governed by the appropriate Bose or Fermi statistics, i.e.

$$n(\mathbf{k}) = \{\exp[\beta(E(\mathbf{k}) - \mu)] \pm 1\}^{-1} \qquad [5.35]$$

where the $\pm$ refers to fermion or boson Landau quasiparticles. Since $E(\mathbf{k})$ is itself dependent upon $n(\mathbf{k})$, this is a self-consistent relationship.

There is one more simplification which can be made; all the major properties of solids involve changes from the equilibrium distribution located close to the Fermi surface, thus

$$f(\mathbf{k},\mathbf{k}') \approx f(\mathbf{k} - \mathbf{k}') \approx f(k_F,\theta) \qquad [5.36]$$

where θ is the angle between the two wave vectors. It is normal to separate the interaction term into a direct and exchange term

$$f(k_F,\theta) = f^d(k_F,\theta) + f^e(k_F,\theta) \qquad [5.37]$$

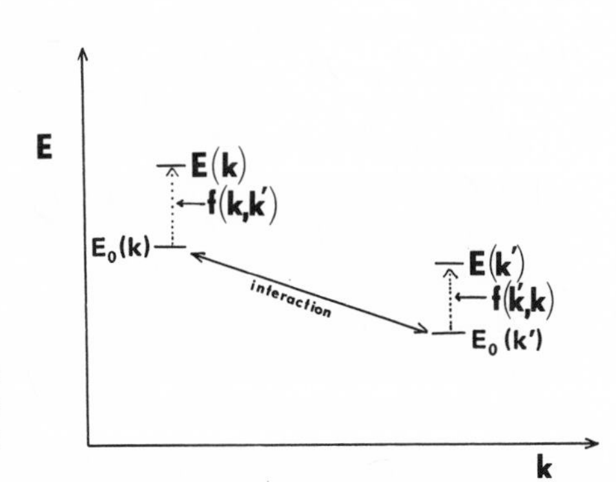

FIGURE 5.1. Effect of the interaction on the Landau quasiparticle energies for two particles introduced at momenta **k**, **k**′.

and then further separate these into

$$\begin{aligned} f^d + \tfrac{1}{2}f^e(k_F,\theta) &= a(k_F,\theta) \\ \tfrac{1}{2}f^e(k_F,\theta) &= b(k_F,\theta) \end{aligned} \qquad [5.38]$$

Summations involved in the calculation of the interaction terms will be of the form

$$\sum_{\mathbf{k}'} f(\mathbf{k},\mathbf{k}')\,\Delta n(\mathbf{k}') \Rightarrow \int d\theta\, f(k_F,\theta)\,\Delta n(k_F,\theta) \qquad [5.39]$$

so that it is natural to expand f in terms of its expansion in spherical harmonics.

$$\begin{Bmatrix} a(k_F,\theta) \\ b(k_F,\theta) \end{Bmatrix} = \sum_l (2l+1) \begin{Bmatrix} a_l \\ b_l \end{Bmatrix} P_l(\cos\theta) \qquad [5.40]$$

The a_l, b_l then form a set of parameters in which the details of the interaction term are subsumed. Suppose, for instance, we wished to calculate the effective mass m^* of the quasiparticle. We can do this by way of the general property of a *uniform system;* that is, the momentum of a unit volume of material must equal the flow of mass.

Consider, for instance, how we would view a system if we were traveling with respect to it at a constant velocity $\mathbf{v}$. This would be equivalent to changing the kinetic energy term in the Hamiltonian by replacing each coordination $\mathbf{p}_i$ by $\mathbf{p}_i - \delta\mathbf{p}$ ($\delta\mathbf{p} = m\mathbf{v}$). The total energy of the system would, therefore, appear to change and become a function of $\mathbf{v}$, i.e.,

$$E(\mathbf{v}) = \langle N|H|N\rangle = E_0 - \sum_i \left\langle N \middle| \frac{\hbar}{m}\delta\mathbf{p}\cdot\mathbf{p}_i \middle| N \right\rangle \qquad [5.41]$$

where $|N\rangle$ is the ground-state wave function.

We can define $\mathbf{p}_i/m$ as a velocity operator so that

$$E(\mathbf{v}) = E_0 - \left\langle N \middle| \sum_i \hbar\,\delta\mathbf{p}\cdot\mathbf{v}_i \middle| N \right\rangle \qquad [5.42]$$

or

$$E(\delta\mathbf{p}) = E_0 - \left\langle N \middle| \sum_i \hbar\, \delta\mathbf{p} \cdot \mathbf{v}_i \middle| N \right\rangle \qquad [5.43]$$

If we take a functional derviative of $E(\delta\mathbf{p})$ (i.e., consider $\delta\mathbf{p}$ as a variable) then

$$\begin{aligned} \frac{1}{\hbar}\frac{\partial E(\delta\mathbf{p})}{\partial(\delta\mathbf{p})} &= -\left\langle N \middle| \sum_i \mathbf{p}_i/m \middle| N \right\rangle \\ &= -\left\langle N \middle| \sum_i \mathbf{v}_i \middle| N \right\rangle \qquad [5.44] \\ &= -\left\langle N | \mathbf{V} | N \right\rangle \end{aligned}$$

This is trivial for a one-particle noninteracting state where

$$\mathbf{V} = \frac{\mathbf{p}}{m} = -i\frac{\hbar}{m}\frac{\partial}{\partial\mathbf{r}} \qquad [5.45]$$

so that in a state $|\mathbf{k}\rangle$, $\langle\mathbf{v}\rangle$ is just $\hbar\mathbf{k}/m$. The energy as a function of $\delta\mathbf{p}$ is

$$E(\delta\mathbf{p}) = \frac{\hbar^2}{2m}(\mathbf{k} - \delta\mathbf{p})^2 \qquad [5.46]$$

and

$$\frac{1}{\hbar}\frac{\partial E(\delta\mathbf{p})}{\partial(\delta\mathbf{p})} = -\frac{\hbar\mathbf{k}}{m} = \langle\mathbf{k}|\mathbf{v}|\mathbf{k}\rangle \qquad [5.47]$$

When we have more than one particle, in particular in the case of a large system, it is better to split the velocity operator and consider the contribution from each state; then we can write

$$E(\delta\mathbf{p}) = \sum_{\mathbf{k}} E(\mathbf{k}; \delta\mathbf{p}) \qquad [5.48]$$

$$\mathbf{V} = \sum_k \mathbf{v}_k \qquad [5.49]$$

$$\langle \mathbf{v}_k \rangle = -\frac{1}{\hbar}\frac{\partial E(\mathbf{k};\delta\mathbf{p})}{\partial(\delta\mathbf{p})} \qquad [5.50]$$

where $E(\mathbf{k})$ is the energy contribution from the state $\mathbf{k}$ and, in the Landau theory, will be given by Ex. (5.34). There are, thus, two contributions to the velocity expectation value

$$\langle \mathbf{v}_k \rangle = -\frac{1}{\hbar}\frac{\partial E_0(\mathbf{k};\delta\mathbf{p})}{\partial(\delta\mathbf{p})} + \frac{\partial}{\partial(\delta\mathbf{p})}\left(\sum_{k'} f(\mathbf{k},\mathbf{k}')\,\Delta n(k')\right) \qquad [5.51]$$

Considering the last term first, if one looks at the system from a moving reference frame it is equivalent to giving an added $-\delta\mathbf{p}$ to the momentum, so we essentially move the whole Fermi sphere by an amount $-\delta\mathbf{p}$ as illustrated in Fig. 5.2. This means that compared to the equilibrium Fermi surface there is an extra distribution of quasiparticles above and below the original Fermi level. We can obtain this by considering those particles whose energy has crossed the Fermi surface. From Eq. (5.34) we have to first order

$$\begin{aligned}\delta E(\mathbf{k}) &= \frac{\partial E_0(\mathbf{k})}{\partial(\delta\mathbf{p})}\cdot\delta\mathbf{p} \\ &= \hbar\,\langle \mathbf{v}_k^0 \rangle \cdot \delta\mathbf{p}\end{aligned} \qquad [5.52]$$

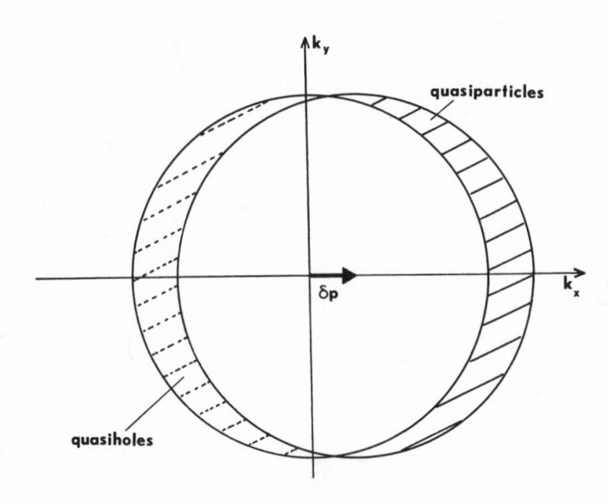

FIGURE 5.2. Change in the Fermi sphere due to the change of momentum showing the equivalent distribution of added quasiparticles and quasiholes.

Thus

$$\Delta n(\mathbf{k}') = \hbar \langle \mathbf{v}_{k'}^0 \rangle \cdot \delta\mathbf{p}\, \delta\{E(k') - \mu\} \tag{5.53}$$

since we are only interested in the limit of $\delta\mathbf{p} \to 0$. Thus, the expression for $\langle v_k \rangle$ becomes

$$\langle \mathbf{v}_k \rangle = \langle \mathbf{v}_k^0 \rangle + \sum_{k'} f(k,k')\langle \mathbf{v}_{k'}^0 \rangle\, \delta\{E(k') - \mu\} \tag{5.54}$$

In the second term summation we have

$$\frac{1}{(2\pi)^3} \int k'^2\, dk'\, d\theta\, d\phi\, |\langle \mathbf{v}_k^0 \rangle| \cos\theta\, \delta\{E(k') - \mu\} f(k_F,\theta)$$

Thus, $\langle \mathbf{v}_k \rangle$ contains two terms, the second of which is proportional to the second term in the interaction expansion, i.e., a. Therefore

$$\langle \mathbf{v}_k \rangle = \langle \mathbf{v}_k^0 \rangle + \langle \mathbf{v}_k^0 \rangle \alpha a_1 \tag{5.55}$$

It is usual to define an effective mass for the quasiparticle through

$$\frac{k_F}{m^*} = \frac{dE_0(\mathbf{k})}{d\mathbf{k}} \tag{5.56}$$

then we have

$$\frac{1}{m} = \frac{1}{m^*} + \frac{\alpha a_1}{m^*} \tag{5.57}$$

$$\frac{m^*}{m} = (a + \alpha a_1) \tag{5.58}$$

where keeping track of all of the numerical factors gives

$$\alpha = \frac{k_F m^*}{\pi^2} \tag{5.59}$$

It is also usual to amalgamate α with a_1 (and in fact all the a's and b's) so that we obtain a dimensionless parameter A_1, which can be derived from experiment. That is

$$\frac{m^*}{m} = 1 + A_1 \qquad [5.60]$$

Consider how we can physically interpret this: There is a great difference between an interacting and noninteracting system here. As the particle moves, it will tend to drag along with it other quasiparticles. One cannot build up a single localized wave packet to represent the motion. The change in effective mass represents this, since the extra dragged flow can contribute to the mass transport. As well as the effective mass, one can derive expressions for a whole series of equilibrium properties, but Landau theory has the great advantage that it can also treat the nonequilibrium properties by utilizing, essentially, the Boltzmann equation.

Supposing the system is disturbed on a macroscopic scale so that the energy, position, time, etc., can all be treated in a quasiclassical way (just as was done in the Thomas–Fermi example). It is natural to extend Eqs. (5.33) and (5.34) so that we have

$$\Delta E(\mathbf{k},\mathbf{r}) = \sum_{\mathbf{k}} E_0(\mathbf{k},\mathbf{r})\, \delta n(\mathbf{k},\mathbf{r}) + \frac{1}{2}\sum_{\mathbf{k},\mathbf{k}'} \int d\mathbf{r}\, d\mathbf{r}'\, f(\mathbf{k}',\mathbf{k}',\mathbf{r},\mathbf{r}')\, \delta n(\mathbf{k},\mathbf{r})\, \delta n(\mathbf{k}',\mathbf{r}') \qquad [5.61]$$

with

$$E(\mathbf{k},\mathbf{r}) = E_0(\mathbf{k},\mathbf{r}) + \sum_{\mathbf{k}'} \int d\mathbf{r}'\, f(\mathbf{k},\mathbf{k}',\mathbf{r},\mathbf{r}')\, \delta n(\mathbf{k}',\mathbf{r}') \qquad [5.62]$$

As it stands this equation is intractable, since the functions $E_0(\mathbf{k},\mathbf{r})$, $f(\mathbf{k},\mathbf{k},\mathbf{r},\mathbf{r}')$ will be different for each inhomogeneity, and though the result may be accurate, it would not serve as a useful approximation to most systems. It is general, therefore, to make the approximation

$$E_0(\mathbf{k},\mathbf{r}) \rightarrow E_0(\mathbf{k})$$
$$f(\mathbf{k},\mathbf{k}',\mathbf{r},\mathbf{r}') \rightarrow f(\mathbf{k},\mathbf{k}')$$

the same functions as in the homogeneous system. It corresponds to the Landau theory for the homogeneous system applying to each small real space element of the inhomogeneous system independently. We can see that this would be a reasonable thing to do if the quasiparticle was localized so that it did not extend over the same range as the inhomogeneity, and also, the form of the interaction did not change, i.e., the system was subject to only a small change. One can think of quite a wide range of situations where this might not hold; for example, at boundaries between different types of systems, i.e., surfaces and interfaces. Leaving out these quite specialized cases, we have

$$E(\mathbf{k},\mathbf{r}) = E_0(\mathbf{k}) + \sum_k f(\mathbf{k},\mathbf{k}')\,\delta n(\mathbf{k}',\mathbf{r}) \qquad [5.63]$$

so that the energy is a local function of the density. Now supposing we wish to consider the properties of the system under such conditions; there will be a nonequilibrium distribution

$$\delta n(\mathbf{k},\mathbf{r}) = n(\mathbf{k},\mathbf{r}) - n^0(\mathbf{k}) \qquad [5.64]$$

Since we have a one-to-one correspondence between model single-particle systems and the quasiparticles, one can use the idea of a gas of quasiparticles described by a classical Hamiltonian $E(\mathbf{k},\mathbf{r})$ so that

$$\frac{d\mathbf{r}}{dt} = \frac{\partial E(\mathbf{k},\mathbf{r})}{\partial \mathbf{k}}, \qquad \frac{d\mathbf{k}}{dt} = -\frac{\partial E(\mathbf{k},\mathbf{r})}{\partial \mathbf{r}} \qquad [5.65]$$

and Liouville's equation holds for the conservation of quasiparticles

$$\frac{\partial n(\mathbf{k},\mathbf{r})}{\partial t} + \frac{\partial n(\mathbf{k},\mathbf{r})}{\partial \mathbf{r}} = \frac{d\mathbf{r}}{dt} + \frac{\partial n(k,r)}{\partial \mathbf{k}} \cdot \frac{d\mathbf{k}}{dt} = 0 \qquad [5.66]$$

In terms of the nonequilibrium distribution we obtain

$$\frac{\partial[\delta n(\mathbf{k},\mathbf{r},t)]}{\partial t} + \frac{\partial}{\partial \mathbf{r}}[\delta n(\mathbf{k},\mathbf{r},t)] \cdot \frac{\partial E_0(\mathbf{k})}{\partial \mathbf{k}} - \frac{\partial n_0(\mathbf{k})}{\partial \mathbf{k}} \sum_{\mathbf{k}'} f(\mathbf{k},\mathbf{k}') \frac{\partial}{\partial \mathbf{r}}[\delta n(\mathbf{k}',\mathbf{r},t)] = 0$$

[5.67]

Now

$$\frac{\partial n_0(\mathbf{k})}{\partial \mathbf{k}} \approx -\left(\frac{\partial E_0(\mathbf{k})}{\partial \mathbf{k}}\right) \delta(E_0(\mathbf{k}) - \mu) \qquad [5.68]$$

i.e., the changes in the ground state take place at the Fermi surface. The first two terms are then the classical transport equations for a gas of independent particles and the last term describes the interaction between those particles. In the same way, one can extend Eq. (5.67) to include the effects of external forces, but this will take us too far from our present interest and we will not consider it here. Suffice to say that the one-to-one correspondence between the quasiparticle and what amounts to a quasiclassical gas of particles enables one to obtain a number of properties of the many-body system with relatively little work in terms of the interaction $f(\mathbf{k},\mathbf{k}')$.

BIBLIOGRAPHY

LANDAU, L. D., and LIFSHITZ, E. M., *Statistical Physics,* Pergamon, Elmsford: NY, 1980.

PINES D., and NOZIÈRES P., *The Theory of Quantum Liquids,* Benjamin, Menlo Park: CA, 1966.

PROBLEMS

1. Close to the Fermi surface, the major decay process is by way of electron–hole pair production. By using Fermi's golden rule, show that energy and momentum conservation produces a scattering time of the form $\tau^{-1}\alpha(\omega - \mu)^2$, where ω is the energy above the Fermi surface and hence the imaginary part of the energy goes as

$$|E_i| \ \alpha(\omega - \mu)^2$$

2. Consider the case of transverse oscillations on a string of fixed length subject to a small damping force. Show explicitly that if a wave packet (much smaller than the length of the string) is propagated along the string, in either direction, the wave packet is damped.

3. Calculate from Eqs. (5.28) and (5.26) the effective mass in the Hartree–Fock approximation as a function of momentum.

4. By considering the effect of the compression of a simple metal on the Fermi surface show that the compressibility K is given by

$$\frac{K}{K_0} = \frac{m^*}{m}(1 + A_0)^{-1}$$

where K_0 is the noninteracting compressibility and A_0 is the Landau parameter.

CHAPTER SIX

MANY-BODY GREEN'S FUNCTIONS

When considering how to describe an interacting system, we should be very clear what it is we require. First of all, it will not be useful to consider the ground state; it will almost certainly be a highly complicated object and, apart perhaps from its formation energy from some other state, of very little interest. What we are normally interested in is what happens to the system when it is disturbed, by some experimental probe for instance. What is required then, are expectation values of operators, transitions between states, time development in particular circumstances, etc. In the sense of the previous chapter, it is the group of quasiparticles and their properties that will be of prime importance, since these relate to the excited states of the system.

6.1. DEFINITION OF THE MANY-BODY GREEN'S FUNCTION

Perhaps the simplest perturbation one can think of applying to a system of N particles is to add one particle or take one away, producing a system of $N \pm 1$ particles. Suppose we start with the ground state of N particles ($|N\rangle$), then add a particle of spin α to the system at $\mathbf{r}$, t. The new state will be $\hat{\psi}_\alpha^+(\mathbf{r},t)|N\rangle$. We can construct a type of Green's function by considering the propagation of the extra particle. If we remove a particle of spin β at $\mathbf{r}'$, t' from the new state, then the overlap with the ground state should tell us something about the probability that the system is left undisturbed or, alternatively, the probability that the particle propagated from $\mathbf{r}$, t to $\mathbf{r}'$, t' with a change of spin from α to β on the way. That is, we envisage the following train of events.

(i) $|N\rangle$: ground state of system,
(ii) $\hat{\psi}^+_\alpha(\mathbf{r},t)|N\rangle$: ground state of system with particle added at $\mathbf{r}$, t with spin α,
(iii) $\hat{\psi}_\beta(\mathbf{r}',t')\hat{\psi}^+_\alpha(\mathbf{r},t)|N\rangle$: particle removed at $\mathbf{r}'$, t' with spin β from $N + 1$ state, and
(iv) $\langle N|\hat{\psi}_\beta(\mathbf{r}',t')\hat{\psi}^+_\alpha(\mathbf{r},t)|N\rangle$: overlap of system after perturbation with the original ground state.

In a classical noninteracting system we would expect an overlap of either 0 or 1. With an interacting classical system we would expect that the collisions would leave the system changed from the ground state, but we would still have to remove the original particle at $\mathbf{r}'$, t'. In a quantum system we would be allowed to remove any particle and still get overlap because of indistinguishability, but since this applies whether the system is interacting or not, a function such as we have written above is still useful.

If we wished to describe the system in the absence of one of its particles (or the presence of a "hole") the sequence would be

(i) $|N\rangle$,
(ii) $\hat{\psi}_\alpha(\mathbf{r},t)|N\rangle$,
(iii) $\hat{\psi}^+_\beta(\mathbf{r}',t')\hat{\psi}_\alpha(\mathbf{r},t)|N\rangle$, and
(iv) $\langle N|\hat{\psi}^+_\beta(\mathbf{r}',t')\hat{\psi}_\alpha(\mathbf{r},t)|N\rangle$.

Both of these sequences have little physical meaning unless the time t' is later than t; otherwise they become noncausal. It is very inconvenient, however, to have two functions to work with; instead we define a time-ordering operator T such that

$$T[\hat{\psi}_\alpha(\mathbf{r},t)\hat{\psi}^+_\beta(\mathbf{r}',t')] = \hat{\psi}_\alpha(\mathbf{r},t)\hat{\psi}^+_\beta(\mathbf{r}',t'), \qquad t > t' \tag{6.1a}$$

$$= \pm\hat{\psi}^+_\alpha(\mathbf{r}',t')\hat{\psi}_\beta(\mathbf{r}, t), \qquad t < t' \tag{6.1b}$$

The $\pm$ signs are for bosons ($+$) and fermions ($-$) and are designed to take care of the commutators and anticommutators, which appear in later equations in the development of the theory. The two expressions for an added and subtracted particle can now be incorporated into one expression for our definition of the *many-body Green's function.*

$$G_{\alpha\beta}(\mathbf{r},t, \mathbf{r}',t') = -i\langle N|T[\hat{\psi}_\alpha(\mathbf{r},t)\hat{\psi}^+_\beta(\mathbf{r}',t')]|N\rangle \tag{6.2}$$

For $t > t'$

$$G_{\alpha\beta}(\mathbf{r},t,\mathbf{r}',t') = -i\langle N|\hat{\psi}_\alpha(r,t)\hat{\psi}^+_\beta(r',t')|N\rangle \qquad [6.3a]$$

will describe the motion of an added particle from $\mathbf{r}'$, t' to $\mathbf{r}$, t.

$$G_{\alpha\beta}(\mathbf{r},t,\mathbf{r}',t') = \mp i\langle N|\hat{\psi}^+_\beta(r',t')\hat{\psi}_\alpha(\mathbf{r},t)|N\rangle, \qquad t < t' \qquad [6.3b]$$

will describe the motion of an added hole from $\mathbf{r}$, t to $\mathbf{r}'$, t'. As a matter of convenience, it is better to include the spin index as one of the variables, so we write $x = \mathbf{r}, \alpha$; $x' = \mathbf{r}', \beta$ and then

$$\sum_{\text{spin}} \int d\mathbf{r} \Rightarrow \int dx \qquad [6.4]$$

and

$$\begin{aligned} G_{\alpha\beta}(\mathbf{r},t,\mathbf{r}',t') &= -i\langle N|T[\hat{\psi}(x,t)\hat{\psi}^+(x',t')]|N\rangle \\ &= G(x,t,x',t') \end{aligned} \qquad [6.5]$$

6.2. RELATIONSHIP TO SINGLE-PARTICLE GREEN'S FUNCTION

Equation (6.2) is our basic definition but it looks very little like the Green's functions we met in Chap. 2. Consider, however, the noninteracting system. In order to be specific we will restrict ourselves to fermions occupying states up to some maximum level m_F. The ground state and field operators are given by

$$|N\rangle = \prod_{\substack{m \\ m<m_F}} \hat{c}^+_m|0\rangle \qquad [6.6]$$

$$\hat{\psi}(x,t) = \sum_m \phi_m(x)\, e^{(-iE_m t)/\hbar}\, \hat{c}_m \qquad [6.7]$$

Substituting into Eq. (6.2) we have

$$G(x,t,x',t') = -i\sum_{l,m} \phi_l(x)\phi^*_m(x')\, e^{[-i(E_l t - E_m t')]/\hbar}$$
$$[\langle N|\hat{c}_l\hat{c}^+_m|N\rangle\theta(t-t') - \langle N|\hat{c}^+_m\hat{c}_l|N\rangle\theta(t'-t)] \qquad [6.8]$$

Now for a fermion system we have

(i) $t > t'; l = m$ and $m > m_F$

(ii) $t < t'; l = m$ and $m < m_F$

to ensure a nonzero result, so that

$$G(x,t,x',t') = -i \sum_{m>m_F} \phi_m(x)\phi_m^*(x')\, e^{[-iE_m(t-t')]/\hbar}\, \theta(t - t') + i \sum_{m<m_F} \phi_m(x)\phi_m^*(x')\, e^{[-iE_m(t-t')]/\hbar}\, \theta(t' - t) \quad [6.9]$$

Apart from the factor i, the many-body Green's function reverts to a form of the *single-particle Green's function,* which differentiates between the occupied and unoccupied states. It corresponds, in fact, to a change in the boundary conditions for the system. Whereas before we were forced to introduce the retarded and advanced Green's functions in order to deal with the undefined energy–time transform, we now introduce a new boundary condition which differentiates between the occupied and unoccupied states—the *time-ordered Green's function.* From Eq. (6.9) we see that there is a simple relationship between the time-ordered, advanced, and retarded Green's functions.

$$G(x,t,x',t') = i[G_A(x,t,x',t') + G_R(x,t,x',t')] \quad [6.10]$$

6.3. ENERGY STRUCTURE AND THE GREEN'S FUNCTION

One of the main attributes of the Green's function of Chap. 2 is that its poles are the eigenvalues of the governing equation. Let us see how this property carries over to the Green's function of Eq. (6.2). First, we must transform our time-dependent Green's function into an energy-dependent one. To do this we first move to the Schrödinger representation. If $\hat{H}$ is the Hamiltonian of the system, we have

$$\hat{H}|N\rangle = E_N^0|N\rangle \quad [6.11]$$

where E_N^0 is the ground-state energy, and

$$\hat{\psi}(x,t) = e^{(i\hat{H}t)/\hbar}\, \hat{\psi}(x)\, e^{(-i\hat{H}t)/\hbar} \quad [6.12]$$

(assuming that $\hat{H}$ is not an explicit function of time). Substituting into Eq. (6.2)

$$iG(x,t,x',t') = e^{[iE_N^0(t-t')]/\hbar} \langle N|\hat{\psi}(x)\, e^{(-i\hat{H}t)/\hbar}\, e^{(i\hat{H}t')/\hbar}\, \hat{\psi}^+(x')|N\rangle\theta(t-t') \pm e^{[-iE_N^0(t-t')]/\hbar} \langle N|\hat{\psi}^+(x')\, e^{(-iHt')/\hbar}\, e^{(iHt)/\hbar}\, \hat{\psi}(x)|N\rangle\theta(t'-t) \quad [6.13]$$

In order to remove the time operators inside the expectation values, we introduce the complete set of states with M particles $|M, j\rangle$, where j is a general label to describe the possible excited states. Since the states form a complete set

$$\sum_j |M,j\rangle\langle j,M| = 1 \quad [6.14]$$

and also

$$\hat{H}|M,j\rangle = E_M^j|M,j\rangle \quad [6.15]$$

We introduce the identity (6.14) between each of the pairs of exponentials to give

$$G(x,t,x',t') = \frac{1}{i}\sum_j e^{i/\hbar(E_N^0-E_M)(t-t')} \langle N|\hat{\psi}(x)|M,j\rangle\langle j,M|\hat{\psi}^+(x')|N\rangle\theta(t-t') \pm \frac{1}{i}\sum_j e^{-i/\hbar(E_N^0-E_M)(t-t')} \langle N|\hat{\psi}^+(x')|M,j\rangle\langle M,j|\hat{\psi}(x)|N\rangle\theta(t'-t) \quad [6.16]$$

The Green's function is now explicitly only a function of the time difference $t - t'$. We can, therefore, rewrite the Green's function as

$$G(x,t,x',t') \Rightarrow G(x,x',t-t')$$

or, with $t - t' = \tau$, we may write

$$G(x,t,x',t') = G(x,x',\tau)$$

An energy transform follows as

$$G(x,x',\omega) = \frac{1}{2\pi\hbar}\int_{-\infty}^{+\infty} d\tau\, e^{(i\omega\tau)/\hbar}\, G(x,x',\tau) \quad [6.17]$$

Thus, we have

$$G(x,x',\omega) = \sum_j \left\{ \frac{\langle N|\hat{\psi}(x)|M,j\rangle\langle M,j|\hat{\psi}^+(x')|N\rangle}{\omega - (E_M^j - E_N^0) + i\delta} \pm \frac{\langle N|\hat{\psi}^+(x')|M,j\rangle\langle M,j|\hat{\psi}(x)|N\rangle}{\omega + (E_M^j - E_N^0) - i\delta} \right\} \qquad [6.18]$$

where the infinitesimals $\pm i\delta$ reflect the time ordering. Let us consider the two terms separately.

1. The expectation values $\langle N|\hat{\psi}(x)|M,j\rangle$ and $\langle M,j|\hat{\psi}^+(x')|N\rangle$ both require that $|M, j\rangle$ have $N + 1$ particles. Thus, E_M^j is an energy level of the $(N + 1)$-particle state (E_{N+1}^j). If we write the ground-state energy as E_{N+1}^0, then we have

$$\omega - (E_M^j - E_N^0) = \omega - (E_{N+1}^j - E_{N+1}^0) - (E_{N+1}^0 - E_N^0) \qquad [6.19]$$

The expression $E_{N+1}^0 - E_N^0$ is the minimum possible energy to add one particle to the N-particle state, and so is the chemical potential or Fermi energy μ. (In a Fermi system the chemical potential has a familiar meaning in interacting and noninteracting systems. For bosons the chemical potential is always set to zero in noninteracting systems but may be finite for interacting ones; i.e., because of the interaction, the energy required to put a particle into the lowest state will depend upon the occupancy of that state.)

$$E_{N+1}^j - E_{N+1}^0 = \epsilon_{N+1}(j)$$

is the jth excitation energy of the $(N + 1)$-particle state. Equation (6.19) becomes

$$\omega - (E_M^j - E_N^0) = \omega - \epsilon_{N+1}(j) - \mu \qquad [6.20]$$

2. The expectation values require that $|M,j\rangle$ be an $(N - 1)$-particle state. Following the same reasoning as in (1) we have

$$\begin{aligned} \omega + E_{N-1}^j - E_N^0 &= \omega + (E_{N-1}^j - E_{N-1}^0) + (E_{N-1}^0 - E_N^0) \\ &= \omega + \epsilon_{N-1}(j) - \mu \end{aligned} \qquad [6.21]$$

Note that since $\epsilon_{N-1}(j)$ is the excited state of the $(N-1)$-particle state it is still inherently positive. We can now write our energy-dependent Green's function as

$$G(x,x',\omega) = \sum_j \left[\frac{\langle N|\hat{\psi}(x)|N+1,j\rangle\langle j,N+1|\hat{\psi}^+(x')|N\rangle}{\omega - \epsilon_{N+1}(j) - \mu + i\delta} \pm \frac{\langle N|\hat{\psi}^+(x')|N-1,j\rangle\langle N-1,j|\hat{\psi}(x)|N\rangle}{\omega + \epsilon_{N-1}(j) - \mu - i\delta} \right] \quad [6.22]$$

Thus, the N-particle Green's function has poles at the excited states of the $(N \pm 1)$-particle system, i.e.

$$\omega = \mu + \epsilon_{N+1}(j) - i\delta \quad [6.23a]$$

and

$$\omega = \mu - \epsilon_{N-1}(j) + i\delta \quad [6.23b]$$

This is in complete accord with our interpretation of the function as describing, for different time ordering, the behavior of the particle system with one particle added or removed. The resulting energy structure is indicated in Fig. 6.1. We see that the chemical potential provides a clear

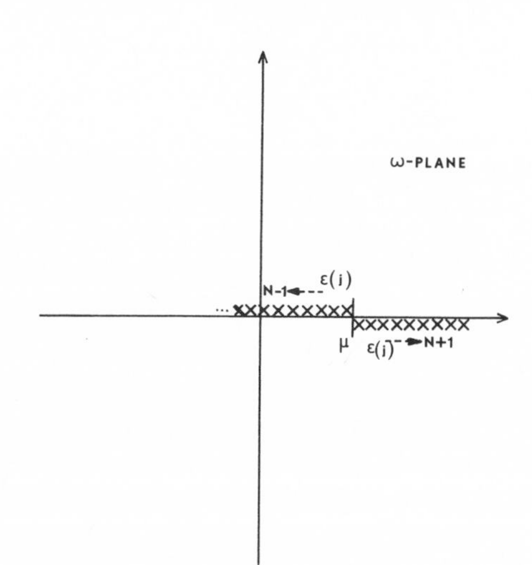

FIGURE 6.1. Energy structure of the Green's function in the complex energy plane showing the position of the poles.

separation point for the system, above which the poles are below the real axis and correspond to the $(N + 1)$-particle system, while below this energy only the $(N - 1)$-particle excitation energy poles appear and these are above the real axis. The residues of the Green's function at the poles are complicated functions of two variables and do not immediately have a clear physical interpretation, but in the case of a uniform system, they do have a very simple and useful meaning.

6.4. THE LEHMAN REPRESENTATION AND QUASIPARTICLES

If the system is completely uniform (as for instance in the jellium solid) then all states are eigenstates of the momentum operator $\hat{p}$ even though the system is interacting. A useful basis for the field operators are the plane-wave states:

$$\hat{\psi}(x) = \sum_k e^{ikx}\, \hat{b}_k \qquad [6.24a]$$

$$\hat{\psi}^+(x) = \sum_k e^{-ikx}\, \hat{b}_k^+ \qquad [6.24b]$$

$$\hat{p} = \sum_k \hbar k \hat{b}_k^+ b_k \qquad [6.25]$$

[with $k \equiv (\mathbf{k}, \alpha)$ in the same sense as $x = (\mathbf{r}, \alpha)$]

We can now write

$$\hat{p}|N,0\rangle = 0 \qquad [6.26]$$

which defines the ground state and in general

$$\hat{p}|N + 1,j\rangle = \mathbf{p}_j|N + 1,j\rangle \qquad [6.27]$$

(Note that *many* interacting states j may have the same momentum.)

It is usual to rewrite the field operator in terms of the momentum operator by analogy with the Heisenberg transformation

$$\hat{\psi}(x) = e^{(-i\hat{p}x)/\hbar}\, \hat{\psi}(0)e^{(+i\hat{p}x)/\hbar} \qquad [6.28]$$

Substituting into Eq. (6.22) we have

$$G(x - x',\omega) = \sum_j \left[\frac{e^{ip_j(x-x')} \, |\langle N,0|\hat{\psi}(0)|N + 1,j\rangle|^2}{\omega - \epsilon_{N+1}(j) - \mu + i\delta} \pm e^{-ip_j(x-x')} \frac{|\langle N,0|\hat{\psi}^+(0)|N - 1,j\rangle|^2}{\omega + \epsilon_{N-1}(j) - \mu - i\delta} \right] \quad [6.29]$$

If we define the momentum transformation as

$$G(k,\omega) = \int G(x - x',\omega) \, e^{-ik(x-x')} \, d(x - x') \quad [6.30]$$

we have

$$G(k,\omega) = \sum_j \frac{|\langle N,0|\hat{\psi}(0)|N + 1,j,k\rangle|^2}{\omega - \epsilon_{N+1}(j,k) - \mu + i\delta} \pm \frac{|\langle N,0|\hat{\psi}^+(0)|N - 1,j,-k\rangle|^2}{\omega + \epsilon_{N-1}(j,-k) - \mu - i\delta} \Bigg] \quad [6.31]$$

where we have added the extra quantum number $\pm k$ to indicate that we are interested only in those excited states j with momentum $\pm\mathbf{k}$.

Consider first the noninteracting (fermion) state; then

$$|N + 1,j,k\rangle = \hat{b}_k^+|N,0\rangle \qquad (k > k_F) \quad [6.32a]$$

$$|N - 1,j,-k\rangle = \hat{b}_k|N,0\rangle \qquad (k < k_F) \quad [6.32b]$$

Since

$$|N,0\rangle = \prod_{k<k_F} \hat{b}_k^+|0,0\rangle \quad [6.33]$$

Also we have

$$\hat{\psi}(0) = \sum_k \hat{b}_k \quad [6.34]$$

$$\hat{H} = \sum_k E(k)\hat{b}_k^+\hat{b}_k \quad [6.35]$$

The energy denominator simplifies to

$$\omega - E(k) \pm i\delta$$

so that

$$G(k,\omega) = \frac{\theta(k - k_F)}{\omega - E(k) + i\delta} - \frac{\theta(k_F - k)}{\omega - E(k) - i\delta} \qquad [6.36]$$

The Green's function has two types of poles, one for the energy of a particle added to the empty states above the Fermi sea and one for the hole obtained by removing a particle. The residues around these poles are both unity, reflecting the unit probability of the particle being in and remaining in that single-particle state. Apart from the infinitesimals, we are again back to the single-particle Green's function.

Once we have introduced interactions there will be many ways of dividing a momentum k among the particles; i.e., once the particle is introduced, the interactions will ensure that the momentum is spread among the other particles. Thus, the state label j will be more of a continuous variable than a discrete set of states. Suppose we label all the states with momentum k by their excitation energy. Then we can replace the expectation values in the two parts of the Green's function by functions of k and the energy $\epsilon_{N\pm1}(j,k)$ ($\equiv \omega'$). Thus

$$\sum_j \frac{|\langle N+1,j,\mathbf{k}|\hat{\psi}^+(0)|N,0\rangle|^2}{\omega - \epsilon_{N+1}(j,\mathbf{k}) - \mu + i\delta} \Rightarrow \int_0^\infty d\omega' \frac{A(k,\omega')}{\omega - \omega' - \mu + i\delta} \qquad [6.37]$$

and we write

$$G(k,\omega) = \int_0^\infty d\omega' \left[\frac{A(k,\omega')}{\omega - \omega' - \mu + i\delta} \pm \frac{B(k,\omega')}{\omega + \omega' - \mu - i\delta}\right] \qquad [6.38]$$

This is the *Lehman representation* of the Green's function. The A and B coefficients are called the *spectral weight functions.* In this form the noninteracting system has

$$A(k,\omega) = \delta(\omega' - E(k) + \mu) \qquad [6.39a]$$
$$B(k,\omega) = \delta(\omega' + E(k) - \mu) \qquad [6.39b]$$

That is, A and B are delta functions at the energy of the single-particle state.

This is very convenient, for if we were able to define a set of single-particle states, they would result in delta functions in the Lehman functions A and B. It follows, then, that if we evaluate the Green's function in the Lehman representation and find something which resembles a delta function, an isolated peak for instance, then it would seem to represent a single-particlelike state—a quasiparticle. Consider the situation shown in Fig. 6.2, where we have a hypothetical curve for $A(k,\omega)$. This obviously splits into two parts, one is the quasiparticle peak $A_1(k,\omega)$ and the other some general background value. For energies near the peak ω_0 we can write, since A_1 will dominate

$$G(k,\omega) \approx \frac{1}{\omega - \omega_0 - \mu + i\delta} \int d\omega' \, A_1(k,\omega') + \text{background} \quad [6.40]$$

Since the states form a complete set, from (6.37)

$$\int_0^\infty A(k,\omega')\, d\omega' = \int_0^\infty B(k,\omega')\, d\omega' = 1 \quad [6.41]$$

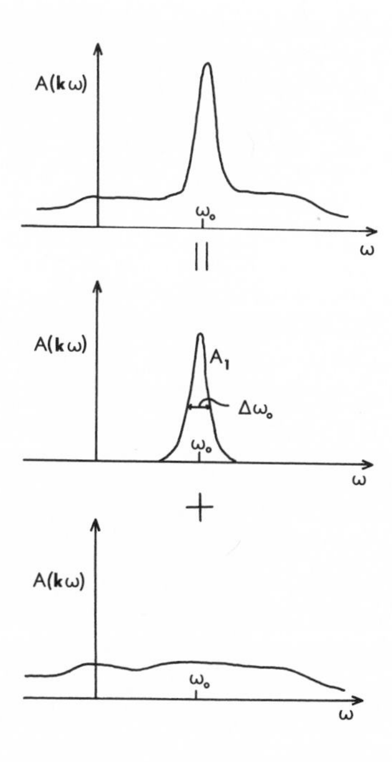

FIGURE 6.2. Splitting of the Lehman spectral weight function into quasiparticle (A_1) and background.

so that

$$\int A_1(k,\omega')\, d\omega' = Z < 1 \qquad [6.42]$$

and we have

$$G(k,\omega) = \frac{Z}{\omega - \omega_0 - \mu + i\delta} + \text{background} \qquad [6.43]$$

Thus, the quasiparticle will have a normalization, or "spectral weight," which is less than unity, reflecting the fact that, because of the interactions, the state is not an isolated complete state but an integral part of a coupled system.

It is also interesting to see what the quasiparticle peak looks like in terms of its time variation. We have, leaving out the background

$$\begin{aligned} G(k,\tau) &= \int d\omega\, e^{(-i\omega\tau)/\hbar}\, G(k,\omega) \\ &\approx \int_{\omega' \approx \omega_0} \frac{A_1(k\omega')\, e^{(-i\omega\tau)/\hbar}}{\omega - \omega' - \mu + i\delta}\, d\omega'\, d\omega \\ &\approx \int A_1(k,\omega' - \mu)\, e^{(-i\omega'\tau)/\hbar}\, d\omega' \end{aligned}$$

The width of the quasiparticle peak now becomes crucial, since it gives a decay component to the Green's function in time. Taking $A(k,\omega' - \mu)$ as a Lorentzian we have

$$G(\mathbf{k},\tau) \simeq e^{(-i\omega_0\tau)/\hbar}\, e^{(-\Delta\omega_0\tau)/\hbar} \qquad [6.44]$$

the broader the width ($\Delta\omega_0$), the faster the decay. Thus, the quasiparticle also decays or (perhaps better) has a finite lifetime. If the process to be considered has a time scale longer than that lifetime, the quasiparticle will not be a good effective single-particle state.

The usefulness of the Lehman representation is that it gives a way of looking at the Green's function for an interacting system which automatically isolates the quasiparticle states. It does not, of course, guarantee that such single-particle states will exist. The spectral weight may turn out to be a completely smooth function with no quasiparticle states at all—only calculation can decide that—but the very success of the single-particle theory of solids suggests that well-defined quasiparticles will exist.

6.5. EXPECTATION VALUES

Having shown that the Green's function contains propagation and energy excitation values, the next property to look for is the ability to calculate expectation values. If we have a single-particle operator $\hat{J}(x)$, then, in terms of the field operators

$$\hat{J}(t) = \int dx\, \hat{\psi}^{+}(x,t)\hat{J}(x)\hat{\psi}(x,t) \qquad [6.45]$$

This suggests an operator density

$$\hat{\mathcal{J}}(x,t) = \hat{\psi}^{+}(x,t)\hat{J}(x)\hat{\psi}(x,t) \qquad [6.46]$$

For an N-particle system, the expectation value for the operator density becomes

$$\langle \mathcal{J}(x,t)\rangle = \langle N|\hat{\mathcal{J}}(x,t)|N\rangle \qquad [6.47]$$

and

$$\langle J\rangle = \int dx\, \langle \hat{\mathcal{J}}(x,t)\rangle \qquad [6.48]$$

In terms of the field operators

$$\begin{aligned}\langle \mathcal{J}(x,t)\rangle &= \langle N|\hat{\psi}^{+}(x,t)\hat{J}(x)\hat{\psi}(x,t)|N\rangle \\ &= \lim_{x\to x'} [\hat{J}(x)\langle N|\hat{\psi}^{+}(x',t)\hat{\psi}(x,\, t)|N\rangle]\end{aligned} \qquad [6.49]$$

since $\hat{J}(x)$ does not operate on $|N\rangle$ and $\hat{\psi}^{+}(x',t)$ does not operate on $\hat{J}(x)$. Using the commutator rules we also have

$$\begin{aligned}\langle \mathcal{J}(x,t)\rangle &= \pm\lim_{x\to x'} [\hat{J}(x)\langle N|\hat{\psi}(x,t)\hat{\psi}^{+}(x',t)|N\rangle \\ &= \pm i \lim_{\substack{x\to x' \\ t\to t'+\delta}} [\hat{J}(x)G(x,t,x',t')]\end{aligned} \qquad [6.50]$$

where the $\pm$ sign refers to Bose or Fermi particles. From this we can write

$$\langle J\rangle = \pm i \lim_{\substack{t\to t'+\delta \\ x\to x'}} [\int \hat{J}(x)G(x,t,x',t')\, dx] \qquad [6.51]$$

or more elegantly

$$\langle J \rangle = \pm i \lim_{\substack{t \to t' + \delta \\ x \to x'}} \mathrm{Tr}[\hat{J}(x) G(x,t,x',t')] \qquad [6.52]$$

Thus, in knowing the Green's function we also have all of the expectation values for the single-particle operators (see Sec. 2.5.4).

6.6. EQUATION OF MOTION FOR THE GREEN'S FUNCTION

Having defined the function we are interested in and found that it has very desirable properties, it simply remains to calculate it. We must start from the Hamiltonian for the system. The form of that Hamiltonian will determine the properties of the Green's function. In order to progress, we take the specific and very important case of the interacting Fermi gas. We have

$$\hat{H}(x_1,t_1) = \hat{H}_0(x_1) + v(x_1,x_2)\,\delta(t_1 - t_2) \qquad [6.53]$$

where $\hat{H}_0(x_1)$ is the one-electron part of the Hamiltonian and $v(x_1,x_2)$ ($\equiv e^2/|\mathbf{r}_1 - \mathbf{r}_2|$) is the coulomb interaction. (Note that the interaction does not depend upon spin, but we have not explicitly separated out the spatial coordinate.) In the Heisenberg representation the field operator ($\hat{\psi}(x_1,t_1)$) has the equation of motion

$$\frac{\hbar}{i}\frac{\partial}{\partial t_1}\hat{\psi}(x_1,t_1) = [\hat{H},\hat{\psi}(x_1,t_1)] \qquad [6.54]$$

where

$$\hat{H} = \int \hat{\psi}^+(x,t)\hat{H}_0(x)\hat{\psi}(x,t)\,dx + \int \hat{\psi}(x,t)\hat{\psi}^+(x',t')v(x,x')\,\delta(t - t')\hat{\psi}(x',t')\hat{\psi}(x,t)\,dx\,dx' \qquad [6.55]$$

This leads to

$$i\hbar\frac{\partial}{\partial t_1}\hat{\psi}(x_1,t_1) = \left[\hat{H}_0(x_1) + \int v(x_1,x_3)\hat{\psi}^+(x_3,t_1)\hat{\psi}(x_3,t_1)\,dx_3\right]\hat{\psi}(x_1,t_1) \qquad [6.56]$$

Since

$$G(x_1,t_1,x_2,t_2) = \frac{1}{i}\langle N|T[\hat{\psi}(x_1,t_1)\hat{\psi}^+(x_2,t_2)]|N\rangle$$

we multiply by the field operator $\hat{\psi}^+(x_2,t_2)$ and then take expectation values, remembering the time ordering, to obtain

$$\left[i\hbar\frac{\partial}{\partial t_1} - \hat{H}_0(x_1)\right]G(x_1,t_1,x_2,t_2) + i\int v(x_1,x_3)\langle N|T[\hat{\psi}^+(x_3,t_1)\hat{\psi}(x_3,t_1)\hat{\psi}(x_1,t_1)\hat{\psi}^+(x_2,t_2)]N\rangle\, dx_3 = \hbar\delta(x_1 - x_2)\delta(t_1 - t_2) \quad [6.57]$$

If we look at the structure of this equation, we have

$$\boxed{\text{noninteracting one-particle equation}} + \boxed{\text{interaction terms}} = \hbar\delta(x_1 - x_2)\delta(t_1 - t_2) \quad [6.58]$$

Since we are specifically interested in the interaction effects, we will assume that the noninteracting part of the equation is always exactly soluble, i.e.

$$\left[i\hbar\frac{\partial}{\partial t_1} - \hat{H}_0(x_1)\right]G_0(x_1,t_1,x_2,t_2) = \hbar\delta(x_1 - x_2)\delta(t_1 - t_2) \quad [6.59]$$

For a given $\hat{H}_0(x_1)$ this *defines* a noninteracting Green's function $G_0(x_1,t_1,x_2,t_2)$. There are two points worth mentioning in this context.

(i) The calculation of G_0 is nontrivial for a solid, other than the jellium solid, as the vast literature on band structure shows.

(ii) The Hamiltonian $\hat{H}_0(x_1)$ is not necessarily unique nor the best one to start with. It is normal to take some of the larger interaction effects, which produce an effective one-particle potential, and include them within $\hat{H}_0(x_1)$. The Hartree potential, or approximations to it, is an obvious example. For this reason one must be careful when comparing calculations of the effects of the interaction on the "noninteracting" Green's function.

If we now look at the interaction term, it contains the function

$$\langle N | T[\hat{\psi}^+(x_3,t_1)\hat{\psi}(x_3,t_1)\hat{\psi}(x_1,t_1)\hat{\psi}^+(x_2,t_2)] | N\rangle$$

The presence of two creation and two annihilation field operators means that it must describe the propagation of two particles. It is in fact a special case of the *two-particle Green's function* defined as

$$G_2(x_1,t_1,x_2,t_2,x_3,t_3,x_4,t_4) \\ = (i)^2\langle N | T[\hat{\psi}(x_1,t_1)\hat{\psi}(x_3,t_3)\hat{\psi}^+(x_4,t_4)\hat{\psi}^+(x_2,t_2)] | N\rangle \quad [6.60]$$

(Note the arrangement of the indices.) According to the time ordering, this function will describe the transport of two particles, two holes or one electron, and one hole. Thus, the equation of motion for the single-particle Green's function contains the two-particle Green's function.

Following the same procedure, we could now write down the equation of motion for the two-particle Green's function. Not surprisingly, in view of the two field operators in the interaction term, this equation would involve a three-particle Green's function. Thus, we could proceed *ad infinitum* and be left with an infinite series of coupled equations essentially telling us of progressively more complicated interactions which would be generated by and affect the propagation of our original added particle. What is required is a way of either summing or terminating this series such that the results are in agreement with experiment and hence contain the important physics.

The strange property which makes many-body theory in solids so fascinating is that it is often *not* the most mathematically complete solution which is the most accurate. Often physical properties must be invoked before any decision can be made on which are the most important terms in the series.

6.7. HARTREE AND HARTREE–FOCK APPROXIMATIONS

Before going on to quite complicated considerations of the solution to Eq. (6.57), it is useful to go back and see how the simplest many-body solutions may be derived. Consider the two-particle Green's function which appears in Eq. (6.57). Given the ordering of the field operators required, this must be written as ($t_1^+ = t_1 + \delta$)

$$G_2(x_1,t_1,x_2,t_2,x_3,t_1,x_3,t_1^+)$$

For the two possible time orderings we have

(i) $t_1 > t_2$: G_2 describes the propagation of a particle from x_2, t_2 to x_1, t_1 *or* x_3, t_1^+ and the propagation of a hole from x_1, t_1 *or* x_3, t_1^+ to x_3, t_1 and

(ii) $t_1 < t_2$: propagation of two holes from $(x_1,t_1),(x_3,t_1)$ to $(x_2,t_2),(x_3,t_1^+)$ *or* $(x_3,t_1^+),(x_2,t_2)$

These cases are illustrated in Fig. 6.3. As it stands we do not know how these propagate, but the obvious first choice is to allow each particle to propagate independently according to the single-particle Green's function. Since there are two particles involved, we are making the separation

$$G_2 \Rightarrow G_1 \times G_1 \qquad [6.61]$$

Allowing for all of the processes in (i) and (ii) we have

$$G_2(x_1,t_1,x_2,t_2,x_3,t_1,x_3,t_1^+) = G(x_1,t_1,x_2,t_2)G(x_3,t_1,x_3,t_1^+) + G(x_1,t_1,x_3,t_1^+)G(x_3,t_1,x_2,t_2) \qquad [6.62]$$

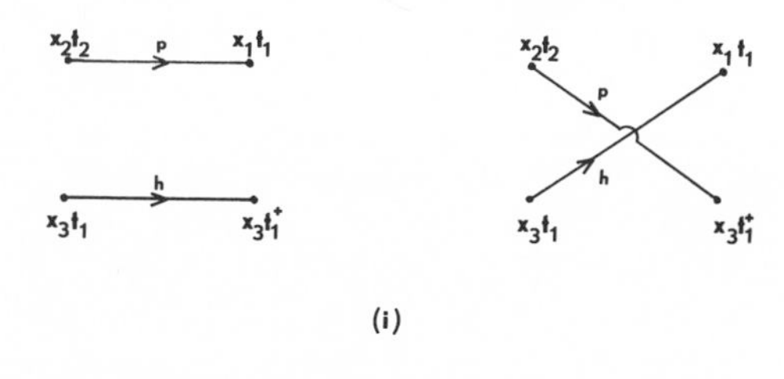

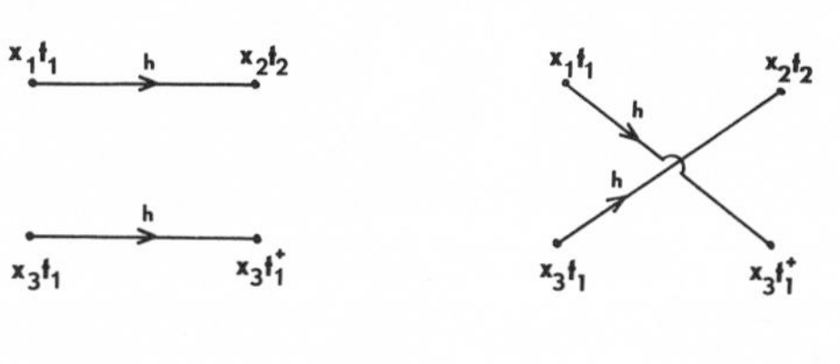

FIGURE 6.3. Possible "Hartree-Fock" reductions of the two-particle Green's function for the time orderings described in the text.

The first term is called the direct and the second the exchange, though obviously each is the exchange of the other. Consider the first term only as an approximation; we have, substituting into Eq. (6.57)

$$\left\{\left[i\hbar\frac{\partial}{\partial t_1} - \hat{H}_0(x_1)\right] + \left[i\int v(x_1,x_3)G(x_3,t_1,x_3,t_1^+)\,dx_3\right]\right\}G(x_1,t_1,x_2,t_2) = \hbar\delta(x_1 - x_2)\,\delta(t_1 - t_2) \quad [6.63]$$

The equation has degenerated into a simple independent-particlelike equation with an added potential given by

$$\begin{aligned} -V(x_1,t_1) &= i\textstyle\int v(x_1,x_3)G(x_3,t_1,x_3,t_1^+)\,dx_3 \\ &= -\textstyle\int v(x_1,x_3)\langle N|\hat{\psi}^+(x_3,t_1)\hat{\psi}(x_3,t_1)|N\rangle\,dx_3 \end{aligned} \quad [6.64]$$

The expectation value is recognizable as the electron density $\rho(x_3,t_1)$, so we have

$$\begin{aligned} -V(x_1,t_1) &= -\textstyle\int v(x_1,x_3)\rho(x_3,t_1)\,dx_3 \\ &= -(\text{Hartree potential}) \\ V(x_1,t_1) &\equiv +V_H(x_1,t_1) \end{aligned} \quad [6.65]$$

Thus, we can derive the Hartree approximation as one-half of the lowest order of approximation to the equation of motion. The next step is to take both terms for the two-particle Green's function—the first obviously gives the Hartree term again—so we have

$$\left[i\hbar\frac{\partial}{\partial t_1} - \hat{H}_0(x_1) - V_H(x_1,t_1)\right]G(x_1,t_1,x_2,t_2) + i\int v(x_1,x_3) \times G(x_1,t_1,x_3,t_1^+)G(x_3,t_1,x_2,t_2)\,dx_3 = \hbar\delta(x_1 - x_2)\delta(t_1 - t_2) \quad [6.66]$$

The interaction term is now a nonlocal operator [i.e., it does not allow the removal of a simple $G(x_1,t_1,x_2,t_2)$ term]. It is, in fact, the Green's function variation of the exchange interaction appearing in the Hartree–Fock equation [Eq. (1.30)]. In order to see this explicitly it is necessary at this stage to revert to the independent-particle Green's function. We will leave that as an exercise.

Referring back to the original discussion of the Hartree and Hartree–Fock equations in Chap. 1 we see that we have an example of two approximations to the solution of Eq. (6.57) at the same level, one apparantly *mathematically* less justified than the other yet, for solids at least, giving a much better description of many properties.

6.8. THE SELF-ENERGY

Having dealt with the two-particle Green's function in this way, the obvious next stage is to carry on up the hierarchy, but this becomes rapidly far too complicated. For instance, considering the third- and fourth-order Green's functions, we need

$$G_3 \rightarrow \text{all combinations of } G_2 \cdot G_1 \quad \text{and} \quad G_1 \cdot G_1 \cdot G_1$$

with G_2 then described by all combinations of $G_1 \cdot G_1$

$$G_4 \rightarrow \text{all combinations of } G_3 \cdot G_1 G_2 \cdot G_2 \quad \text{and} \quad G_1 \cdot G_1 \cdot G_1 \cdot G_1$$

then we replace G_3, G_2 by their expansions, and so on. Keeping track of all of the possible terms is very difficult and not, in view of the behavior of the lowest-order solution—i.e., Hartree–Fock—necessarily justified. The need is for a systematic way of developing the solution to the series.

Suppose we assume that we have solved the infinite series of equations and look for a solution in the form

$$\left[i\hbar \frac{\partial}{\partial t_1} - \hat{H}_0(x_1) - V(x_1,t_1) \right] G(x_1,t_1,x_2,t_2) - \int \Sigma(x_1,t_1,x_3,t_3) G(x_3,t_3,x_2,t_2) dx_3\, dt_3 = \hbar \delta(x_1 - x_2)\delta(t_1 - t_2) \qquad [6.67]$$

where

$$V(x_1,t_1) = \phi(x_1,t_1) + V_H(x_1,t_1) \qquad [6.68]$$

is the Hartree potential plus any external potential such as an experimental probe. The operator $\Sigma(x_1,t_1,x_3,t_3)$ is called the *self-energy operator* and includes all of the interaction effects within itself. This trick of introducing a function to help formally solve the integro-differential equation is well known in mathematical theory—it is a bonus in this context that it turns out to have a clear physical meaning.

To see why it is termed the self-energy, we transform to the energy-dependent form of Eq. (6.67). As long as the Hamiltonian $H_0(x_1)$ has no explicit time dependence (which is normally the case) we can write

$$[\omega - \hat{H}_0(x_1) - V(x_1,\omega)]G(x_1,x_2,\omega) - \int\Sigma(x_1,x_3,\omega)G(x_3,x_2,\omega)\,dx_3 = \delta(x_1 - x_2) \quad [6.69]$$

If we write this in the matrix form, which we introduced when considering perturbations to the single-particle Green's function, we have

$$(\omega\mathbf{1} - \mathsf{H}_0 - \mathsf{V})\mathsf{G} - \Sigma\mathsf{G} = 1 \quad [6.70]$$

or formally

$$\mathsf{G}^{-1} = (\omega 1 - \mathsf{H}_0 - \mathsf{V} - \Sigma) \quad [6.71]$$

The noninteracting (i.e., without Σ) equation has a solution

$$\mathsf{G}_0^{-1} = \omega 1 - \mathsf{H}_0 - \mathsf{V} \quad [6.72]$$

and so

$$\mathsf{G}^{-1} = \mathsf{G}_0^{-1} - \Sigma \quad [6.73]$$

The poles of the Green's function G (or zeros of the inverse G^{-1}) are, from (6.72) and (6.73), moved in energy by Σ from those of the noninteracting Green's function G_0. Thus, the self-energy may be considered as a contribution to the energy of the excitations of the system from the interaction effects.

There is a question of terminology here; many authors define the self-energy to include the Hartree potential. We do not do this here because in most respects the Hartree potential is a one-electron effective potential and in almost all practical calculations is, approximately at least, contained in the one-electron part of the Hamiltonian.

The calculation of the Green's function now reduces to the calculation of the kernel of the integral equation (6.67), the self-energy. There are two basic ways of tackling this problem, which comes out as a perturbative series, one uses a theorem concerning combinations of operators (Wick's theorem), while the other, the one we shall use, develops the idea of the self-energy and the relationship to the equation of motion. Which is most appropriate is a matter of taste. Both result in the same answers, of course, but the equation of motion method does appear to give a more consistent perturbation series.

BIBLIOGRAPHY

HEDIN, L., and LUNDQVIST, S., *Solid State Phys.*, **23**(1), 1969. (Review article.)

KADANOFF, L., and BAYM, G., *Quantum Statistical Mechanics*, Benjamin, Menlo Park: CA, 1971.

GALITSKII, V. M., and MIGDAL, A. B., *Sov. Phys. JETP*, **7**(96), 1958.

KLEIN, A., and PRANGE, R., *Phys. Rev.*, **112**(994), 1958.

LEHMAN, H., *Nuovo Climento*, **11**(342), 1954.

PROBLEMS

1. Consider the inverse transform to Eq. (6.17)

$$G(x,x',\tau) = \frac{1}{2\pi}\int_c G(x,x',\omega)\, e^{(-i\omega\tau)/\hbar}\, d\omega$$

Define a suitable contour and consider the analytic continuation of $G(x, x', \omega)$ in terms of the Lehman representation of the Green's function [Eq. (6.38)]. How would you then interpret the appearance of the quasi-particle states?

2. Show that, for the noninteracting Green's function, Eq. (6.52) reduces to the sum over the occupied states.

3. Show that the Green's function does indeed satisfy Eq. (6.57).

4. Confirm that Eq. (6.66) is equivalent to the Hartree–Fock equation.

5. Write the next approximation to the two-particle Green's function corresponding to the particles interacting once during their propagation and hence obtain an expression for the self-energy to second order in the interaction.

CHAPTER SEVEN

THE SELF-ENERGY AND PERTURBATION SERIES

No matter how long one carries on in a descriptive manner or how many new definitions one introduces, at some point it is necessary to actually calculate the solution. The derivation of the Thomas–Fermi, Hartree, and Hartree–Fock solutions are useful, but we need to be able to calculate systematically the Green's function or, alternatively, the *self-energy*.

7.1. FUNCTIONAL DERIVATIVES AND THE CALCULATION OF G AND Σ

With an external field $\phi(x_1,t_1)$, the basic equation of motion is

$$\left[i\hbar \frac{\partial}{\partial t_1} - H_0(x_1) - \phi(x_1,t_1)\right] G(x_1,t_1,x_2,t_2) + i\int v(x_1,x_3)\langle N|T[\hat{\psi}^+(x_3,t_1)\hat{\psi}(x_3,t_1)\hat{\psi}(x_1,t_1)\hat{\psi}^+(x_2,t_2)]|N\rangle\, dx_3 = \hbar\delta(x_1 - x_3)\,\delta(t_1 - t_3) \quad [7.1]$$

where both the ground state $|N\rangle$ and the field operators $\hat{\psi}$ contain implicitly the effect of the external potential. It is convenient to extract this dependence and make it explicit. To do this we make use of the relationships derived in the interaction representation. We also need to require that, for $t \to \pm\infty$, the external potential tends to zero and also that the potential turns on and off slowly so that the system is adiabatic and returns to its original ground state [cf. Eq. (4.26)]. This enables the state $|N\rangle$ to be uniquely defined. From Eq. (4.29) we have

$$\hat{\psi}_H(t) = \left[T \exp\left(\frac{i}{\hbar}\int_0^t \hat{\phi}(t_3)\,dt_3\right)\right]\hat{\psi}_0\left[T \exp\left(\frac{-i}{\hbar}\int_0^t \hat{\phi}(t_3)\,dt_3\right)\right] \quad [7.2]$$

where

$$\hat{\phi}(t) = \int dx_3\, \phi(x_3,t)\hat{\psi}_0^+(x_3,t)\hat{\psi}_0(x_3,t) \quad [7.3]$$

and $\hat{\psi}_0(x,t)$ are the field operators defined in the interaction representation: i.e., they are solutions of the unperturbed ($\phi = 0$) equation

$$\frac{\hbar}{i}\frac{\partial}{\partial t}\hat{\psi}_0 = [\hat{H}_{(\phi=0)},\hat{\psi}_0] \quad [7.4]$$

They are, therefore, also the unperturbed Heisenberg operators. The ground state is given by

$$|N\rangle = \frac{T\exp\left(\frac{-i}{\hbar}\int_{-\infty}^{0}\hat{\phi}(t_3)\,dt_3\right)|N_I,-\infty\rangle}{\langle N_I,-\infty|N\rangle} \quad [7.5a]$$

$$= \frac{T\exp\left(\frac{+i}{\hbar}\int_{0}^{\infty}\hat{\phi}(t_3)\,dt_3\right)|N_I,+\infty\rangle}{\langle N_I,+\infty|N\rangle} \quad [7.5b]$$

where $|N_I,-\infty\rangle$ ($=|N_I,+\infty\rangle$) is the ground state without the application of the perturbation. In the interaction picture this is identical to the Heisenberg ground state $|N_H^0\rangle$. We can now rewrite the Green's function, for either time ordering, as

$$G(x_1,t_1,x_2,t_2) = \frac{1}{i}\frac{\langle N_H^0|T[\hat{S}\hat{\psi}_0(x_1,t_1)\hat{\psi}_0^+(x_2,t_2)]|N_H^0\rangle}{\langle N_H^0|\hat{S}|N_H^0\rangle} \quad [7.6]$$

where

$$\hat{S} = T\exp\left(\frac{-i}{\hbar}\int_{-\infty}^{+\infty}\hat{\psi}(t_3)\,dt_3\right) \quad [7.7]$$

For example, take $t_1 \geq t_2$

$$\begin{aligned}
&\langle N_H^0 | T[\hat{S}\hat{\psi}_0(x_1,t_1)\hat{\psi}_0^+(x_2,t_2)] | N_H^0 \rangle \\
&= \left\langle N_H^0 \right| \exp\left[\frac{-i}{\hbar}\int_{t_1}^{\infty} \hat{\phi}(t_3)\, dt_3\right] \hat{\psi}_0(x_1,t_1) \\
&\quad \times \exp\left[\frac{-i}{\hbar}\int_{t_2}^{t_1} \hat{\phi}(t_3)\, dt_3\right] \hat{\psi}_0^+(x_2,t_2) \exp\left[\frac{-i}{\hbar}\int_{-\infty}^{t_2} \hat{\phi}(t_3)\, dt_3\right] \left| N_H^0 \right\rangle \\
&= \langle N | \hat{\psi}(x_1,t_1)\hat{\psi}^+(x_2,t_2) | N \rangle
\end{aligned} \tag{7.8}$$

and so on. The denominators ensure that the phase of the Green's function and wave functions remain well defined and are a result of the Gell-Mann and Lowe theorem (see reference at end of chapter).

Suppose we now take the functional derivative based upon the variation

$$\phi(x,t) \to \phi(x,t) + \delta\phi(x,t) \tag{7.9}$$

and simplify the notation, since indices will proliferate enormously from now on, so that

$$(x_1,t_1) \to (1), \qquad (x_2,t_2) \to (2) \tag{7.10}$$

then

$$\begin{aligned}
\frac{\delta G(1,2)}{\delta\phi(3)} &= \frac{1}{i} \frac{\left\langle N_H^0 \right| T\left[\dfrac{\delta \hat{S}}{\delta\phi(3)} \hat{\psi}_0(1)\hat{\psi}_0^+(2)\right] \left| N_H^0 \right\rangle}{\langle N_H^0 | \hat{S} | N_H^0 \rangle} \\
&\quad - \frac{1}{i} \frac{\langle N_H^0 | T[\hat{S}\hat{\psi}_0(1)\hat{\psi}_0^+(2)] N_H^0 \rangle}{[\langle N_H^0 | \hat{S} | N_H^0 \rangle]^2} \left\langle N_H^0 \left| \frac{\delta \hat{S}}{\delta\phi(3)} \right| N_H^0 \right\rangle
\end{aligned} \tag{7.11}$$

Now

$$\begin{aligned}
-\frac{\delta \hat{S}}{\delta\phi(3)} &= \frac{i}{\hbar} \hat{S} \left[\frac{\delta}{\delta\phi(3)} \int_{-\infty}^{+\infty} d[4]\phi(4)\hat{\psi}_0^+(4)\hat{\psi}_0(4)\right] \\
&= \frac{i}{\hbar} \hat{S}\hat{\psi}_0^+(3)\hat{\psi}_0(3)
\end{aligned} \tag{7.12}$$

Substituting into Eq. (7.11) gives

$$-\frac{\hbar\delta G(1,2)}{\delta\phi(3)} = G_2(1,2,3,3^+) - G(1,2)G(3,3^+) \qquad [7.13]$$

where the 3^+ is necessary to conserve the order of the field operators. Thus, the two-particle Green's function can be written in terms of two single-particle Green's functions and a functional derivative. This is not really surprising since, in an interacting system, the introduction of a particle (described by the single-particle Green's function) is simply one of a set of possible perturbations. The reaction of the system to a perturbation must be described by essentially the same set of functions, in particular the second- and higher-order Green's functions. Equation (7.13) expresses that relationship. Furthermore, we see that the single-particle Green's function pair is just the one we need to extract the Hartree potential, i.e.

$$\begin{aligned} i\int v(x,x_3)G(x_1,t_1,x_2,t_2,x_3,t_3,x_3,t_3^+)\,dx_3 \\ = +i\int v(x_1,x_3)G(x_3,t_3,x_3,t_3^+)\,dx_3 \\ - i\hbar\int v(x_1,x_3)\delta(t_1-t_3)\frac{\delta G(x_1,t_1,x_2,t_2)}{\delta\phi(x_3,t_3)}\,dx_3\,dt_3 \\ = -V_H - i\hbar\int v(x_1,x_3)\delta(t_1-t_3)\frac{\delta G(x_1,t_1,x_2,t_2)}{\delta\phi(x_3,t_3)}\,dx_3\,dt_3 \qquad [7.14] \end{aligned}$$

Comparing this with our definition of the self-energy in Eq. (6.67), we now have

$$\begin{aligned} \int \Sigma(x_1,t_1,x_3,t_3)G(x_3,t_3,x_2,t_2)\,dx_3\,dt_3 \\ = i\hbar\int v(x_1,x_3)\delta(t_1-t_3)\frac{\delta G(x_1,t_1,x_2,t_2)}{\delta\phi(x_3,t_3)}\,dx_3\,dt_3 \qquad [7.15a] \end{aligned}$$

or

$$\int \Sigma(1,3)G(3,2)\,d[3] = \hbar i\int v(1,3)\frac{\delta G(1,2)}{\delta\phi(3)}\,d[3] \qquad [7.15b]$$

If we now define the inverse of the Green's function through the identity

$$\int G^{-1}(1,3)G(3,2)\,d[3] = \delta(1-2) \qquad [7.16a]$$

or

$$\mathsf{G}^{-1}\mathsf{G} = 1 \qquad [7.16b]$$

then we can write

$$\frac{\delta(\mathsf{G}^{-1}\mathsf{G})}{\delta\phi} = 0 \qquad [7.17]$$

or

$$\frac{\delta \mathsf{G}^{-1}}{\delta\phi}\mathsf{G} + \mathsf{G}^{-1}\frac{\delta \mathsf{G}}{\delta\phi} = 0 \qquad [7.18]$$

That is,

$$\frac{\delta \mathsf{G}}{\delta\phi} = -\mathsf{G}\frac{\delta \mathsf{G}^{-1}}{\delta\phi}\mathsf{G} \qquad [7.19]$$

In full form this is

$$\frac{\delta G(1,2)}{\delta\phi(3)} = -\int G(1,4)\frac{\delta G^{-1}(4,5)}{\delta\phi(3)}G(5,2)\,d[4]\,d[5] \qquad [7.20]$$

Substituting into equation (7.15) we have

$$\Sigma(1.2) = -i\hbar\int v(1,4)G(1,3)\frac{\delta G^{-1}(3,2)}{\delta\phi(4)}\,d[3]\,d[4] \qquad [7.21]$$

This constitutes a *formal* solution to the problem when taken with Equation (6.67).

$$\left[i\hbar\frac{\partial}{\partial t_1} - H(1) - V(1)\right]G(1,2) - \int\Sigma(1,2)G(3,2)\,d[3] = \hbar\delta(1,2) \qquad [7.22]$$

and

$$V(1) = \phi(1) - i\int v(1,3)G(3,3^+)\, d[3] \qquad [7.23]$$

These equations are normally solved by iteration.

It is interesting to note that we were able to show that the change in the Green's function was the difference between a two-particle and two single-particle Green's functions. From Eq. (7.13) we have

$$\hbar\,\delta G(1,2) = [G_2(1,2,3,3^+) - G(1,2)G(3,3^+)]\,\delta\phi(3) \qquad [7.24]$$

If we substitute for G_2 from Eq. (6.62) (the Hartree–Fock approximation) we have

$$\begin{aligned}\hbar\,\delta G(1,2) &= G(1,3^+)G(3,2)\,\delta\phi(3) \\ &= G(1,3^+)\phi(3)G(3,2)\end{aligned} \qquad [7.25]$$

Thus, in the approximation in which the two-particle Green's function reverts to two single-particle Green's functions, the change in the Green's function is simply the first term in the perturbation series (2.24) as expected in a noninteracting system.

7.2. ITERATIVE SOLUTION FOR THE GREEN'S FUNCTION AND SELF-ENERGY

Suppose we start with the noninteracting Green's function G_0

$$\left[i\hbar\frac{\partial}{\partial t_1} - H_0(x_1) - \phi(x_1,t_1)\right] G_0(x_1,t_1,x_2,t_2) = \hbar\,\delta(x_1 - x_2)\,\delta(t_1 - t_2) \qquad [7.26]$$

Comparing this with the definition for $G_0^{-1}(x_1,t_1,x_2,t_2)$

$$\int G_0^{-1}(x_1,t_1,x_3,t_3)G_0(x_3,t_3,x_2,t_2)\, dx_3\, dt_3 = \delta(x_1 - x_2)\,\delta(t_1 - t_2) \qquad [7.27]$$

we see that

$$G_0^{-1}(x_1,t_1,x_2,t_2) = \frac{1}{\hbar}\left[i\hbar\frac{\partial}{\partial t_1} - H_0(x_1) - \phi(x_1,t_1)\right]\delta(x_1 - x_2)\,\delta(t_1 - t_2) \tag{7.28}$$

Using this as the start of the iteration series we have

$$-\frac{\delta G_0^{-1}(x_1,t_1,x_2,t_2)}{\delta\phi(x_3,t_3)} = \frac{1}{\hbar}\,\delta(x_1 - x_3)\,\delta(t_1 - t_3)\,\delta(x_1 - x_2)\,\delta(t_1 - t_2) \tag{7.29}$$

It now follows from Eqs. (7.21) and (7.23), that

$$\Sigma^{(1)}(x_1,t_1,x_2,t_2) = iv(x_1,x_2)G_0(x_1,t_1,x_2,t_2) \tag{7.30}$$

$$\begin{aligned} V^{(1)}(x_1,t_1) &= \phi(x_1,t_1) - i\int v(x_1,x_3)G_0(x_3,t_1,x_3,t_1^+)\,dx_3 \\ &= \phi(x_1,t_1) + V_H^{(1)}(x_1,t_1) \end{aligned} \tag{7.31}$$

Since

$$\left[i\hbar\frac{\partial}{\partial t_1} - H_0(x_1) - V(x_1,t_1)\right]G(x_1,t_1,x_2,t_2)$$
$$\int \Sigma(x_1,t_1,x_3,t_3)G(x_3,t_3,x_2,t_2)\,dx_3\,dt_3 = \hbar\,\delta(x_1 - x_2)\,\delta(t_1 - t_2) \tag{7.32}$$

we have, from the definition of $G_0(x_1,t_1,x_2,t_2)$ and Eq. (6.70),

$$\mathbf{G}_0^{-1}\mathbf{G} - (\mathbf{V}_H + \Sigma)\mathbf{G} = 1 \tag{7.33}$$

or

$$\mathbf{G} = \mathbf{G}_0 + \mathbf{G}_0(\mathbf{V}_H + \Sigma)\mathbf{G} \tag{7.34}$$

In "full" form this becomes

$$\begin{aligned} G(x_1,t_1,x_2,t_2) = {} & G_0(x_1,t_1,x_2,t_2) + \int G_0(x_1,t_1,x_3,t_3) \\ & \times [V(x_3,t_3)\,\delta(x_4 - x_3)\,\delta(t_4 - t_3) + \Sigma(x_3,t_3,x_4,t_4)] \\ & \times G(x_4,t_4,x_2,t_2)\,dx_3\,dt_3\,dx_4\,dt_4 \end{aligned} \tag{7.35}$$

Substituting for $\Sigma^{(1)}$ and $V_H^{(1)}$, we obtain the first approximation to the Green's function $G_{(1)}(x_1,t_1,x_2,t_2)$ in the form of an integral equation. Alternatively, substituting directly into Eq. (7.22) we see that we obtain Eq. (6.65). Thus, for a third time we have derived the Hartree–Fock equation. The difference now is that it is the first term in a logical iterative sequence rather than being an intuitive guess.

Suppose we now carry on the iteration. Fortunately, we do not have to solve for $G_{(1)}(x_1,t_1,x_2,t_2)$; instead we can immediately propose an expression for $G_{(1)}(x_1,t_1,x_2,t_2)$ from the general expression [Eq. (6.73)]

$$\begin{aligned} G^{-1}(x_1,t_1,x_2,t_2) &= G_0^{-1}(x_1,t_1,x_2,t_2) \\ &\quad - V_H(x_1,t_1)\,\delta(x_1 - x_2)\,\delta(t_1 - t_2) - \Sigma(x_1,t_1,x_2,t_2) \end{aligned} \tag{7.36}$$

So, to continue the iteration, we need

$$\begin{aligned} G_{(1)}^{-1}(x_1,t_1,\ x_2,t_2) &= G_0^{-1}(x_1,t_1x_2,t_2) \\ &\quad - V_H^{(1)}(x_1,t_1)\,\delta(x_1 - x_2)\,\delta(t_1 - t_2) - \Sigma^{(1)}(x_1,t_1,x_2,t_2) \end{aligned} \tag{7.37}$$

Taking the functional derivative

$$\begin{aligned} \frac{\delta G_{(1)}^{-1}(x_1,t_1,x_2,t_2)}{\delta\phi(x_3,t_3)} &= \frac{\delta G_0^{-1}(x_1,t_1,x_2,t_2)}{\delta\phi(x_3,t_3)} \\ &\quad - \frac{\delta V_H^{(1)}(x_1,t_1)}{\delta\phi(x_3,t_3)}\,\delta(x_1 - x_2)\,\delta(t_1 - t_2) - \frac{\delta\Sigma^{(1)}(x_1,t_1,x_2,t_2)}{\delta\phi(x_3,t_3)} \end{aligned} \tag{7.38}$$

The first term simply repeats the first approximation (but with $G_{(1)}$ in place of G_0 in Eqs. (7.30) and (7.31)), the second and third are the corrections to it. In both the latter terms it is necessary to use Eqs. (7.19) and (7.29) to evaluate the term $\delta G_0/\delta\phi$. We then obtain

$$\frac{\delta V_H^{(1)}(x_1,t_1)}{\delta\phi(x_3,t_3)} = \int v(x_1,x_3)G_0(x_4,t_1,x_3,t_3)G_0(x_3,t_3,x_4,t_1^+)\,dx_4 \tag{7.39}$$

and

$$\frac{\delta\Sigma^{(1)}(x_1,t_1,x_2,t_2)}{\delta\phi(x_3,t_3)} = iv(x_1,x_2)G_0(x_1,t_1,x_3,t_3)G_0(x_3,t_3,x_2,t_2) \tag{7.40}$$

Substituting these expressions into Eq. (7.38) and then using the resulting functional derivative in Eq. (7.21), we have

$$\begin{aligned}\Sigma^{(2)}(x_1,t_1,x_2,t_2) &= iv(x_1,x_2)G_{(1)}(x_1,t_1,x_2,t_1^+)\\ &+ i\int v(x_1,x_4)v(x_3,x_5)G_{(1)}(x_1,t_1,x_3,t_3)G_0(x_5,t_3,x_4,t_1)G_0(x_4,t_4,x_5,t_3^+)\,dx_3\,dt_3\,dx_4\,dx_5\\ &- \int v(x_1,x_4)G_{(1)}(x_1,t_1,x_3,t_2)v(x_3,x_2)G_0(x_3,t_2,x_4,t_1)G_0(x_4,t_1,x_2,t_2)\,dx_3\,dx_4\end{aligned} \quad [7.41]$$

Now, it is not really necessary to go into these terms in detail but obviously we can continue on in the series. We see that it is, in fact, an iteration in powers of the coulomb interaction. Thus, the second correction is of order v^2, the third will be of order v^3, and so on. The first term in the series was Hartree–Fock, so subsequent terms are corrections to this theory.

There is a dilemma here; the first approximation is wrong for solids, yet corrections to Hartree–Fock are of the order of the coulomb interaction. Since this interaction is very strong and long range, we would not expect the subsequent terms to die off very quickly. Thus, if we are to get realistic corrections at all, we have to go to very high orders in the series. In fact, in a different form, this problem was recognized early in the development of the subject, and it has been shown by a number of authors that the series needs to be treated very carefully if correct answers are to be obtained. Rather than repeat the arguments which were used then, it is better to review physically what the iteration series really means.

The solution to the equation of motion depends upon the formula for the two-particle Green's function in terms of the single-particle Green's function, i.e.

$$-\hbar\,\frac{\delta G(1,2)}{\delta\phi(3)} = G_2(1,2,3,3^+) - G(1,2)G(3,3^+) \quad [7.13]$$

Now, we could consider this as relating the change in the properties of the system (δG) under an external perturbation ($\delta\phi$) to the properties of G_2 and G. If we think of G as representing the motion of a single particle, we see that there is a real physical problem with Eq. (7.13) because the particle cannot distinguish between an external perturbation and any internal potential, produced by the response to that perturbation, from the rest of the system.

Taking a simple example, if we apply a large electric field to the surface of a metal, the induced charge on the surface will screen the interior of the metal from the field; for almost all of the metal the perturbation caused by the field is very weak. Using Eq. (7.13), however, would give an enormous change in the Green's function, in first order, which would not be physically realistic. Only if we went to high orders in the perturbation series might we get the correct small answer.

Just as in the case of Thomas–Fermi theory, any particle reacts not to the external potential, but to the total potential, including the response of all of the other particles. Except for special cases, the result of allowing the rest of the system to respond is to reduce drastically the effective potential. Thus, we should consider the change in the Green's function with respect to the total potential, not the external potential. This should give a more realistic answer in low order and hopefully the resulting series will be more convergent.

7.3. SCREENING AND THE PERTURBATION SERIES

Consider the potential function

$$\begin{aligned} V(x_1,t_1) &= \phi(x_1,t_1) - i\int v(x_1,x_2)G(x_2,t_1,x_2,t_1^+)\,dx_2 \\ &= \phi(x_1,t_1) + \int v(x_1,x_2)\rho(x_2,t_1)\,dx_2 \end{aligned} \qquad [7.42]$$

where ρ is the charge density. In terms of this function, a change in the external potential $\delta\phi$ will result in

$$\delta V(x_1,t_1) = \delta\phi(x_1,t_1) + \int v(x_1,x_2)\,\delta\rho(x_2,t_2)\,dx_2 \qquad [7.43]$$

where $\delta\rho$ is the resulting change in the charge density. Thus, we see that the change in the potential consists of the external potential change $\delta\phi$ less the induced potential due to the charge density change. We conclude, therefore, on physical grounds, that it is this change in the total single-particle potential δV which we should consider. It is usual and useful to define some functions by way of analogies to classical electrostatics.

7.3.1. Dielectric Response Function. Classically, if we apply a field to a solid, the screening of the solid is defined in terms of a dielectric constant. Thus, we would write

$$V_{\text{INT}} \approx \frac{\phi_{\text{EXL}}}{\epsilon}$$

In many-body theory we *define* the *inverse* of the dielectric response as

$$\epsilon^{-1}(x_1,t_1,x_2,t_2) = \frac{\delta V(x_1,t_1)}{\delta\phi(x_2,t_2)} \qquad [7.44]$$

Substituting from Eq. (7.43) this gives

$$\epsilon^{-1}(x_1,t_1,x_2,t_2) = \delta(x_1 - x_2)\,\delta(t_1 - t_2) + \int v(x_1,x_3)\,\frac{\delta\rho(x_3,t_1)}{\delta\phi(x_2,t_2)}\,dx_3 \qquad [7.45]$$

7.3.2. Screened Interaction. Following the discussion on Thomas–Fermi response, in which we argued that the interaction was reduced by the screening, we define a screened coulomb interaction as

$$W(x_1,t_1,x_2,t_2) = \int\epsilon^{-1}(x_1,t_1,x_3,t_3)v(x_3,x_2)\,\delta(t_3 - t_2)\,dx_3\,dt_3 \qquad [7.46]$$

or formally

$$\mathsf{W} = \epsilon^{-1}\mathsf{v} \qquad [7.47]$$

7.3.1. Polarization Propagator. By analogy with the classical polarization, we define a *polarization propagator* through the relationship

$$\epsilon(x_1,t_1,x_2,t_2) = \delta(x_1 - x_2)\,\delta(t_1 - t_2) - \int v(x_1,x_3)\,\delta(t_1 - t_3)\,P(x_3,t_3,x_2,t_2dx_3\,dt_3 \qquad [7.48]$$

or

$$\epsilon = 1 - \mathsf{vP} \qquad [7.49]$$

These functions can be manipulated to produce a number of relationships. From Eq. (7.47) we have

$$\epsilon\mathsf{W} = \mathsf{v} \qquad [7.50]$$

Substituting from Eq. (7.49) and rearranging gives

$$\mathsf{W} = \mathsf{v} + \mathsf{vPW} \tag{7.51}$$

an integral equation for the screened interaction in terms of the polarization. The importance of the polarization propagator is that it is immediately obtainable in terms of the partial derivatives. We can write Eq. (7.45) as

$$\begin{aligned} \epsilon^{-1} &= 1 + \mathsf{v}\frac{\delta\rho}{\delta\phi} \\ &= 1 + \mathsf{v}\frac{\delta\rho}{\delta V}\cdot\frac{\delta V}{\delta\phi} \end{aligned} \tag{7.52}$$

From (7.44)

$$\epsilon^{-1} = 1 + \mathsf{v}\frac{\delta\rho}{\delta V}\epsilon^{-1}$$

or

$$\epsilon = 1 - \mathsf{v}\frac{\delta\rho}{\delta V} \tag{7.53}$$

Thus, we have

$$\mathsf{P} = +\frac{\delta\rho}{\delta V} \tag{7.54}$$

or

$$\begin{aligned} P(x_1,t_1,x_2,t_2) &= +\frac{\delta\rho(x_1,t_1)}{\delta V(x_2,t_2)} \\ &= -i\frac{\delta G(x_1,t_1,x_1,t_1^+)}{\delta V(x_2,t_2)} \end{aligned} \tag{7.55}$$

Thus, we can write

$$P(1,2) = i\int G(1,3)\,\frac{\delta G^{-1}(3,4)}{\delta V(2)}\,G(4,1^+)\;d[3]\;d[4] \tag{7.56}$$

The function

$$\Gamma(x_1,t_1,x_2,t_2,x_3,t_3) = -\frac{\hbar\,\delta G^{-1}(x_1,t_1,x_2,t_2)}{\delta V(x_3,t_3)} \qquad [7.57]$$

appears so often that it is given a special name—*the vertex function.*

Having defined these functions we return to the self-energy

$$\Sigma(1,2) = -i\hbar \int v(1,4)G(1,3)\,\frac{\delta G^{-1}(3,2)}{\delta\phi(4)}\,d[3]\,d[4] \qquad [7.58]$$

If we convert to the derivative with respect to the total potential

$$\begin{aligned}\Sigma(1,2) &= -i\hbar \int v(1,4)G(1,3)\,\frac{\delta G^{-1}(3,2)}{\delta V(5)}\,\frac{\delta V(5)}{\delta\phi(4)}\,d[3]\,d[4]\,d[5] \\ &= -i\hbar \int v(1,4)\,\epsilon^{-1}(5,4)G(1,3)\,\frac{\delta G^{-1}(3,2)}{\delta V(5)}\,d[3]\,d[4]\,d[5]\end{aligned} \qquad [7.59]$$

Combining the coulomb potential with the dielectric response we have

$$\Sigma(1,2) = i\hbar \int W(1,4)G(1,3)\,\frac{\delta G^{-1}(3,2)}{\delta V(4)}\,d[3]\,d[4] \qquad [7.60]$$

This expression is formally identical to Eq. (7.21) except that the bare coulomb interaction has been replaced by the screened interaction, and the functional derivative with respect to the external potential by the derivative with respect to total screened interaction. The formal solution to the interaction problem consists now of the following set of equations:

$$\Gamma(1,2,3) = -\hbar\,\frac{\delta G^{-1}(1,2)}{\delta V(3)} \qquad [7.61a]$$

$$P(1,2) = -i/\hbar \textstyle\int G(1,3)\Gamma(3,4,2)G(4,1^+)\,d[3]\,d[4] \qquad [7.61b]$$

$$\epsilon(1,2) = \delta(1-2) - \textstyle\int v(1,3)P(3,2)\,d[3] \qquad [7.61c]$$

$$W(1,2) = \textstyle\int \epsilon^{-1}(1,3)v(3,2)\,d[3] \qquad [7.61d]$$

$$\Sigma(1,2) = i\textstyle\int W(1,4)G(1,3)\Gamma(3,2,4)\,d[3]\,d[4] \qquad [7.61e]$$

$$V(1) = \phi(1) - i\textstyle\int v(1,3)G(3,3^+)\,d[3] \qquad [7.61f]$$

In order to start the iteration off we need a first-order Green's function. We use the best *Hartree* solution to define the noninteracting Green's function

$$\left[i\hbar\frac{\partial}{\partial t_1} - H(x_1) - V(x_1,t_1)\right] G_0(x_1,t_1,x_2,t_2) = \hbar\delta(x - x_1)\delta(t - t_1) \quad [7.62]$$

which is solved self-consistently with

$$V(x_1,t_1) = -i\int v(x_1,x_3)G_0(x_3,t_1,x_3,t_1^+)\, dx_3 \quad [7.63]$$

The iteration cycle is then closed by

$$G^{-1}(1,2) = G_0^{-1}(1,2) - \Sigma(1,2) \quad [7.64]$$

as before. From Eq. (7.62) we have

$$G_0^{-1}(x_1,t_1,x_2,t_2) = \frac{1}{\hbar}\left[i\hbar\frac{\partial}{\partial t_1} - H(x_1) - V(x_1,t_1)\right] \delta(x_1 - x_2)\,\delta(t_1 - t_2) \quad [7.65]$$

so that

$$-\hbar\delta\frac{G_0^{-1}(x_1,t_1,x_2,t_2)}{\delta V(x_3,t_3)} = \delta(x_1 - x_2)\,\delta(t_1 - t_2)\,\delta(x_1 - x_3)\,\delta(t_1 - t_3) \quad [7.66a]$$

or

$$\Gamma(1,2,3) = \delta(1 - 2)\,\delta(1 - 3) \quad [7.66b]$$

It follows that

$$P_0(1,2) = \frac{-i}{\hbar}G_0(1,2)G_0(2,1^+) \quad [7.67]$$

In terms of the polarization we obtain the response function ϵ and the screened interaction W. The solution of Eq. (7.51), or equivalently the inversion of the response function $\epsilon \rightarrow \epsilon^{-1}$, is *not* a trivial calculation, however, except in the special case of a uniform system (we will return to this later). But we will assume that this has been done and we have the screened interaction. The first approximation to the self-energy is then

$$\Sigma^{(1)}(1,2) = iW_0(1,2)G_0(1,2) \qquad [7.68]$$

This is equivalent to Eq. (7.30) except for the substitution of the screened interaction for the bare coulomb interaction. Thus, we have a modification to Hartree–Fock in the first iteration. Further, we see that, since it is the screened interaction that is now involved, we will not have the problems associated with the long-range part of the coulomb interaction.

The second iteration proceeds from

$$G_{(1)}^{-1}(1,2) = G_0^{-1}(1,2) - \Sigma^{(1)}(1,2) \qquad [7.69]$$

The term G_0 is again the solution of the Hartree equation, which is now assumed to be

$$\left[i\hbar \frac{\partial}{\partial t_1} - H(x_1) - V_H^{(1)}(x_1,t_1) \right] G_0(x_1t_1x_2t_2) = \hbar\, \delta(x - x_1)\, \delta(t - t_1) \qquad [7.70]$$

with

$$V_H^{(1)}(x_1,t_1) = -i\int v(x_1,x_3)G_{(1)}(x_3,t_1,x_3,t_1^+)\, dx_3 \qquad [7.71]$$

That is, G_0 is still the solution to the "best" Hartree equation. Thus, G_0 *changes* implicitly from iteration to iteration as does the interaction. The next stage of the iteration starts with the new vertex function

$$\Gamma(1,2,3) = \delta(1 - 3)\, \delta(1 - 2) - iW_0(1,2)G_0(1,3)G_0(3,2)$$

Introducing this approximation into the series of Eqs. (7.61) we have

$$\Sigma^{(2)}(1,2) = iW_0(1,2)G_{(1)}(1,2) - i\int W_{(1)}(1,2)G_{(1)}(1,3)W_0(3,2)G_0(3,4)G_0(4,2)\, d[3]\, d[4]$$

The new interaction $W_{(1)}$ corresponds to the new polarization in the second iteration. At this stage the series becomes complicated because we now have two interactions from different iterations as well as an implicit change in G_0, all appearing in the second level of iteration. One could hope that by using the second iteration interaction for all expressions at this level we might have a better calculation of the self-energy. We might also hope that using the best Hartree approximation [Eq. (7.70)], i.e., including the change in the Green's function, will give us still better results. Unfortunately this is not in general true. As we have said before, it is much more important to get the physics correct than the mathematics.

Ignoring the differences between interactions, we see that schematically the expansion of the self-energy now behaves like

$$\Sigma = \alpha_1 W + \alpha_2 W^2 + \alpha_3 W^3 + \cdots \qquad [7.72]$$

Thus, we have an expansion in orders of the weaker-screened interaction. There is hope, therefore, that this series will converge much more rapidly, so that we might need to consider only the first few terms at most.

7.4. THE SCREENED INTERACTION AND SELECTIVE SUMMATIONS

It is obvious that it is the introduction of the screened interaction that is the main change in the expansion series we are now considering. It is as well, therefore, to briefly look at the relationship this implies between the self-energies of Eqs. (7.21) and (7.60). Combining Eqs. (7.51) and (7.67) to give an equation for the first approximation to the screened interaction

$$W_0(1,2) = v(1,2) + \int v(1,3)P_0(3,4)W_0(4,2)\, d[3]\, d[4] \qquad [7.73]$$

This integral equation (a Dyson equation) can be solved by repeated substitution in exactly the same way as for the single-particle Green's function of Eq. (2.24). That is, formally

$$\begin{aligned} \mathrm{W}_0 &= \mathrm{v} + \mathrm{vP}_0\mathrm{W}_0 \qquad [7.74] \\ &= \mathrm{v} + \mathrm{vP}_0\mathrm{v} + \mathrm{vP}_0\mathrm{vP}_0\mathrm{v} + \mathrm{vP}_0\mathrm{vP}_0\mathrm{vP}_0\mathrm{v} + \cdots \end{aligned}$$

Substituting into the self-energy expression and replacing the polarization by the Green's functions of Eq. (7.67), we obtain an expansion in terms of the bare interaction

$$\Sigma^{(1)}(x_1,t_1,x_2,t_2) = iv(x_1,x_2)G_0(x_1,t_1,x_2,t_2) + i\int v(x_1, x_3)G_0(x_3,t_1,x_4,t_2^+)G_0(x_4,t_2^+,x_1,t_1) \times v(x_4x_2)G_0(x_1,t_1,x_2,t_1^+)\, dx_3\, dx_4 + \cdots \qquad [7.75]$$

The first approximation to the self-energy in terms of the screened interaction contains also second and higher terms in the expansion of the self-energy in the bare interaction. If we were to compare, term for term, the expansion of the "bare interaction" self-energies with the expansion of the first approximation for the self-energy in the screened interaction, we would find, not surprisingly, that each term in the expansion of the screened interaction corresponded to a term in the bare interaction self-energy. Thus, the use of the screened interaction in the first self-energy term sums a *selected set* of terms, in the bare interaction, to infinite order.

As we expand the self-energy to higher orders in the screened interaction, we thus automatically include a large number of terms each time, which we would have had to calculate in the bare interaction. Looking at Eq. (7.74), we see why this will be a large improvement in solids. In essence it is the expansion of the expression

$$W = \frac{v}{1 - vP_0} = \frac{v}{\epsilon} \qquad [7.76]$$

If the dielectric constant is large, as it is in solids, vP_0 is large and the infinite series converges slowly. For atoms or very dilute interacting systems, however, vP_0 can be small. The first term (v) would then be a good approximation and Hartree–Fock would work.

In conclusion, mathematically the use of the screened interaction selectively sums an infinite set of terms in the expansion of the self-energy, while physically it corresponds to accepting the principle that any particle in an interacting system will not respond to the external potential but rather to the total potential, including any induced potential due to the response of the system.

7.5. THE UNIFORM SYSTEM

We have deliberately kept the discussion at the level of greatest generality. The case of a uniform system is, however, a very special case. Translational invariance allows enormous simplifications and many of the integral equations revert to simple algebraic relationships.

In a uniform system a function $f(x_1,t_1,x_2,t_2)$ can only be a function of the position difference $x_1 - x_2$, since absolute positions have no relevance. In addition, for the systems we normally consider, the time difference is also the only significant variable. Thus

$$f(x_1,t_1,x_2,t_2) \Rightarrow f((x_1 - x_2),(t_1 - t_2))$$

Putting

$$x_1 - x_2 = x, \qquad t_1 - t_2 = \tau$$

we can define a Fourier transform to momentum energy variables through

$$f(k,\omega) = \frac{1}{(2\pi)^4\hbar} \int f(x,\tau)\, e^{-i[kx-(\omega/\hbar)\tau]}\, dx\, d\tau \qquad [7.77]$$

Thus, we replace all the functions G, P, ϵ, W, Σ by their momentum–energy transforms. It is unnecessary to go through the full derivation of the resulting equations. However, in a uniform system the vertex function transforms in the following way:

$$\Gamma(x_1,t_1,x_2,t_2,x_3,t_3) \Rightarrow \Gamma(x_1 - x_2, x_2 - x_3, t_1 - t_2, t_2 - t_3)$$

or

$$\Gamma(x_1 - x_2, t_1 - t_2;\ x_2 - x_3, t_2 - t_3)$$

On Fourier transforming, we have two sets of variables, i.e.

$$\Gamma \rightarrow \Gamma(k,\omega;\ q,\Omega)$$

The set of Eqs. (7.61) become

$$P(q,\omega) = \frac{-i}{(2\pi)^4} \int G(q+k,\omega+\omega')G(k,\omega') \times \Gamma(q+k,\, \omega+\omega';\, -k,-\omega')\, dk\, d\omega' \qquad [7.78a]$$

$$\epsilon(q,\omega) = 1 - v(q)P(q,\omega) \qquad [7.78b]$$

$$W(q,\omega) = \frac{v(q)}{\epsilon(q,\omega)} = \frac{v(q)}{1 - v(q)P(q,\omega)} \qquad [7.78c]$$

$$\Sigma(q,\omega) = \frac{i}{(2\pi)^4} \int G(q-k,\omega-\omega')W(k,\omega') \times \Gamma(q-k,\omega-\omega';k,\omega')\, dk\, d\omega' \qquad [7.78d]$$

The Green's function equation becomes

$$G(q,\omega) = G_0(q,\omega) + G_0(q,\omega)\Sigma(q,\omega)G(q,\omega) \qquad [7.79]$$

or

$$G^{-1}(q,\omega) = G_0^{-1}(q,\omega) - \Sigma(q,\omega) \qquad [7.80]$$

In the uniform system the self-energy may be usefully connected to the Lehman representation and the spectral weight. Writing the Green's function in the Lehman representation we have

$$G(q,\omega) = \frac{1}{\omega - E(q) - \Sigma(q,\omega)} = \int_0^\infty d\omega' \left[\frac{A(q,\omega')}{\omega - \omega' - \mu + i\delta} + \frac{B(q,\omega')}{\omega + \omega' - \mu - i\delta} \right] \qquad [7.81]$$

We expect the self-energy to be, in general, complex, so writing

$$\Sigma(q,\omega) = \Sigma_R(q,\omega) + i\Sigma_I(q,\omega) \qquad [7.82]$$

and remembering that $A(q,\omega')$, $B(q,\omega')$ are positive definite quantities we can equate imaginary parts to give

$$\frac{i\Sigma_I(q,\omega)}{[\omega - E(q) - \Sigma_R(q,\omega)]^2 + \Sigma_I^2(q,\omega)} = \begin{cases} -\pi i A(q,\omega - \mu)\theta(\omega - \mu) \\ +\pi i B(q,\mu - \omega)\theta(\mu - \omega) \end{cases} \qquad [7.83]$$

from which we have

$$A(q,\omega - \mu) = \frac{-\Sigma_I(q,\omega)}{[\omega - E(q) - \Sigma_R(q,\omega)]^2 + \Sigma_I^2(q,\omega)} \qquad [7.84a]$$

$$B(q,\mu - \omega) = \frac{+\Sigma_I(q,\omega)}{[\omega - E(q) - \Sigma_R(q,\omega)]^2 + \Sigma_I^2(q,\omega)} \qquad [7.84b]$$

For a small value of the imaginary part of the self-energy, we see that we have a Lorentzian form for the spectral weight functions around the excitation energy

$$\omega = E(q) - \Sigma_R(q,\omega) \qquad [7.85]$$

But in general the spectral weight function will be smeared out over a range of energies governed to a large extent by the value of $\Sigma_I(q,\omega)$.

We can now proceed with the iterative solution in terms of the single-particle Green's function

$$G_0(q,\omega) = \frac{1}{\omega - E(q) + i\delta \operatorname{sgn}(E(q) - \mu)} \qquad [7.86]$$

The first and second iterations are, as one might expect, simply Fourier transforms of the equations we have already derived, and all of the remarks we have made apply equally to these solutions. The only saving is in the algebraic relationships between the interacting and noninteracting Green's functions and the polarization, response function, and screened interaction. As before, we have a rapid increase in the complexity of the terms in the series.

There are a large number of situations where it is necessary to identify the physical process taking place so that, in a situation where we expect it to be important in the calculation of the self-energy, we can be sure to include all the terms which depend upon it. This is not immediately possible in the formulation we now have because of the algebraic complexity. What is needed is a *physical* interpretation of the terms in the perturbation expansion akin to the propagator diagrams we used for the single-particle Green's function in Chap. 2.

BIBLIOGRAPHY

Functional Derivative Techniques

HEDIN, L., and LUNDQVIST, S., *Solid State Phys.*, **23**, 1, 1969.

KADANOFF, L., and BAYM, G., *Quantum Statistical Mechanics*, Benjamin, Menlo Park: CA, 1976.

BAYM, G., *Ann. Phys., (N.Y.)*, **14**, 1, 1961.

HEDIN, L., *Phys. Rev.*, **139**, A769, 1965.

GELL-MANN, M., and LOWE, F., *Phys. Rev.*, **84**, 350, 1951.

Dielectric Response

NOZIÈRES, P., and PINES, D., *Nuovo Cimento*, **9**, 470, 1958.

PINES, D., *Elementary Excitations in Solids*, Benjamin, Menlo Park: CA, 1964.

PROBLEMS

1. Show by direct substitution that the Green's function of Eq. (7.6) is a solution to the "perturbed" equation of motion.

2. A one-dimensional insulator is described by two bands separated by an energy gap such that the eigenstates and eigenvalues are given by

$$|k,A\rangle = \frac{1}{\sqrt{N}}\sum_l e^{ikl}\,A(r-l), \qquad E_A(k) = \alpha k^2$$

$$|k,B\rangle = \frac{1}{\sqrt{N}}\sum_l e^{ikl}\,B(r-l), \qquad E_B(k) = E_g + \beta k^2$$

where l is the site index and $A(r)$, $B(r)$ are localized functions so that the overlap between sites is small:

(i) Obtain the first-order polarization propagator $P_0(r,r',\omega)$ and show that it may be written in the form

$$P_0(r,r',\omega) \approx [A(r-l)B^*(r-l)]\tilde{P}_0(l,l',\omega)[A(r'-l')B^*(r'-l')]^*$$

and find $\tilde{P}_0(l,l',\omega)$.

(ii) For the case of $\alpha = \beta = 0$ show that the response consists of a set of independent polarizable centers and calculate the response function $\epsilon(r,r',\omega)$ in this limit.

3. The particles in a uniform gas interact by way of a contact energy-dependent interaction

$$W(x - x',\omega) = \frac{A\delta(x - x')}{(\omega - \omega_p + i\delta)(\omega - \omega_p - i\delta)}$$

Obtain expressions for the complex self-energy in first order and show that the quasiparticle energies are real for energies within $\pm\omega_p$ of the Fermi energy.

4. If the matrix $A(a,b)$ may be written in the form

$$A(a,b) = \delta(a,b) - \sum_{p} X(a,p)Y(p,b)$$

show that

$$A^{-1}(a,b) = \delta(a,b) + \sum_{p,q} X(a,p)S(p,q)Y(q,b)$$

where

$$S^{-1}(p,q) = 1 - \sum_{a} Y(p,a)X(a,q)$$

Apply this to the result of Prob. 2(ii) and obtain an expression for the inverse response function $\epsilon^{-1}(r,r',\omega)$.

5. Show that Eqs. (7.78) are indeed the transform of Eqs. (7.61).

CHAPTER EIGHT

DIAGRAMMATIC INTERPRETATION OF THE GREEN'S FUNCTION SERIES

The introduction of the screened interaction can be considered, as in Sec. 7.4, as the sum of a particular set of terms in the iteration series. It may very well turn out that some other sum of terms corresponds to an important physical process, but the complexity of the mathematics of Chap. 7 would obscure rather than illuminate the connection.

Chapter 2 contains a simple example of this in the occurrence of bound states in the perturbation expansion of the Green's function. The appearance of a bound state corresponded to a divergence of the infinite perturbation series. With the diagrammatic description of the Green's function as a propagator, this divergence had a simple visual explanation—it was becoming favorable for the particle to remain in the vicinity of the scattering potential.

In an interacting system with a major increase of complexity it is even more advantageous to have an equivalent descriptive facility because the number of possible processes increases rapidly. For this reason, diagrams that illustrate the terms in the perturbation expansion of the Green's function and self-energy are very important in practical applications.

8.1. DIAGRAMMATIC INTERPRETATION OF THE PERTURBATION SERIES

We can continue to iterate the series of Eqs. (7.61) as far as we like, but their complexity tends to obscure the physical processes that are taking place. A number of shorthand notations for the equations have already been used, matrix methods being the most obvious. The shortcoming of the matrix method is that, although it facilitates manipulation enormously, it does not give any physical insight, nor is it obvious how we can treat the vertex function or the self-energy terms as matrices.

The visual method used is based, not unnaturally, upon the Green's function as a propagator. We have already used these diagrams in the single-particle Green's function scattering problem, but in order to treat the interacting system we need a number of symbols. We write

$G(1,2) \rightarrow$ 1 ——→—— 2 [8.1]

$G_0(1,2) \rightarrow$ 1 - - →- - - 2 [8.2]

$v(1,2) \rightarrow$ 1 ·········· 2 [8.3]

$W(1,2) \rightarrow$ 1 ∿∿∿∿∿ 2 [8.4]

$\Gamma(1,2,3) \rightarrow$ 1, 2, 3 [8.5]

$P(1,2) \rightarrow$ 1, 2 [8.6]

$\Sigma(1,2) \rightarrow$ 1, 2 [8.7]

The shape of the last three symbols are related to their "standard shape" in the second (for Γ) and first iterations of the self-energy equations.

If we have the general rule that all internal vertices are integrated over, then Eqs. (7.61) become (neglecting constants)

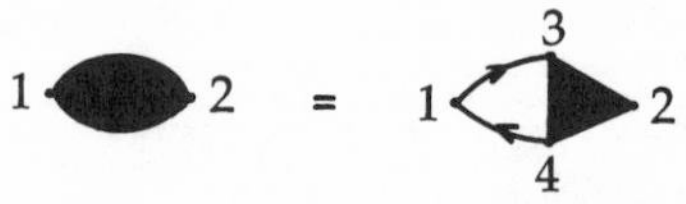

1, 2 = 1, 3, 4, 2 [8.8]

1 ∿∿∿∿∿ 2 = 1 ·········· 2 + 1 ·········· 3, 4 ∿∿∿∿∿ 2 [8.9]

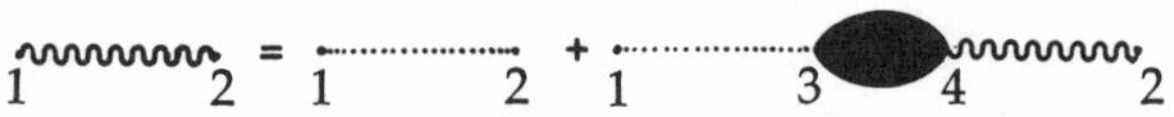

1 2 = 1 3 4 2 [8.10]

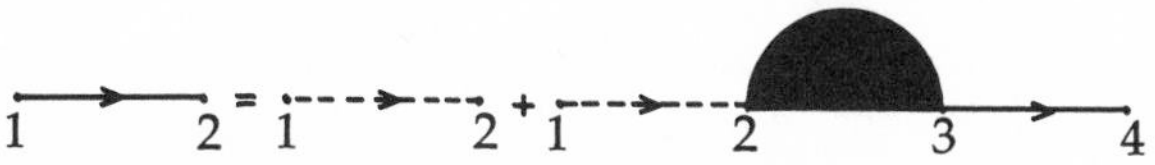

1 2 = 1 2 + 1 2 3 4 [8.11]

On this basis the first iteration [Eqs. (7.66)–7.69)] is drawn as

$$\Gamma = \frac{1}{2} \bullet 3 \; [\equiv \delta(1,3)\, \delta(1,2)] \tag{8.12}$$

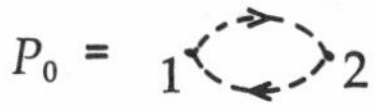

P_0 = 1 2 [8.13]

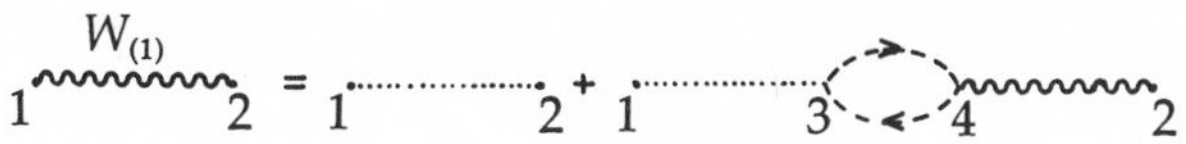

$W_{(1)}$

1 2 = 1 2 + 1 3 4 2 [8.14]

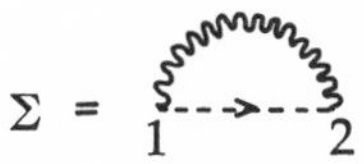

Σ = 1 2 [8.15]

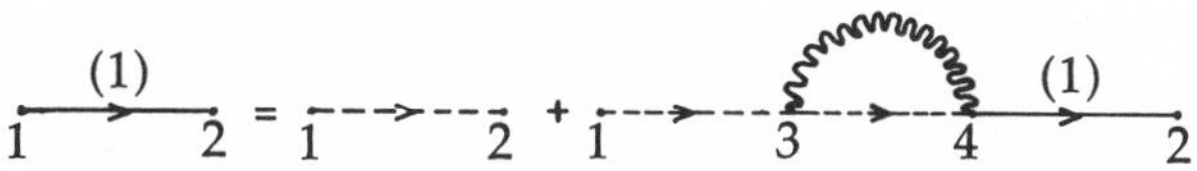

(1)

1 2 = 1 2 + 1 3 4 (1) 2 [8.16]

The second iteration results from [cf. Eq. (7.70)]

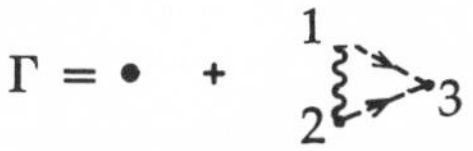

$\Gamma = \bullet$ + 1 2 3 [8.17]

Substituting this "picture" of Γ into the series gives

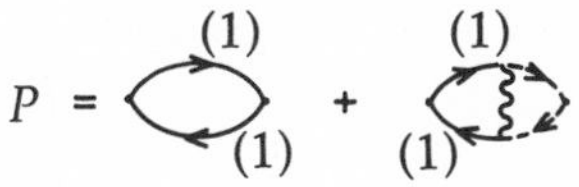

P = (1) (1) + (1) (1) [8.18]

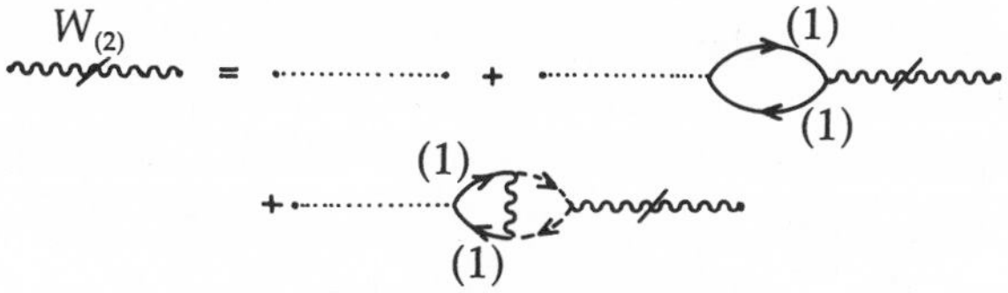

$W_{(2)}$

= + (1) (1)

+ (1) (1) [8.19]

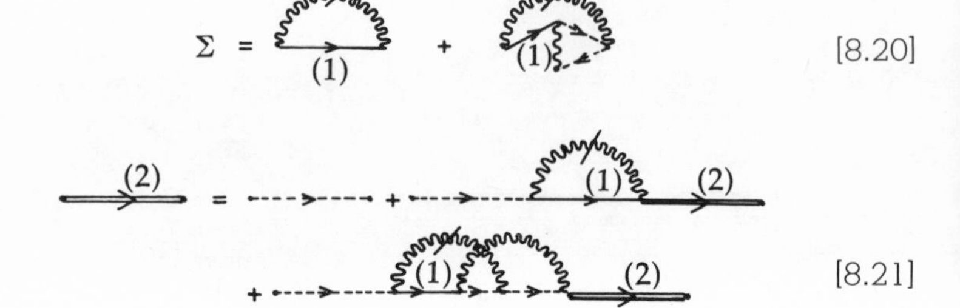

and so on.

Even in second order the diagrams are becoming complicated. Usually one assumes that some terms are known as accurately as possible for the purposes of the perturbation series. The interaction is a case in point. Instead of keeping track of first- and second-order corrections one simply puts the *best* interactions one can into all of the terms. The noninteracting Green's function is another case where keeping track of repeated changes in the Hartree potential is not often considered—one assumes that one already has the "best" noninteracting Green's function. This is dangerous, as we shall see, but not normally fatal.

This is, in fact, a rigorous procedure when one considers the iteration carried to infinite orders. It is obvious, for instance, that at each stage the Green's function in the evaluations of P, Σ, W, etc., is one stage behind the final calculated Green's function of that iteration. Thus, the first term of the self-energy is of the form

$$\Sigma = iWG \qquad [8.22]$$

in each iteration, but G and W are the *best* at that iteration. Thus, in practice, in an infinite expansion of the self-energy in terms of the Green's function, the series becomes

Σ = [diagram] + [diagram] + ... [8.23]

where G, W are the *true* Green's functions and interaction. But one must be careful in actual calculations. The substitution of G_0 for G, for instance, may give a *better* calculation (in terms of being accurate) than the use of $G_{(1)}$, for example, in the second iteration. The series tend to be asymptotic rather than simply convergent. Before going on, let us see how we may physically interpret these diagrams in the first and second iteration.

8.1.1. Green's Function. If we interpret

1•----➤----•2

as the transport of a particle from point 1 to point 2 then it follows that

1•----◄----•2

must correspond to the transport of a hole between 1 and 2 for the *same* time ordering of 1 and 2.

8.1.2. Polarization.

The lowest-order polarization diagram consists of the simultaneous creation of an electron and a hole at point 1, their transport to point 2, and their subsequent annihilation. This fits into our usual classical ideas of the polarization process as the production of positive and negative charges in the medium and the resulting reduction of the applied field.

8.1.3. Interaction. The diagrams for the interaction can best be thought of in the second quantization form:

1•··········•2

represents the transport of photons from 1 to 2; i.e., the interaction of charged particles via the coulomb field is equivalent to the exchange of photons. The screened interaction equation, expanded as a perturbation series, now becomes

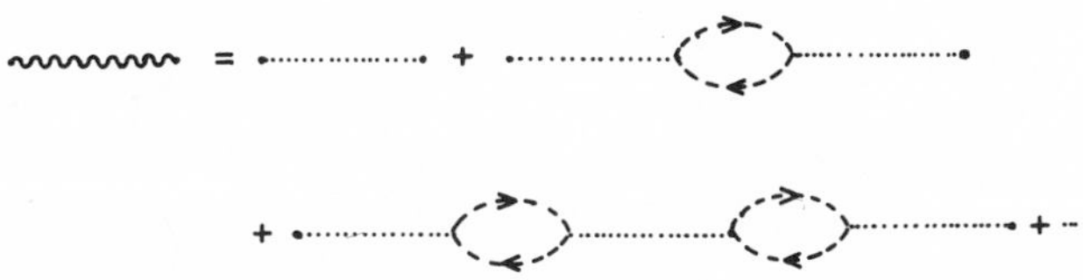

[8.24]

A term like

can be interpreted as the photon creating an electron–hole pair which then propagates and annihilates at a later time producing another photon, which then excites another electron–hole pair somewhere else, and so on. So the screened interaction between any two points consists of a whole series of excitations involving many electron–hole pairs, i.e., the system as a whole is involved. We will later see that we can also interpret the screened interaction line as an exchange of an elementary excitation between the interacting particles, the plasmon.

8.1.4. Self-Energy. Diagrams like

must be considered as the interaction of a particle with itself; that is, the particle creates a disturbance in the system (represented by the interaction) which is then reabsorbed by the particle as it propagates. Thus, the motion of the particle through the system consists of a combination of its free motion and all of its interactions with the rest of the system. Thus, the self-energy incorporates all of the effects of the self-interaction, as its name implies.

8.2. DIAGRAMMATIC EXPANSION

Having an interpretation of the perturbation expansion is of benefit. There is, however, a way of actually developing the perturbation series using the diagrams. Suppose we start with the basic first-order vertex function

$$\Gamma_{(0)}(1,2,3) = -\hbar\frac{\delta G_0^{-1}(1,2)}{\delta V(3)} = \delta(1-2)\delta(2-3) \Rightarrow {}^{1}_{2}\bullet 3 \qquad [8.25]$$

Each subsequent iteration starts with

$$G^{-1}(1,2) = G_0^{-1}(1,2) - \Sigma(1,2) \qquad [8.26]$$

so that

$$\Gamma(1,2,3) = \delta(1-2)\,\delta(2-3) + \frac{\delta\Sigma(1,2)}{\delta V(3)} \qquad [8.27]$$

in the nth order, for instance, we would have

$$\Gamma_n(1,2,3) = \delta(1-3)\delta(2-3) + \frac{\delta\Sigma^{(n-1)}(1,2)}{\delta V(3)} \qquad [8.28]$$

The self-energy at each iteration will consist of a series of connected Green's function and interaction lines which must be differentiated to calculate the next-order vertex function. We have, however

$$\frac{\delta G(1,2)}{\delta V(3)} = -\int G(1,4)\frac{\delta G^{-1}(4,5)}{\delta V(3)}G(5,2)\,d[4]\,d[5] \qquad [8.29]$$

In diagram terms this may be expressed as

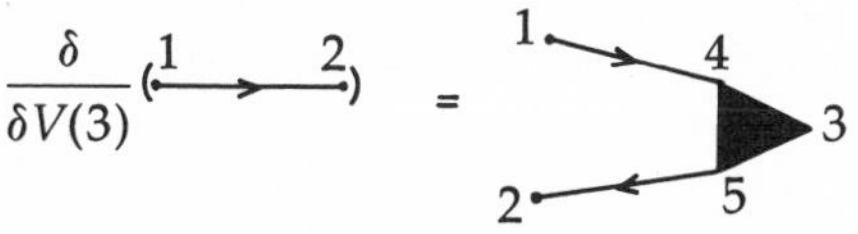

[8.30]

so that the action of differentiation simply inserts the previous vertex into each Green's function line in the self-energy. Thus, in first order

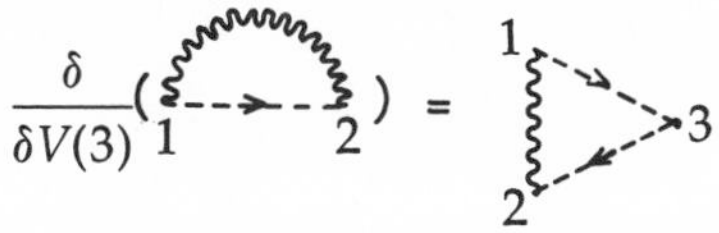

[8.31]

In addition to this, however, there will be a contribution from the interaction lines. From Eq. (7.51) we can obtain the relation

$$\frac{\delta W(1,2)}{\delta V(3)} = \int W(1,4)\frac{\delta P(4,5)}{\delta V(3)}W(5,2)\,d[4]\,d[5] \qquad [8.32]$$

The polarization propagator contains, at each iteration, Green's functions and lower-order interaction lines, so that the differential will contain a number of terms. For instance, in first order again

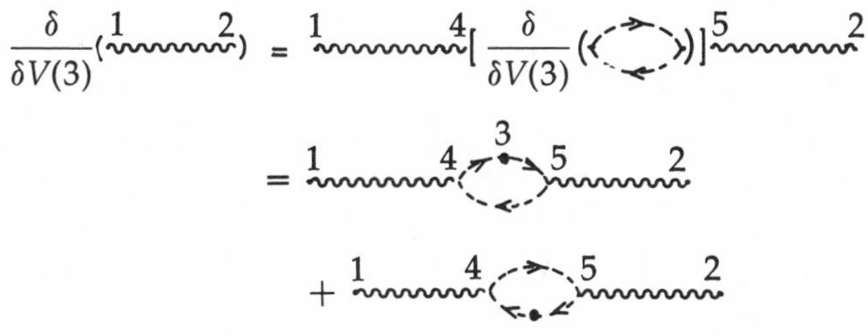

[8.33]

Notice that the derivatives of the interaction jump an order (since $\delta W/\delta V \approx W^2$). In the next order we would have to consider the differential

$$\frac{\delta}{\delta V(3)}\left[\ \overset{(1)}{\underset{(1)}{\bigcirc}} + \overset{(1)}{\underset{(1)}{\bigcirc}}\ \right]$$

where both the vertex insertions (8.25) and (8.30) could enter the Green's function lines and (8.33) could be inserted in the interaction line. Obviously the process becomes rapidly more and more complicated as the order progresses. Figure 8.1 illustrates the procedure up to the third iteration. As before we have replaced each Green's function by the full Green's function, since this would be the result if we had a *complete* solution.

If we are actually calculating the Green's function, the complete solution will depend upon the level of iteration we reach. For instance, suppose we wish to develop the series in terms of G_0, the noninteracting Green's functions, only. To first order

$$\Sigma^{(1)} = \quad [8.34]$$

and

$$G_{(1)} = \quad + \quad [8.35]$$

To second order the self-energy becomes [Eq. (7.71)]

$$\Sigma^{(2)} = \underset{G_{(1)}}{} + \underset{G_{(1)}}{} \quad [8.36]$$

If we now expand $G_{(1)}$ in terms of the noninteracting Green's function

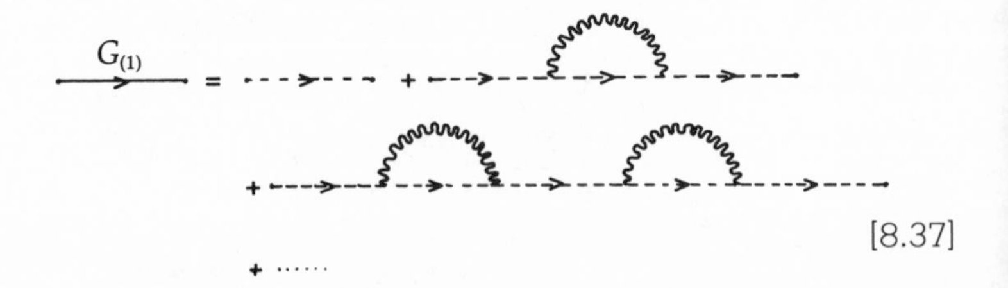

[8.37]

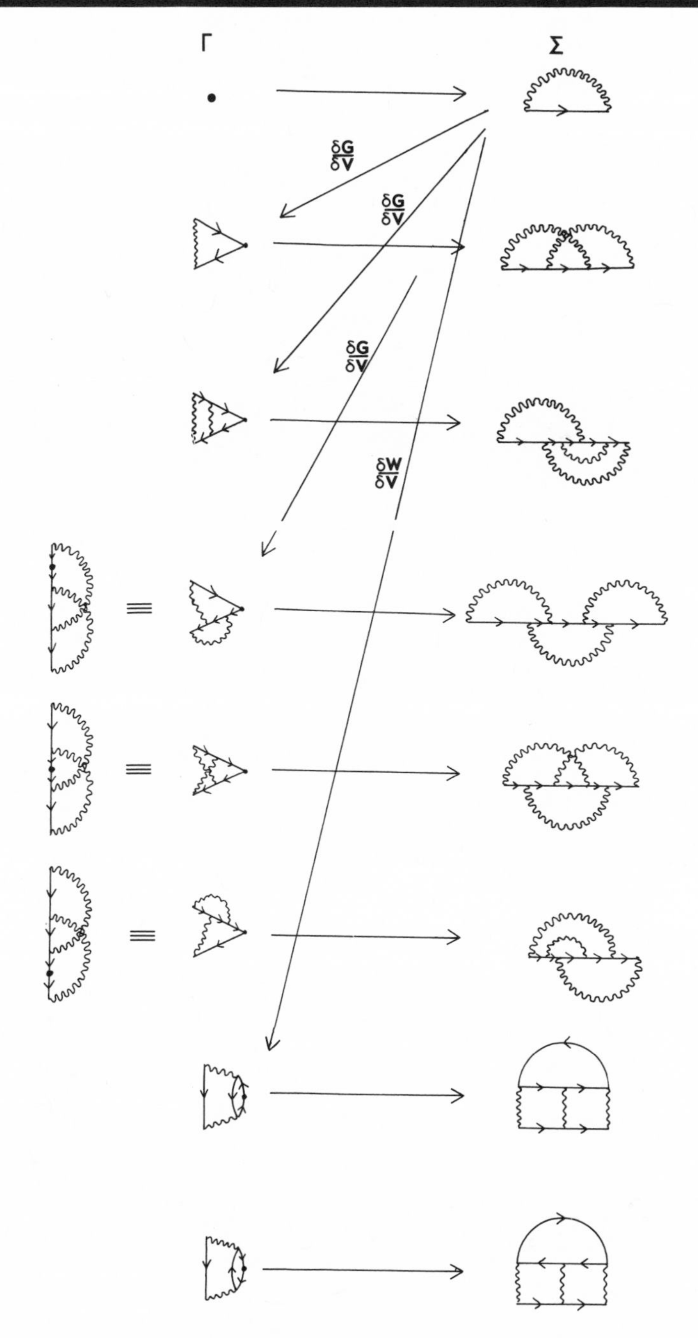

FIGURE 8.1. Diagrammatic iteration of the self-energy and vertex function up to the third order.

The actual full expansion to second order of the self-energy in terms of the noninteracting Green's function is now, after inserting the series for $G_{(1)}$ into Eq. (8.36)

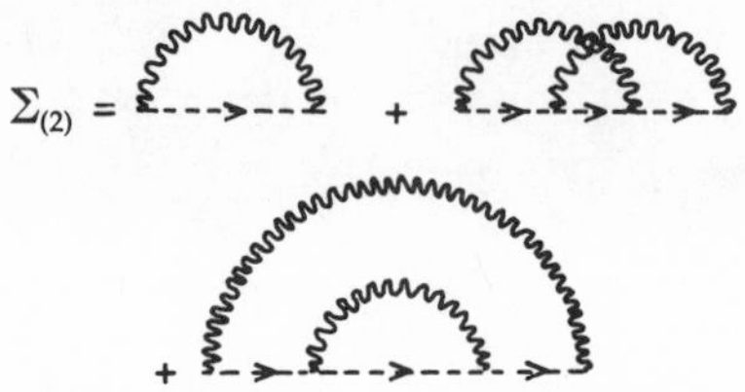

$\Sigma_{(2)}$ = + + [8.38]

In a similar manner the full expansion of the interaction to second order in corrections to the polarization is

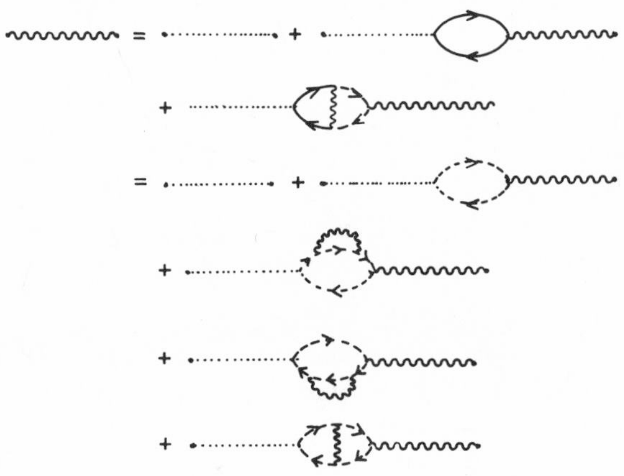

[8.39]

The second, third, and fourth terms come from the expansion of the Green's function, while the fifth term comes from the second-order vertex [Eq. (8.17)]. Other terms have more interaction lines.

We are thus again and again led to a value judgment of which terms to keep. The most we can hope for is to obtain reasonable approximations based upon the physics of the situation.

8.3. INFINITE SERIES AND IRREDUCIBLE DIAGRAMS

We have up to now based our discussions firmly upon the perturbation series. There is, however, another form of many-body theory which replaces the equation of motion method by an expansion based upon Wick's theorem. We are not going to go into this method, but a number of concepts arise in this approach which, although strictly irrelevant to the expansion series, need to be considered since they occur so frequently.

Consider the interaction; diagrammatically we have

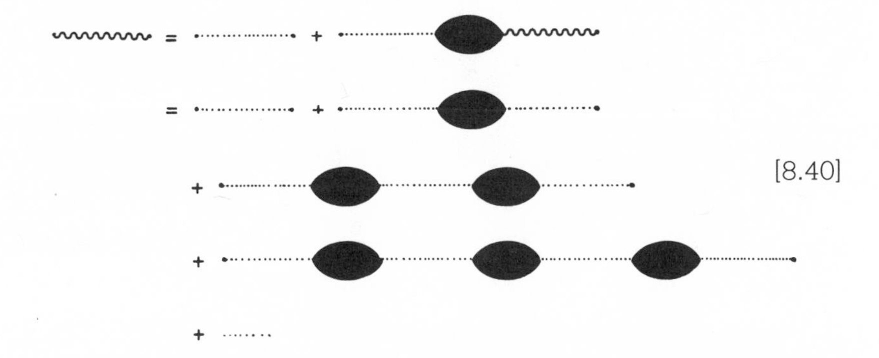

[8.40]

The polarization propagator can be expanded in turn to give a series of diagrams which, following the expansion of the vertex, will be of the form

[8.41]

These polarization diagrams have a common form. They consist of "bubbles" of Green's functions coupled together with interaction lines. There are no free Green's function or interaction lines.

The following diagrams are also polarization diagrams in this sense

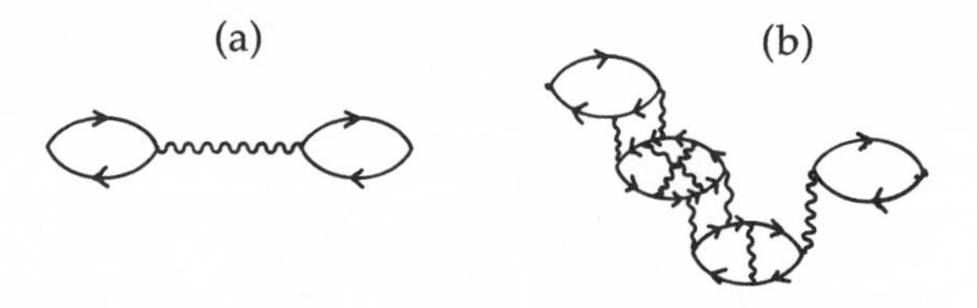

However, diagram (a) may be written as an infinite series, on expanding the interaction according to (8.40).

[8.42]

Each of these terms appears, however, in the expansion of the screened interaction in terms of the approximation for the polarization with

P =

because of the form of the perturbation expansion. Thus, this approximation for the polarization contains many of the terms which would also be present with a polarization of the form

Similarly, the term (b) is included in the interaction at the vertex expansion, of P which includes

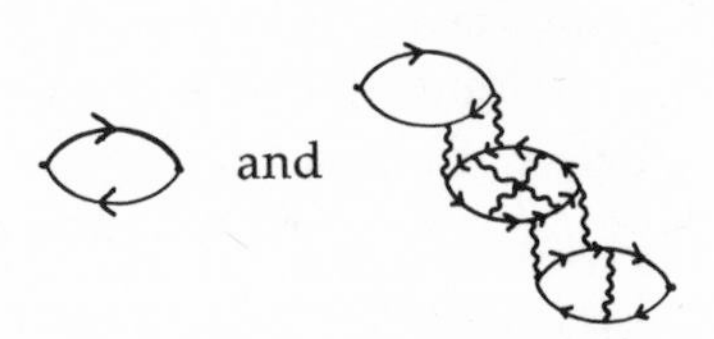

Thus, if we tried to include in our expansion of the polarization terms like (a) and (b), we would be counting many terms twice.

As long as the diagram cannot be separated into two parts joined by a single interaction line, however, there will be no danger of it having been included in the interaction. It will, in fact, be one of the polarization diagrams in the series generated by the vertex expansion. We can thus define two types of polarization diagrams. The first cannot be split into two parts by cutting a single interaction line. These are called *irreducible* polarization diagrams. The second type is of the form discussed above in which two (or more) irreducible diagrams are connected by single interaction lines. These are *reducible* diagrams.

The important point is that the perturbation series for the polarization, generated by the expansion of the vertex function, generates all of the "topologically distinct" (i.e., different) irreducible diagrams—there are no reducible diagrams generated. If, however, one wants to consider a diagrammatic generation of the polarization as separate from the perturbation expansion (since it is easier to draw a set of diagrams than solve an iteration series!) one must be careful only to count the irreducible diagrams.

A similar restriction applies to the self-energy. From the Dyson equation for the Green's function, we have

$$G = G_0 + G_0 \Sigma G \qquad [8.43]$$

or

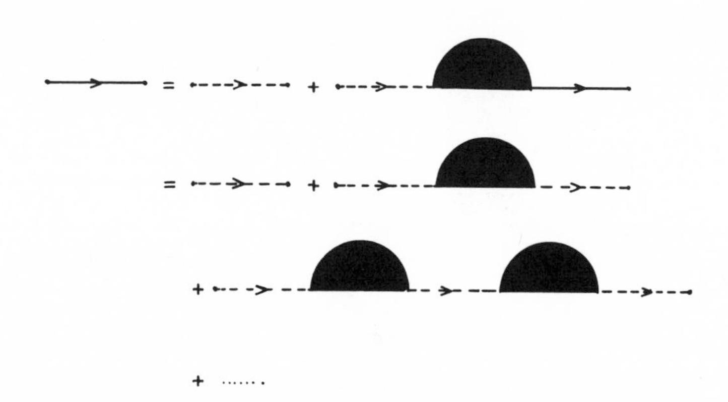

In diagrammatic terms the expansion of the self-energy can only contain irreducible diagrams, i.e., those which cannot be split by cutting a single Green's function line. This is so since any diagram of the form

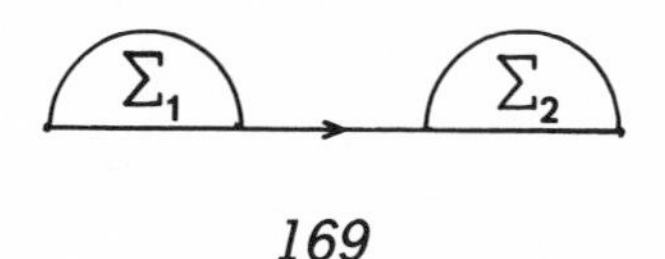

will already be included in the expansion of the Dyson equation, which contains the irreducible self-energies Σ_1 and Σ_2. Thus

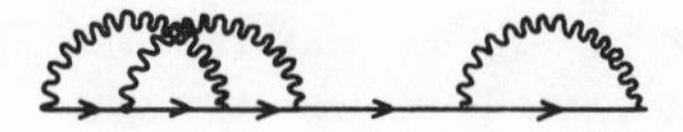

is reducible, but

is not, and neither is

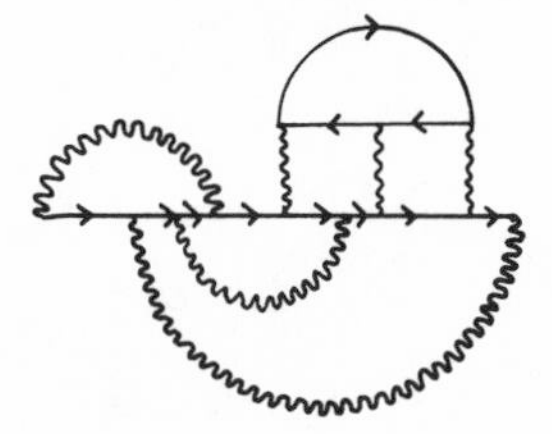

The perturbation series will produce *all* of the irreducible diagrams automatically. It is sometimes necessary, however, to generate selected sets of diagrams to calculate particular effects. In these cases it is a useful check to be able to restrict oneself to the irreducible diagrams.

8.4. THE HARTREE POTENTIAL

The Hartree potential was removed from the self-energy series at the beginning, since it was closely identified with an average effective one-electron potential. This potential is given by

$$V_H(x,t) = -i\int v(x,x_1)\, \delta(t - t_1) G(x_1,t_1,x_1,t_1^+)\, dx_1 \qquad [8.45]$$

and as such it corresponds to the diagram

(x_1,t_1)

(x_1,t)

This is often, for obvious reasons, termed a *tadpole diagram.* Since the Hartree potential contains the full Green's function, various approximations for G gave rise to a series of Hartree terms. In terms of the G_0 expansion for G, we have

V_H = + + + [8.46]

The definition of a Green's function (G_{00}) without the Hartree inclusion would lend to self-energy diagrams including these tadpole diagrams. If we write

$G_{00} \equiv$

then Eq. (7.62) may be written

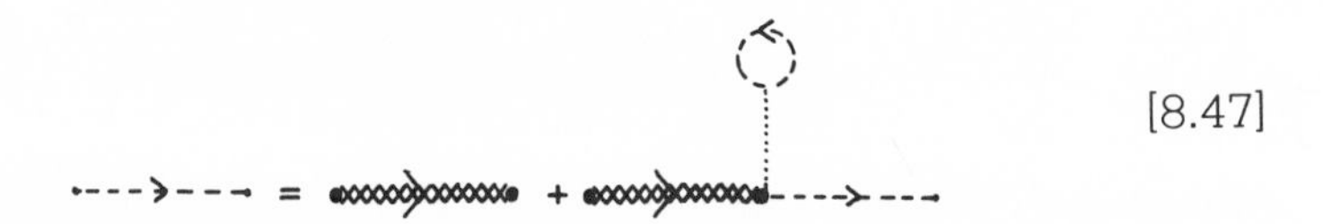

[8.47]

Combining this with the Green's function equation [Eq. (8.11)], the perturbation expansion can be carried out as before, and a typical self-energy diagram in terms of G_{00} would be

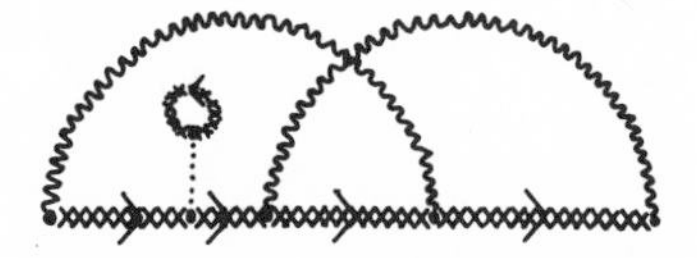

These diagrams are sometimes seen in the literature and indicate, as we have said, a noninteracting Green's function defined *without* the Hartree potential.

8.5. THE UNIFORM SYSTEM

One of the advantages of the diagrammatic interpretation of the iteration series is that it may be used for either the space–time or momentum–energy form used in the uniform system. If we label each of the diagrams by a momentum and energy rather than position time coordinates we can write

$$G_0(q,\omega) \equiv$$ 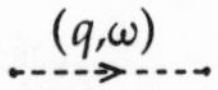

[8.48]

$$G(q,\omega) \equiv$$ 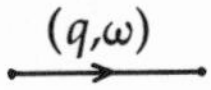

[8.49]

$$v(q,\omega) \equiv$$ 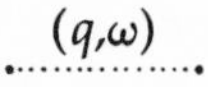

[8.50]

$$w(q,\omega) \equiv$$ 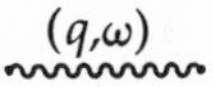

[8.51]

$$\Sigma(q,\omega) \equiv$$ 

[8.52]

$$\vdots$$

The vertex function still has three terminals but only two sets of variables. It is easy to show, from the Fourier transforms, that

$$\Gamma(q,\omega,k,\Omega) \equiv$$ 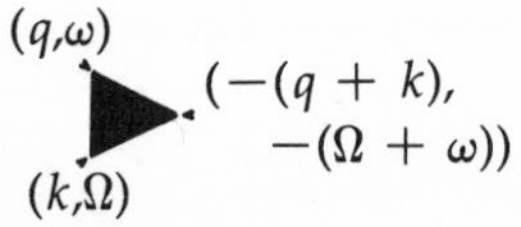

$$=$$ 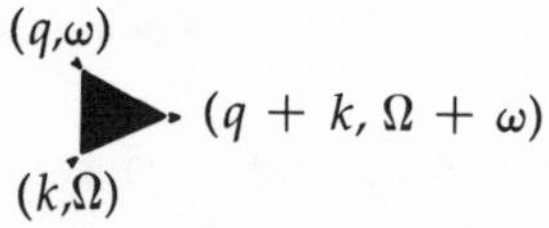

[8.53]

where the arrows show the direction of the momentum and energy flow. If we now draw out the diagrams relating to the iteration series in the Fourier-transformed form [(7.78)–(7.80)] we find exactly the same topological form as in (8.8)–(8.11). There is the added advantage of a conservation of energy and momentum at each vertex if we imagine the q, ω labels on the functions as defining the input and output from the external vertices. Thus

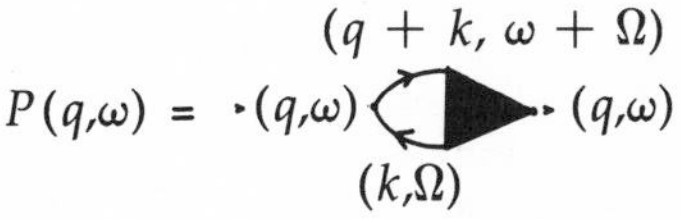

$(q + k, \omega + \Omega)$
$P(q,\omega) = \cdot(q,\omega)$ (q,ω) [8.54]
(k,Ω)

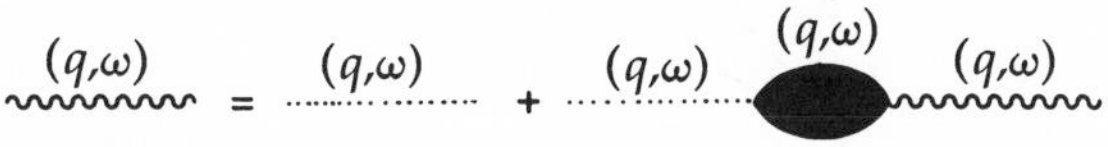

$(q,\omega) = (q,\omega) + (q,\omega)$ (q,ω) (q,ω) [8.55]

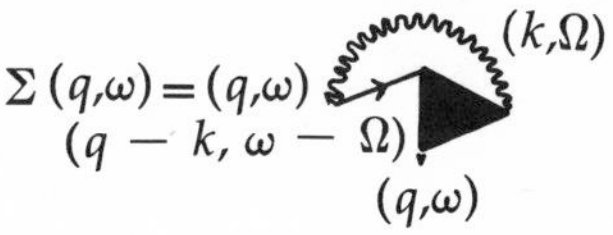

(k,Ω)
$\Sigma(q,\omega) = (q,\omega)$
$(q - k, \omega - \Omega)$ [8.56]
(q,ω) 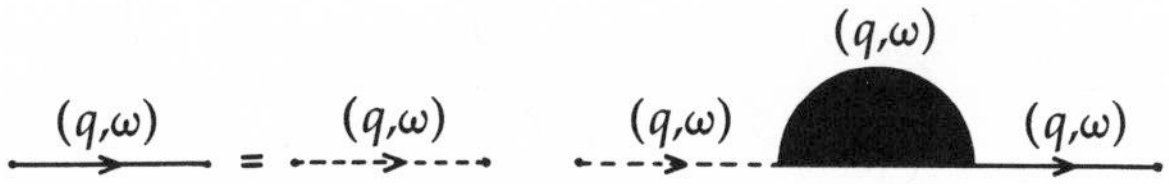

(q,ω)
$(q,\omega) = (q,\omega)$ (q,ω) (q,ω) [8.57]

In each case we integrate over the internal, or "free" momentum and energy variables. The development of the iteration series in terms of vertex insertion to represent the functional derivatives $\delta G^{-1}/\delta V$ also applies in this case (from a simple Fourier transform of the defining equations). Thus, the diagrams in the momentum–energy form are *exactly* the same as for the space–time form, but the conservation of energy and momentum at each vertex helps the interpretation immensely. Since the interaction lines are also carrying energy and momentum, just as the Green's function lines do, this strengthens the identification of the interaction lines as some form of propagator for an "interaction" particle (photon or plasmon).

Everything we have said in relation to the various expansion series applies to the uniform system and the momentum–energy diagrams, but we must stress again that what is important are the defining equations [(7.78)–(7.80)]. The diagrams are a *visual* device for *interpreting* and *extracting physical insight.*

8.6. RULES FOR EVALUATING DIAGRAMS

In those cases where the physical process being considered requires the consideration of a specific set of diagrams, a set of rules for ensuring that the contribution of each diagram is calculated correctly is useful. The rules, set out in this section for completeness, are derivable from the iteration series or, alternatively, they also come directly from Wick's theorem in the standard diagrammatic expansion.

8.6.1. Diagrams in Space-Time

1. Draw the diagram for the process required, remembering that vertices must have conservation of Green's function lines, except for the external vertices; otherwise the number of particles will not be conserved, i.e.,

and  are allowed but  is not.

2. Check that the diagram is not reducible.

3. Decide upon the *order* of the diagram, which is equal to the number of interaction lines.

4. Label vertices.

5. Each Green's function line between vertices A and B becomes $G(A,B)$ or $G(B,A)$ according to the direction assigned. Each interaction line becomes $W(A,B)$.

6. Integrate over all internal variables. We should mention spin—the interaction W cannot reverse spin. The spin, therefore, continues along the Green's function lines and this must be remembered when replacing

$$\int dx \quad \text{by} \quad \sum_{\sigma} \int d\mathbf{r}$$

It will give a factor of 2 in the integration of isolated fermion loops, i.e.

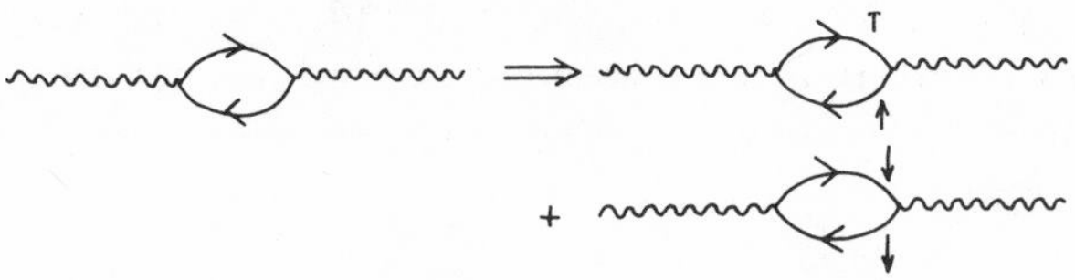

but in a self-energy diagram the spin is the same along the Green's function line

is either

or

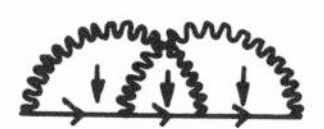

but not both since the spin is determined by the external Green's function lines connecting to the self-energy.

7. Multiply by $(-1)^F$, where F is the number of closed fermion loops.

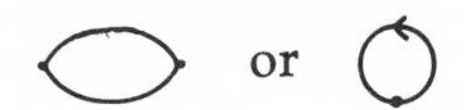

8. For causality

$$G(x_1,t_1,x_2,t_1) \Rightarrow G(x_1,t_1,x_2,t_1^+)$$

9. Multiply by $(i)^n$, where n is the order.

As an example take the second-order self-energy diagram

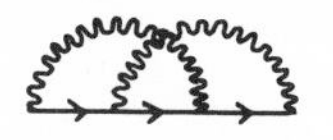

The order is 2, the number of closed Fermion loops (F) is zero, and thus the integral corresponding to the diagram is

$$(i)^2(-1)^0\int d[3]\, d[4]\, G(1,3)G(3,4)G(4,2)W(1,4)W(3,2)$$

8.6.2. Momentum-Energy Diagrams

1. Draw the diagrams as in space–time.
2. Check for reducibility.
3. Decide on the order of diagram.
4. Assign a directed momentum and energy to each line, remembering that there is an input and an output momentum and energy for the diagram.
5. Interaction line describes $W(\mathbf{q},\omega)\ \delta_{\alpha\alpha'}\delta_{\beta\beta'}$

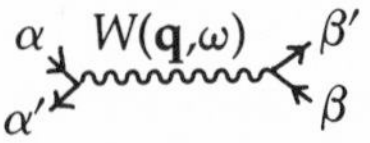

α $W(\mathbf{q},\omega)$ β'
α' β

where α, α', β, β' are the spin indices of the connecting Green's functions and, as before, ensure that none of the spins are flipped.

6. The Green's function line describes $G(\mathbf{q},\omega)\delta_{\alpha,\beta}\ [\equiv G(q,\omega)]$

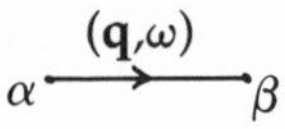

$(\mathbf{q},\omega)$
α β

7. Conserve energy and momentum at each vertex and so reduce the number of free variables.
8. Sum over the free spin, momentum, and energy variables.
9. Multiply by the numerical factor $(-1)^F$ where F is the number of closed Fermion loops.
10. To conserve causality in the diagram, for each Green's function line, either in a closed loop or linked by the *same* interaction line, we replace $G(k,\omega)$ by $G(\mathrm{k},\omega)e^{i\omega\eta}$
11. Multiply by the factor $(i)^n(2\pi)^{-4n}$ to take care of the order factor and the Fourier transforms involved in going from space–time to momentum–energy.

Considering again the diagram for the second-order self-energy, we

draw

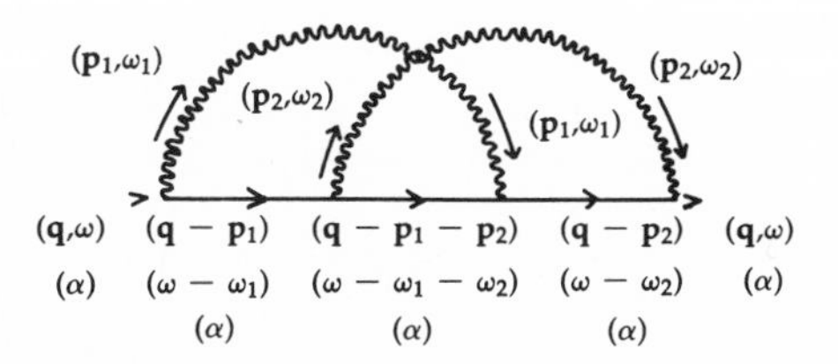

So that

$$\sum_2 (\mathbf{q},\omega) = \frac{(i)^2}{(2\pi)^8} \int d\mathbf{p}_1 d\mathbf{p}_2\, d\omega_1\, d\omega_2\, G(\mathbf{q} - \mathbf{p}_1, \omega - \omega_1)$$
$$\times\, G(\mathbf{q} - \mathbf{p}_1 - \mathbf{p}_2, \omega - \omega_1 - \omega_2) G(\mathbf{q} - \mathbf{p}_2, \omega - \omega_2) W(\mathbf{p}_1 \omega_1) W(\mathbf{p}_2 \omega_2)$$

which is the Fourier transform of the space–time expression.

8.7. SELECTIVE SUMMATIONS

The prime advantage of diagrams is that it allows one to manipulate the iteration equations fairly easily and identify related processes. Consider the screened interaction, for instance. As we have seen in Sec. 8.3, this has an expansion in the form

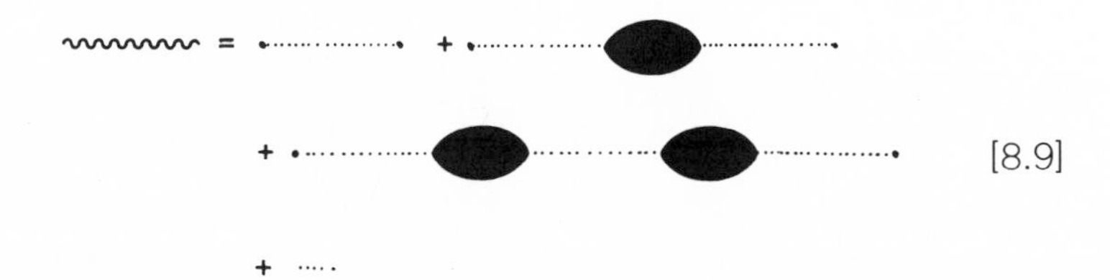

[8.9]

In the lowest order of approximation we can replace the polarization by the bubble diagram

[8.58]

This corresponds to the "electron" and "hole" propagating without interacting with each other. Normally this may be as far as one would wish to go, but consider the physical interpretation of this particular set of diagrams

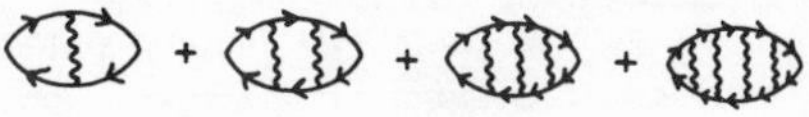

[8.59]

occurring as a subset in the expansion of the polarization propagator. These we would interpret as the electron and hole repeatedly interacting with each other by way of the screened interaction. For an electron and a hole this would be an *attractive* interaction. We might, therefore, expect a bound state to be possible. As we have seen before, such a bound state is characterized by a divergence of the perturbation series. Thus, as we expanded the polarization to higher orders, we would get a divergent result from the set of terms corresponding to these diagrams (there are other terms as well, of course, but this divergence would dominate).

In such a case it is necessary to selectively sum the series so that the divergence is taken care of. Since there are four vertices involved, we could define the sum of all terms involving repeated interactions by a function $S(1,2,3,4)$ and then we would write

$S(1,2,3,4) =$ 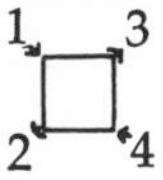 [8.60]

as the series

$S(1,2,3,4) =$ ... + ... [8.61]

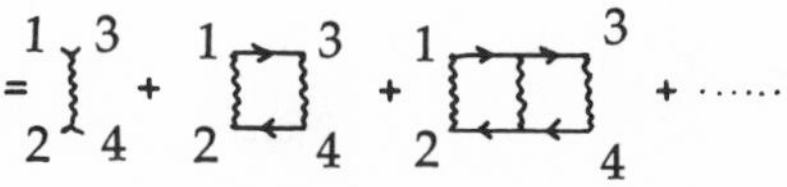

We have seen this type of series before and it is easy to see that it is generated by a repeated substitution solution of the integral equation represented in diagrammatic form by

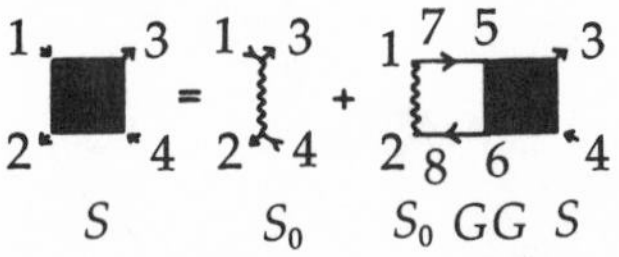

[8.62]

$S \quad S_0 \quad S_0\, GG\, S$

This is, again, a Dyson type of equation, which we can write in full as

$$S(1,2,3,4) = S_0(1,2,3,4) + \int S_0(1,2,7,8)G(7,5)G(6,8)S(5,6,3,4)\, d[5]\, d[6]\, d[7]\, d[8] \quad [8.63]$$

S_0 is the lowest-order interaction [i.e., $W(1,2)\,\delta(1,3)\,\delta(2,4)$]. Having solved this equation to avoid the divergences it would now be possible, in the perturbation series, to replace the diagrams in which the repeated interaction occurred by a single diagram summing the series. Thus, in the present example, the polarization bubbles of (8.58) would become

= + + +........ [8.64]

Physically what we are doing is allowing for the presence of bound electron–hole pairs formed because of the interaction. The presence of the bound state alters the form of the solution so drastically that it must be incorporated from the beginning if it is not to destroy any perturbation series (i.e., the formation of a bound state is *not* a small perturbation).

There would also be equivalent inserts into the self-energy. Thus, we would have to supplement the low-order terms by the terms involved in the "divergence":

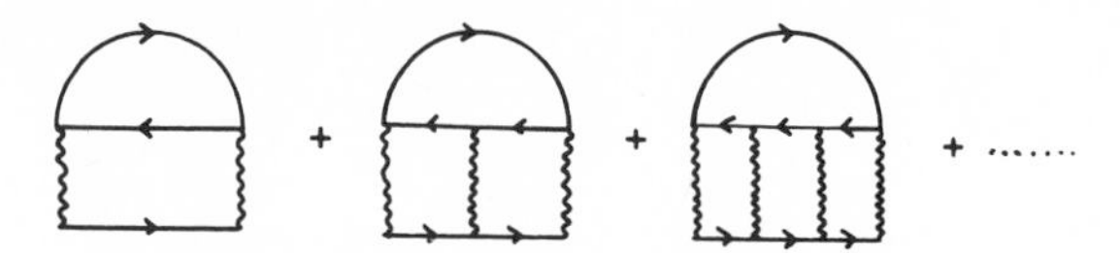

or, summing the series

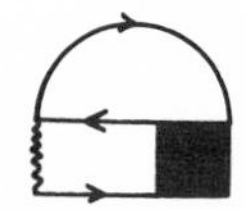

The identification of possible divergences, and then the summation of the selected series to avoid those divergences, is very important in any physical application. As we saw in Sec. 7.4, the replacement of the bare by the screened interaction is just such a case. If we wished to study excitons in semiconductors or insulators, it would be necessary to carry out the kind of summation we have indicated above. In metals the screened interaction is thought to be weak enough to ignore repeated scatterings in most cases. One further important example we will come across is the attractive interaction between electrons brought about by the lattice. Repeated scatterings of the form

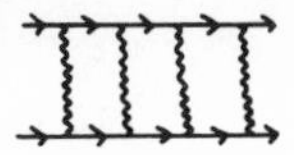

(where the interaction line includes the lattice effect) then result in the formation of Cooper pairs and the phenomena of superconductivity.

In general, we are faced with one of two consequences:

1. The change in the system is small in the sense that it is restricted to a particular range of energy and momentum (as in the case of the exciton). In this case we usually proceed by either avoiding that region or treating it as a special case.
2. The change is a basic one in the properties of the system as a whole (e.g., superconductivity). In this case we have to reformulate our ideas so as to incorporate these changes from the beginning.

8.8. PRACTICAL ASPECTS OF DIAGRAMMATICS

One can obviously have great fun drawing diagrams and seeing how processes become intertwined in the higher-order diagrams, but we must be practical. If one looks at the second-order self-energy diagrams used as examples in Sec. 8.6, even in the simple case of the uniform system one is faced with the evaluation of a complex function in momentum and energy. This requires an eight-dimensional integral (ignoring any previous evaluations necessary to obtain G and W), for each momentum energy point of the self-energy. Thus, in no way can we consider actually calculating any but the simplest diagrams unless we have drastic simplifications, which would reduce the integrations or number of variables.

This restriction to the lowest orders of diagrams is one reason why it is so vitally important that the correct physics be incorporated at this low level. The ways of achieving this are to utilize the selective summations where possible and sacrifice mathematical elegance to physical necessity.

BIBLIOGRAPHY

FEYNMAN, R. P., *Quantum Electrodynamics*, Benjamin, Menlo Park: CA 1961.
FETTER, A. L., and WALECKA, J. D., *Quantum Theory of Many-Particle Systems*, McGraw-Hill, New York, 1971.
MATTUCK, R. D., *A Guide to Feynman Diagrams in the Many-Body Problem*, McGraw-Hill, New York, 1976.
DYSON, F. J., *Phys. Rev.*, **75**, 486, 1949.
WICK, G. C., *Phys. Rev.*, **80**, 286, 1950.

PROBLEMS

1. Extend the argument for the vertex expansion and show that in the iteration development it is correct to replace G_0 by G.

2. Draw the polarization diagrams to the third order on the same basis as the self-energy diagrams of Fig. 8.1.

3. Explain why the following diagrams do not occur in the second-order expansion for the self-energy.

 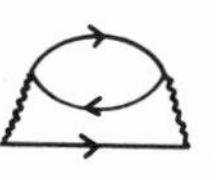 

4. For the system described in Prob. 3 (Chapter 7) in the $\alpha = \beta = 0$ limit, evaluate the ladder sum of Eq. (8.63) (neglecting overlap integrals between different sites). Interpret your result in terms of the Frankel exciton, which is bound on one site. Show that it may propagate by the electron and hole recombining on one site and then the resulting photon exciting a new exciton at another site.

CHAPTER NINE

THE NORMAL SYSTEM

Having derived a formalism to treat an interacting fermion gas, it is now time to see how it may be applied to a real physical system. In this chapter we will consider the interacting electron gas in a solid. What we are interested in is to what extent the first-order terms in the various functions (polarization, self-energy, Green's function) describe the properties of a real solid. To some extent we already know from the success of single-particle solid-state physics that interaction effects do not affect the basic properties of the system. Even the simple Thomas–Fermi theory (Sec. 1.6) gave us an indication of how that might be due to the locking up of the interaction into an inert binding energy, resulting from the coulomb correlation hole.

In applying the self-energy techniques to the solid, we start first with the best single-particle Hamiltonian to produce the Hartree equation

$$\left[i\hbar \frac{\partial}{\partial t_1} - H_0(x_1) - V_H(x_1,t_1) \right] G_0(x_1,t_1,x_2,t_2) = \hbar\delta(x_1 - x_2)\delta(t_1 - t_2) \quad [9.1]$$

In doing so we must recognize that the standard practice of one-electron solid-state physics always includes some, often parametrized, attempt to include exchange and correlation effects as a local potential contribution (Sec. 12.1). This is not serious in itself, but it does mean that the interpretation of the resulting self-energy is made difficult by the presence of such a contribution. It is impossible to avoid this complication, except in the case of the jellium solid or very simple metals, since the omission of an "exchange and correlation potential" leads to a set of eigenfunctions and eigenvalues which do not bear any resemblance to the experimental ones. It is most profitably looked on as a procedure for obtaining the best "noninteracting" Green's function and so reducing the overall interaction effects—in other words, giving a better chance that the first-order corrections will be sufficient.

In the special case of the jellium solid, the Hartree potential and the background ionic potentials are constant, so the above considerations do not apply. We first look at this system as the simplest possible model of a real solid.

9.1. THE JELLIUM SOLID RESPONSE FUNCTION

The jellium solid, being translationally invariant, is best treated in the momentum-energy form so we use Eqs. (7.78)–(7.80) as the basis of our first iteration. For the jellium solid we have

$$G_0(\mathbf{k},\omega) = [\omega - E(\mathbf{k}) + i\delta\,\mathrm{sgn}(E(\mathbf{k}) - \mu)]^{-1} \qquad [9.2]$$

where $E(\mathbf{k})$ is the kinetic energy $[\equiv (\hbar^2k^2/2m)]$ and μ the Fermi energy $[=(\hbar^2k_F^2/2m)]$. If we substitute this into the expression for the polarization (with the factor of 2 for spin)

$$P(\mathbf{q},\omega) = \frac{2i}{(2\pi)^4}\int d\mathbf{k}\, d\omega\, G_0(\mathbf{k}+\mathbf{q},\omega+\omega')G_0(\mathbf{k},\omega') \qquad [9.3]$$

we have

$$P(\mathbf{q},\omega) = \frac{-2i}{(2\pi)^4} \times \int \frac{d\mathbf{k}\, d\omega'\, e^{(i\omega'\eta/\hbar)}}{[\omega' - E(\mathbf{k}) + i\delta\,\mathrm{sgn}(E(\mathbf{k})-\mu)][\omega+\omega'-E(\mathbf{k}+\mathbf{q})+i\delta\,\mathrm{sgn}(E(\mathbf{k}+\mathbf{q})-\mu)]} \qquad [9.4]$$

The exponential is the causality factor and requires the contour for the energy integration to be closed in the upper half of the plane (Fig. 9.1a). The poles of the integration

$$\omega' = E(\mathbf{k}) - i\delta\,\mathrm{sgn}(E(\mathbf{k}) - \mu) \qquad [9.5a]$$
$$\omega' = E(\mathbf{k}+\mathbf{q}) - \omega - i\delta\,\mathrm{sgn}(E(\mathbf{k}+\mathbf{q}) - \mu) \qquad [9.5b]$$

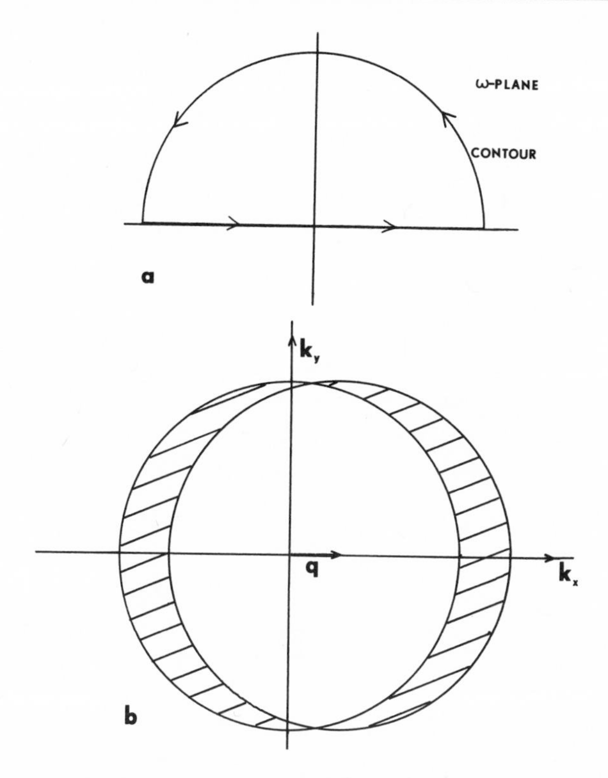

FIGURE 9.1. (a) Integration contour in the complex energy plane for the polarization. (b) Integration region for the polarization is shown hatched and corresponds to the mismatch of the two Fermi spheres.

then correspond to the occupied states since $E(\mathbf{k}) - \mu$ and $E(\mathbf{k} + \mathbf{q}) - \mu$ must both be negative. This simple integration gives

$$P(\mathbf{q},\omega) = -\frac{2}{(2\pi)^3} \int d\mathbf{k} \frac{n(\mathbf{k}) - n(\mathbf{k} + \mathbf{q})}{\omega + E(\mathbf{k} + \mathbf{q}) - E(\mathbf{k}) + i\delta} \qquad [9.6]$$

where

$$\begin{aligned} n(k) &= 1, \qquad |k| < k_F \\ &= 0, \qquad |k| > k_F \end{aligned}$$

is the Fermi factor denoting the occupation of the level.

There are two things to note about this expression:

(i)
$$n(\mathbf{k}+\mathbf{q}) - n(\mathbf{k}) = 0$$

unless $|\mathbf{k}| > \mathbf{k}_F$, $|\mathbf{k}+\mathbf{q}| < k_F$, or $|\mathbf{k}| < \mathbf{k}_F$, $|\mathbf{k}+\mathbf{q}| > k_F$. Contributions to the integral and hence the polarization only comes from excitations *across* the Fermi surface $|\mathbf{k}| = k_F$. A similar situation occurred in Thomas–Fermi theory where it was the Fermi surface which was important. It also agrees with the interpretation of the polarization bubble as an electron–hole pair.

(ii) There remains an $i\delta$ in the denominator. This gives a finite imaginary part to the integrals for the polarization, which must therefore always be considered as a complex function. (This will apply equally to the dielectric function and the self-energy).

The evaluation of the integral is now a standard task first performed by Lindhard. It is usual to consider directly the dielectric function, which we write as

$$\epsilon(\mathbf{q},\omega) = \epsilon_1(\mathbf{q},\omega) + i\epsilon_2(\mathbf{q},\omega) \tag{9.7}$$

rather than the polarization. The range of integration is governed by the Fermi factors and is shown in Fig. 9.1. The result for the real part is

$$\epsilon_1(\mathbf{q},\omega) = 1 + \frac{\lambda^2}{q^2}\left[\frac{1}{2} + \frac{k_F}{4q}\left\{\left[1 - \frac{\left(\omega^2 - \dfrac{\hbar^2 q^2}{2m}\right)^2}{\hbar^2 q^2 v_0^2}\right]\right.\right.$$

$$\times \ln\left|\frac{\omega - \hbar q v_0 - \dfrac{\hbar^2 q^2}{2m}}{\omega + \hbar q v_0 - \dfrac{\hbar^2 q^2}{2m}}\right| + \left[1 + \frac{\left(\omega^2 - \dfrac{\hbar^2 q^2}{2m}\right)^2}{\hbar^2 q^2 v_0^2}\right]$$

$$\left.\left.\times \ln\left[\frac{\omega + \hbar q v_0 + \dfrac{\hbar^2 q^2}{2m}}{\omega - \hbar q v_0 + \dfrac{\hbar^2 q^2}{2m}}\right|\right\}\right] \tag{9.8}$$

where λ is the Thomas–Fermi wave vector and v_0 [$= (\hbar k_F/m)$] is the "Fermi velocity."

For the imaginary part, we use the following relation in Eq. (9.6)

$$\mathrm{Im}\left(\frac{1}{\omega - [E(\mathbf{k}+\mathbf{q}) - E(\mathbf{k})] + i\delta}\right) = -\pi\delta(\omega - [E(\mathbf{k}+\mathbf{q}) - E(\mathbf{k})]) \tag{9.9}$$

It is illuminating to interpret this delta function as the energy requirement for creating an electron–hole pair of net momentum $\mathbf{q}$. The imaginary part of the dielectric function is then associated with the possibility of energy loss mechanisms—a fact well known in normal electromagnetism.

As long as the interaction has not enough energy to excite an electron–hole pair, only virtual processes are available: the electron–hole pair must annihilate rapidly within a time governed by the uncertainty principle and so no energy is dissipated by the interaction. Once the real process is available, the interaction can lose energy to the system by creating electron–hole pairs. It is one of the strengths of the procedure we have been following that both eventualities are included automatically.

The result of the integration is

$$\begin{aligned}
\epsilon_2(\mathbf{q},\omega) &= \frac{\pi}{2}\frac{\hbar\omega}{qv_0}\frac{\lambda^2}{q^2} \qquad \left(\omega \leq \hbar q v_0 - \frac{\hbar^2 q^2}{2m}\right) \\
&= \frac{\pi}{4}\frac{\hbar}{q}\left[1 - \frac{\left(\omega - \dfrac{\hbar^2 q^2}{2m}\right)^2}{\hbar^2 q^2 v_0^2}\right]\frac{\lambda^2}{q^2} \\
&\qquad \left(\hbar q v_0 - \frac{\hbar^2 q^2}{2m} \leq \omega \leq \hbar q v_0 + \frac{\hbar^2 q^2}{2m}\right) \\
&= 0 \qquad \left(\omega \geq \hbar q v_0 + \frac{\hbar^2 q^2}{2m}\right)
\end{aligned} \tag{9.10}$$

the final energy requirement being the maximum energy for electron–hole creation compatible with a momentum $\mathbf{q}$.

Expressions like Eq. (9.8) are difficult to understand or visualize, so it is as well to look at a few limiting forms.

(i) $$\epsilon(\mathbf{q}, 0) = 1 + \frac{\lambda^2}{q^2}, \qquad |q| \to 0 \qquad [9.11]$$

This is the standard Thomas–Fermi expression and is now seen to be the static long-wavelength limit of the accurate response.

(ii) $$\epsilon(0,\omega) = 1 - \frac{\omega_p^2}{\omega^2} \qquad \left(\omega_p^2 = \frac{4\pi Ne^2\hbar^2}{m}\right) \qquad [9.12]$$

This is the Drude response of a classical coulomb plasma with a natural oscillation frequency of $\omega_p/\hbar$. For nonzero wavelengths the high-frequency limit is better described by

$$\epsilon(\mathbf{q},\omega) = 1 + \frac{\omega_p^2}{[\omega^2(q) - \omega_p^2] - \omega^2} \qquad [9.13]$$

where $\omega(q)/\hbar$ is the plasma oscillation frequency at a wave vector $\mathbf{q}$. There is one further aspect of the real part of the response function: at $q = 2k_F$ there is a logarithmic singularity in the derivative due to the change in the nature of the integration volume—at $2k_F$ the spheres of Fig. 9.1b separate. This has very little consequence for the interaction effects but does give rise to long-range oscillations in static screening. The frequency dependance of the dielectric function for a finite $\mathbf{q}$ value is shown in Fig. 9.2.

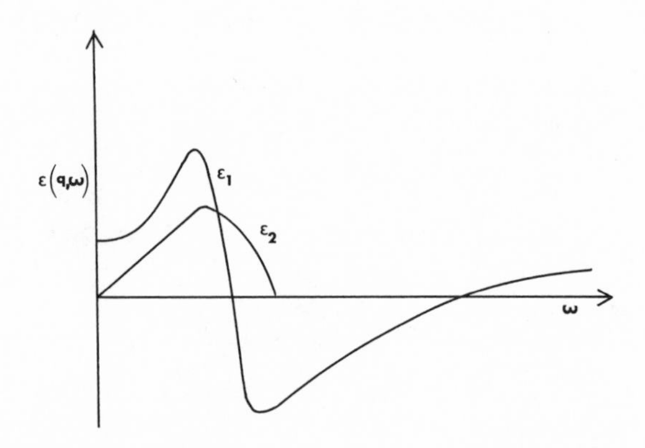

FIGURE 9.2. The dielectric function $\epsilon_1(\mathbf{q},\omega) + i\epsilon_2(q,\omega)$ for a finite momentum value as a function of energy.

The interaction is given now as

$$W(\mathbf{q},\omega) = \frac{v(\mathbf{q})}{\epsilon(\mathbf{q},\omega)} \qquad [9.14]$$

The zeros of $\epsilon(\mathbf{q},\omega)$ appear as poles in the interaction and so will contribute to the self-energy. The energy at which these poles occur will generally be complex, since $\epsilon(\mathbf{q},\omega)$ itself is now complex. The physical meaning of the poles can be understood quite easily. If we write the relationship between the internal and external potential as

$$\delta V(\mathbf{q},\omega) = \frac{\delta\phi(\mathbf{q},\omega)}{\epsilon(\mathbf{q},\omega)} \qquad [9.15]$$

If $\epsilon(\mathbf{q},\omega)$ has a zero [at $\omega(\mathbf{q})$, say] then, as we approach that value, it is possible to have a large internal field with a very small external field. At the pole there is no need for an external field at all. Thus, the zeros of the dielectric function represent a self-sustaining oscillatory field in the sample—the plasma oscillations [which, when quantized, give the plasmon (Chap. 12)]. If $\omega(\mathbf{q})$ is complex, the oscillations are damped so that of the two zeros shown in Fig. 9.2 only the higher energy one will give physically meaningful oscillations. At long wavelengths the pole occurs [Eq. (9.12)] at $\omega/\hbar = \pm\omega_p/\hbar$ the classical plasma frequency.

If we draw the energy structure for the interaction in the complex energy plane, we obtain Fig. 9.3; that is, in Eqs. (9.6) and (9.7), consider ω as complex (and hence subsume the $i\delta$ within it). There are poles at the plasmon energy and a cut along the real axis, due to the logarithms in the response function. The cut is associated with the single-particle excitations and the imaginary part of the response function, while the poles are associated with the collective plasmon oscillations.

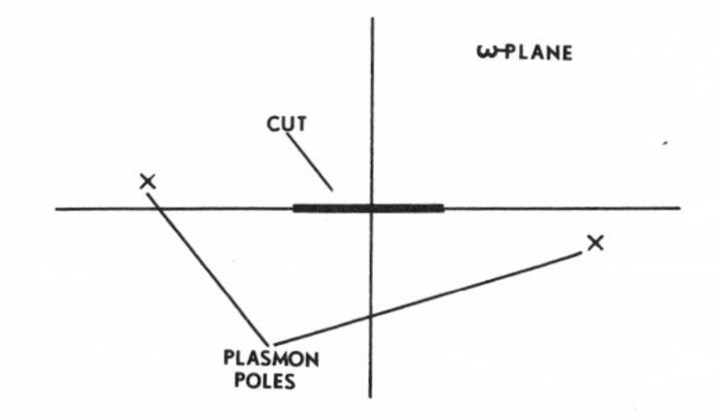

FIGURE 9.3. Energy structure of the dielectric function in the complex energy plane.

9.2. THE SELF-ENERGY (PHYSICAL CONSIDERATIONS)

We are now in a position to calculate the self-energy from Eq. (7.78d). Substituting for the Green's function, we have

$$\Sigma(\mathbf{k},\omega) = \frac{i}{(2\pi)^4} \int \frac{v(\mathbf{q})}{\epsilon(\mathbf{q},\omega)} \frac{e^{-i\eta\omega'}\, d\mathbf{q}\, d\omega'}{\omega - \omega' - E(\mathbf{k}-\mathbf{q}) + i\delta \operatorname{sgn}(E(\mathbf{k}-\mathbf{q}) - \mu)} \tag{9.16}$$

where the exponential is again due to the causality factor. Together with the structure associated with the interaction, we have the poles of Green's function corresponding to the Eigenstate of momentum $\hbar(\mathbf{k} - \mathbf{q})$. We close the energy contour in the lower half of the plane and identify two distinct contributions to the self-energy (Fig. 9.4).

1. Poles of the Green's function:

$$\omega' = \omega - E(\mathbf{k}-\mathbf{q}) - i\delta \tag{9.17}$$

these contribute iff $|\mathbf{k} - \mathbf{q}| < k_F$ and give

$$\sum_{\text{S.E.}} (\mathbf{k},\omega) = \frac{1}{(2\pi)^3} \int \frac{v(\mathbf{q})\, d\mathbf{q}}{\epsilon(\mathbf{q},\omega - E(\mathbf{k}-\mathbf{q}))'} \qquad |\mathbf{k}-\mathbf{q}| < k_F \tag{9.18}$$

The limitation of the integration to the occupied states, plus the form of the self-energy for $\epsilon = 1$, i.e.,

$$\sum_{\text{S.E.}} (\mathbf{k},\omega) = \frac{1}{(2\pi)^3} \int v(q)\, d\mathbf{q} \equiv \sum_{\text{H-F}} (\mathbf{k},\omega), \qquad |\mathbf{k}-\mathbf{q}| < k_F \tag{9.19}$$

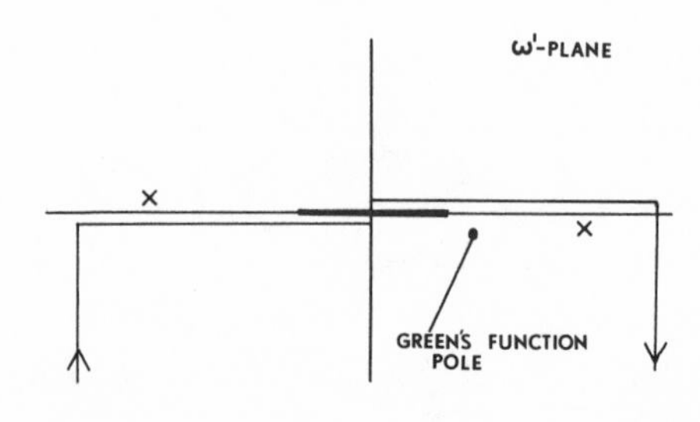

FIGURE 9.4. Contour for the evaluation of the self-energy.

leads to the identification of this term (9.18) as the "dynamic screened exchange." It has the limit of Hartree–Fock exchange if we ignore the screening effect.

2. Poles and cut in the interaction:

$$\sum_{\text{C.H.}}(\mathbf{k},\omega) = \frac{1}{(2\pi)^3}\int \frac{v(\mathbf{q})}{\left.\dfrac{\partial\epsilon(\mathbf{q},\omega)}{\partial\omega}\right|_{\omega=\omega(q)}} \times \frac{d\mathbf{q}}{\omega - \omega(q) - E(\mathbf{k}-\mathbf{q}) + i\delta\,\mathrm{sgn}(E(\mathbf{k}-\mathbf{q})-\mu)} + \frac{i}{(2\pi)^4}\int_{\text{CUT}} \frac{v(\mathbf{q})}{\epsilon(\mathbf{q},\omega)}\cdot\frac{d\mathbf{q}\,d\omega'}{\omega-\omega'-E(\mathbf{k}-\mathbf{q})+i\delta\,\mathrm{sgn}(E(\mathbf{k}-\mathbf{q})-\mu)} \qquad [9.20]$$

For historical reasons, this is termed the coulomb correlation term and in its simplest form corresponds to the Thomas–Fermi coulomb hole as we shall see.

All of these terms need to be evaluated numerically, but for illustrative purposes consider what we are really after—the quasiparticle properties. The new poles in the Green's function will define the quasiparticles energies $\mathscr{E}(\mathbf{k})$ through the equation

$$\mathscr{E}(\mathbf{k}) - E(\mathbf{k}) - \Sigma(\mathbf{k},\mathscr{E}(\mathbf{k})) = 0 \qquad [9.21]$$

In general, since the self-energy will not be real, neither will the energy, thus giving quasiparticles with a *finite* lifetime. If we assume, however, that the change is *small*, the contribution to the self-energy from the plasmon poles may be written as

$$\sum_{\text{C.H.}}(\mathbf{k},\mathscr{E}(\mathbf{k})) = \frac{1}{(2\pi)^3}\int \frac{v(\mathbf{q})}{\left.\dfrac{\partial\epsilon(\mathbf{q},\omega)}{\partial\omega}\right|_{\omega=\omega(q)}} \times \frac{d\mathbf{q}}{[\mathscr{E}(\mathbf{k}) - E(\mathbf{k}-\mathbf{q}) - \omega(q) + i\delta\,\mathrm{sgn}(E(\mathbf{k}-\mathbf{q})-\mu)]} \qquad [9.22]$$

(The cut contribution turns out to be always small so we will ignore it in the discussion.) For all intents and purposes the plasmon energy is very much larger than any difference in energies $\mathscr{E}(\mathbf{k}) - E(\mathbf{k}-\mathbf{q})$ (in fact it is usually larger than the Fermi energy). Since the plasmon energy is large, we shall use Eq. (9.13) to give us the physical content of the response.

If we now substitute both of these approximations into the expression for the self-energy term, we have

$$\sum_{\text{C.H.}} (\mathbf{k},\mathscr{E}(\mathbf{k})) = \frac{-1}{(2\pi)^3} \int \frac{v(\mathbf{q})}{2\omega^2(\mathbf{q})} \omega_p^2 \, d\mathbf{q} \qquad [9.23]$$

In terms of the effective "static" dielectric response

$$\epsilon_{\text{eff}}(\mathbf{q},0) = 1 + \frac{\omega_p^2}{\omega^2(q) - \omega_p^2} \qquad [9.24]$$

resulting from the limiting form of Eq. (9.13), this gives

$$\sum_{\text{C.H.}} (\mathbf{k},\mathscr{E}(\mathbf{k})) = \frac{1}{2(2\pi)^3} \int v(\mathbf{q}) \left(1 - \frac{1}{\epsilon_{\text{eff}}(\mathbf{q},0)}\right) d\mathbf{q} \qquad [9.25]$$

which is half the induced potential due to the coulomb screening hole in a solid characterized by the response function $\epsilon_{\text{eff}}(q,0)$. Thus, this term which is independent of the state concerned, is the one we would seek to connect with the large binding energy term resulting from the correlated motion of the electrons (or in even simpler terms, the screening of the charge by the other electrons) which we identified in the Thomas–Fermi calculation.

The dynamic screened exchange term may also be simplified, since if

$$\mathscr{E}(\mathbf{k}) - E(\mathbf{k} - \mathbf{q}) < \text{plasmon energy} \qquad [9.26]$$

we can write

$$\sum_{\text{S.E.}} (\mathbf{k},\mathscr{E}(\mathbf{k})) = \frac{1}{(2\pi)^3} \int \frac{v(\mathbf{q})}{\epsilon(\mathbf{q},0)} \, d\mathbf{q}, \qquad |\mathbf{k} - \mathbf{q}| < k_F \qquad [9.27]$$

Using the zero-energy long-wavelength limit of the response function we have

$$\sum_{\text{S.E.}} (\mathbf{k},\mathscr{E}(\mathbf{k})) = \frac{1}{(2\pi)^3} \int \frac{4\pi e^2}{q^2 + \lambda^2}, \qquad |\mathbf{k} - \mathbf{q}| < k_F \qquad [9.28]$$

a screened exchange integral. The presence of the screening length both reduces the overall magnitude of the exchange term and removes the anomalous behavior at the Fermi energy. Thus, the simplest investigation of the self-energy gives us terms indicative of the coulomb hole and the screened exchange, which we expected on physical grounds.

A more important aspect of the self-energy we have calculated is, however, that it is complex. Returning to Eqs. (9.18) and (9.20) (ignoring the cut contribution again) we see that the imaginary part of the self-energy will come from the appearance, in the integrals, of

$$\mathrm{Im}\, \epsilon^{-1}(\mathbf{q}, \omega - E(\mathbf{k} - \mathbf{q}) + i\delta) \tag{9.29a}$$

in the screened exchange and

$$\mathrm{Im}\, (\omega - \omega(\mathbf{q}) - \mathrm{E}(\mathbf{k} - \mathbf{q}) + i\delta)^{-1} \tag{9.29b}$$

in the correlation term. The evaluation of the second term is straightforward, using Eq. (9.9). For the first term we have, for small Δ

$$\epsilon(\mathbf{q},\Omega + i\Delta) = \epsilon(\mathbf{q},\Omega) + i\Delta\left(\frac{\partial \varepsilon}{\partial \Delta}\right) \tag{9.30}$$

then

$$\epsilon^{-1}(\mathbf{q},\Omega + i\Delta) = \frac{\epsilon(\mathbf{q},\Omega) - i\Delta\left(\dfrac{\partial \varepsilon}{\partial \Omega}\right)}{\epsilon^2(\mathbf{q},\Omega) + \Delta^2\left(\dfrac{\partial \varepsilon}{\partial \Omega}\right)^2} \tag{9.31}$$

which in the limit $\Delta \to 0$ gives

$$\epsilon^{-1}(\mathbf{q},\Omega + i\delta) = \frac{1}{\epsilon(\mathbf{q},\Omega)} - i\pi\, \delta(\epsilon(\mathbf{q},\Omega)) \tag{9.32}$$

If we use the high-energy limit of the response function, we can simplify this still further to give

$$\epsilon^{-1}(\mathbf{q},\Omega + i\delta) = \frac{1}{\epsilon(\mathbf{q},\Omega)} - \frac{i\pi}{\left.\dfrac{\partial \epsilon}{\partial \Omega}\right|_{\omega(\mathbf{q})}} [\delta(\Omega - \omega(\mathbf{q})) + \delta(\Omega + \omega(\mathbf{q}))] \tag{9.33}$$

we can now substitute into the self-energy to obtain the imaginary part

$$\operatorname{Im} \Sigma(\mathbf{k},\omega) = \frac{\pi}{(2\pi)^3} \int \frac{d\mathbf{q}\, v(\mathbf{q})}{\left.\dfrac{\partial\epsilon(\mathbf{q},\omega)}{\partial\omega}\right|_{\omega(q)}} \{n(\mathbf{k}-\mathbf{q})[\delta(\omega - E(\mathbf{k}-\mathbf{q}))$$
$$+ \delta(\omega - E(\mathbf{k}-\mathbf{q}) + \omega(\mathbf{q}))] - \delta(\omega - E(\mathbf{k}-\mathbf{q}) - \omega(\mathbf{q}))\} \quad [9.34]$$

The first two terms come from the dynamic exchange term and the third term from the coulomb correlation. We can rewrite this as

$$\operatorname{Im} \Sigma(\mathbf{k},\omega) = -\frac{\pi}{(2\pi n)^3} \int \frac{d\mathbf{q}\, v(\mathbf{q})}{\left.\dfrac{\partial\epsilon(\mathbf{q},\omega)}{\partial\omega}\right|_{\omega(\mathbf{q})}} \{[1 - n(\mathbf{k}-\mathbf{q})]$$
$$\times\, \delta(\omega - E(\mathbf{k}-\mathbf{q}) - \omega(\mathbf{q})) - n(\mathbf{k}-\mathbf{q})$$
$$\times\, \delta(\omega - E(\mathbf{k}-\mathbf{q}) + \omega(\mathbf{q}))\} \quad [9.35]$$

There are two components to this imaginary part, identified by the Fermi and energy conservation limitations:

(i) $\omega - E(\mathbf{k}-\mathbf{q}) - \omega(\mathbf{q}) = 0; \qquad n(\mathbf{k}-\mathbf{q}) = 0$

If we consider ω as the initial energy $[\approx \mathscr{E}(\mathbf{k})]$ and $E(\mathbf{k}-\mathbf{q})$ as the final energy, in a real transition of the particle, we have

$$\omega(\mathbf{q}) = \omega - E(\mathbf{k}-\mathbf{q}) \quad [9.36]$$

as the energy and momentum requirement that the transition between ω and $E(\mathbf{k}-\mathbf{q})$ states be able to create a plasmon. The Fermi factor, in addition, ensures that the final state is available.

(ii) $\omega(\mathbf{q}) = E(\mathbf{k}-\mathbf{q}) - \omega; \qquad n(\mathbf{k}-\mathbf{q}) = 1$

The final state is an occupied one. It suggests, then, that the relationship between plasmon initial and final energies should be interpreted as the energy and momentum restriction on the transition of a *hole* being able to produce a plasmon.

Taken together we have the not surprising result that the imaginary part of the self-energy is connected with the ability of the single particle to excite a real transition in the system, in this case a plasmon, and hence lose energy. This is illustrated schematically in Fig. 9.5 for the two cases above.

Coupling the two restrictions of real energy loss and availability of final states, we see that for the energy range

$$\mu - \omega_p < \omega < \omega_p + \mu$$

in the simple model we have used, the self-energy will be real and hence single-particle states will be well-defined entities. We must remember, however, that the actual response function (or interaction) has structure corresponding to the possibility of single-particle excitations [Eq. (9.10)]. These do not have a minimum energy cutoff so the self-energy is complex for all energies. In fact, the lifetime of a state very close to the Fermi level becomes very long because of the two powerful restrictions on the energy loss mechanism: availability of final states and energy–momentum conservation. In terms of the self-energy, this results in

$$|\mathrm{Im}\ \Sigma(\omega)| \approx (\mathrm{consts})(\omega - \mu)^2 \qquad [9.37]$$

as the energy ω tends to the Fermi energy.

9.3. EVALUATION OF THE SELF-ENERGY AND QUASIPARTICLE PROPERTIES

The evaluations of the integrals involved in the self-energy are essentially numerical, as is the solution of the equation

$$G^{-1}(\mathbf{k},\omega) = \omega - E(\mathbf{k}) - \Sigma(\mathbf{k},\omega) = 0 \qquad [9.38]$$

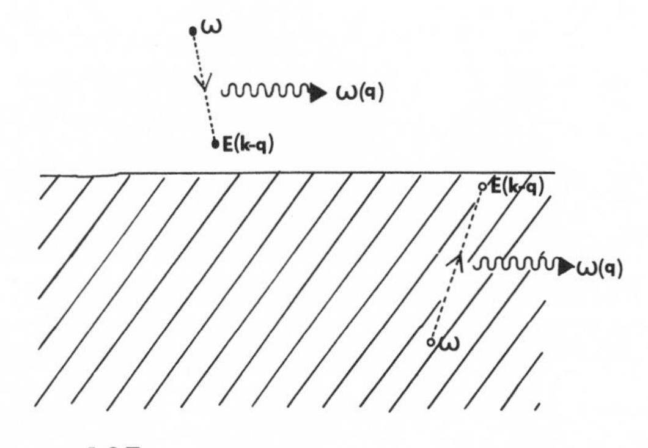

FIGURE 9.5. Interpretation of the imaginary part of the self-energy as the emission of a plasmon.

for the excitation energies of the system. One important consideration is the applicability of the quasiparticle concept to those excitations. We can use the Lehman representation to investigate this. We look for peaks in the spectral weight function. Figures 9.6a, b, and c show the results of a typical calculation of the self-energy $\Sigma(\mathbf{k},\omega)$ as a function of energy for a number of momentum values. The *imaginary* part of the self-energy behaves as we expected from simple considerations. It is very small for energies between a plasmon energy on either side of the Fermi energy, where only the single-particle excitations are possible; but shows a large increase outside this range when plasmon excitations become available. The falloff at high energy is due to limitations on the number of final states imposed by the conservation of energy and momentum, as is the change in the relative size of the peaks as the momentum alters. The *real* part of the self-energy has a similar form, though the peaks in the imaginary part translate into resonancelike behavior at the same energy. This is, of course, what one would expect, since the two terms are simply different aspects of the same complex integral.

If now on the same graph, we draw the straight line representing $\omega - E(\mathbf{k})$ then, for a small imaginary contribution to the self-energy, we obtain a solution to Eq. (9.38). We see that there is apparently more than one solution for each momentum value, but we always find at least one solution close to where we would expect the quasiparticle to be situated. The spectral weight Figs. 9.7a, b, and c gives a rather clearer picture, with at least one well-defined peak and a rather broad background at each momentum value. The presence of more than one peak should not surprise us—after all, this is an interacting system—but let us concentrate on the quasiparticle peak and the quasiparticle properties. For this we

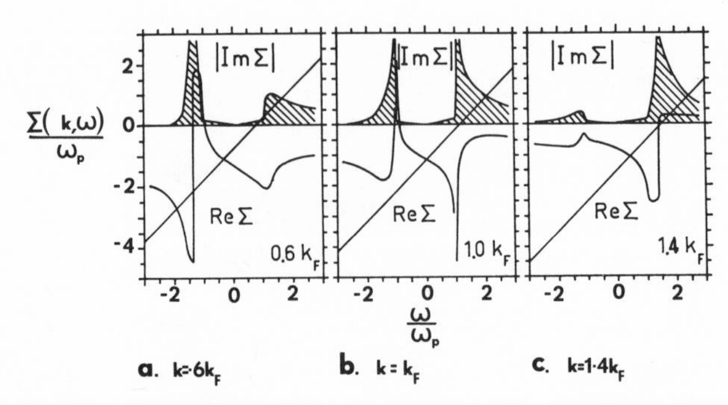

FIGURE 9.6. The self-energy [after B. I. Lundqvist, *Phys. Kondens. Mater.* 7, 117 (1968)].

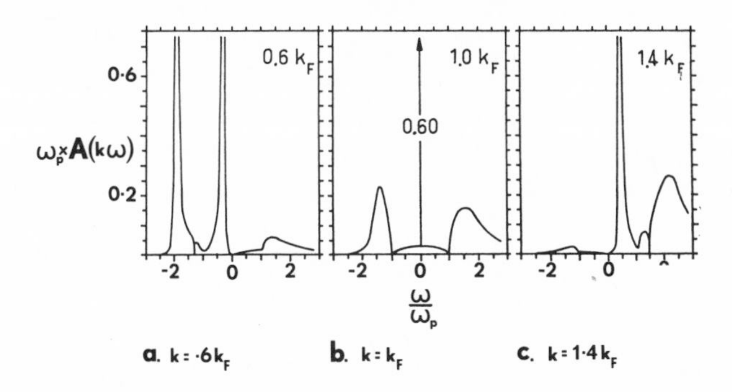

FIGURE 9.7. The spectral weight function (as in Fig. 9.6).

require $\Sigma_R(\mathbf{k},\mathscr{E}(\mathbf{k}))$ and $\Sigma_I(\mathbf{k},\mathscr{E}(\mathbf{k}))$. These are plotted as a function of momentum in Fig. 9.8. We see immediately that:

(i) The real part of the self-energy is larger than the imaginary part, even for quasiparticle states some way away from the Fermi energy.

(ii) The real part of the self-energy is essentially a constant right out to wave vectors of approximately twice the Fermi wave vector (i.e., over an energy range of four times the Fermi energy).

Together these give a remarkable justification of the Hartree model of the solid as a collection of essentially noninteracting particles. The constancy of the real part of the self-energy simply amounts to an energy shift in all of the levels corresponding to the binding energy of the elec-

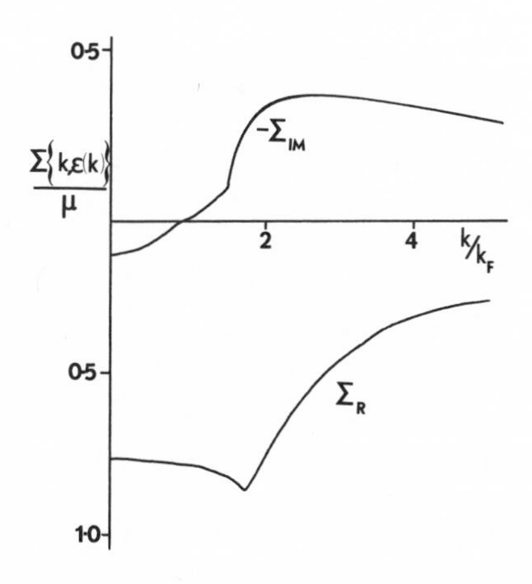

FIGURE 9.8. The real and imaginary part of the self-energy at the quasiparticle peak [after L. Hedin and S. Lundqvist, *Solid State Phys.* **23**,1 (1969)].

trons in the solid, while the small value for the imaginary part of the self-energy allows us to have confidence in the single-particlelike nature of the states. There are problems, however. Consider the resulting Green's function

$$G(\mathbf{k},\omega) = \frac{1}{\omega - E(\mathbf{k}) - \Sigma(\mathbf{k},\omega)} \tag{9.39}$$

At the quasiparticle peak we may write

$$G(\mathbf{k},\omega) \approx \frac{Z}{\omega - \mathscr{E}(\mathbf{k})} + \text{background} \tag{9.40}$$

where Z is the "renormalization constant" and is given by either

$$Z = \left(1 - \frac{\partial \Sigma}{\partial \omega}\right)^{-1} \bigg|_{\omega = \mathscr{E}(\mathbf{k})} \tag{9.41}$$

or alternatively, by integrating the peak in the spectral weight function. For electron densities which correspond to metallic values, Z is in the range 0.5 to 0.7 and varies little over the momentum range $0 < k < 2k_F$.

The fact that the quasiparticles are so "emaciated" does cause problems in the next iteration. If we take the quasiparticle states as our next "best set" of eigenstates in the calculation of, for instance, the polarization and the screened interaction, we have

$$P \approx GG \approx \tfrac{1}{4}G_0G_0 \tag{9.42}$$

(since the energy differences involved do not change if the self-energy is a constant). This in turn would suggest that the response function is now

$$\epsilon \approx 1 - \tfrac{1}{4}\omega P_0 \tag{9.43}$$

where P_0 is the first iteration polarization. The original response function ϵ ($\sim 1 - vP_0$), however, is in very good agreement with experiment for most solids. Thus, for this calculation the original single-particle Green's function would seem to be in better agreement with experiment. This appears to be so for a number of properties that can be calculated, but it is most generally shown in the Landau quasiparticle parameters.

9.4. LANDAU QUASIPARTICLES

We now have the basic justification for the *Landau theory:* Close to the Fermi surface the quasiparticles exist and have a long lifetime. We can now go on to calculate the interaction on the basis of the formalism we have developed. From Eq. (5.34), in the same notation, we have, for a quasiparticle in state $\mathbf{k}$

$$E(\mathbf{k}) = E_0(\mathbf{k}) + \sum_{\mathbf{k}'} f(\mathbf{k},\mathbf{k}')\,\Delta n(\mathbf{k}')$$

from which we have

$$\frac{\delta E(\mathbf{k})}{\delta n(\mathbf{k}')} = f(\mathbf{k},\mathbf{k}') \qquad [9.44a]$$

where $E(\mathbf{k})$ is the total energy of the quasiparticle. In terms of the self-energy theory, the energy $E(\mathbf{k})$ can be replaced by the quasiparticle energy $\mathscr{E}(\mathbf{k})$, provided we consider it to be a function of the occupation $\Delta n(\mathbf{k}')$. (This could be caused through a change in Fermi energy, for instance, by adding a distribution of quasiparticles around the Fermi surface). We then write

$$f(\mathbf{k},\mathbf{k}') = \frac{\delta \varepsilon(\mathbf{k})}{\delta n(\mathbf{k}')} \qquad [9.44b]$$

Since the single-particle energies will be independent of occupation, only the self-energy at the quasiparticle peak needs to be considered, i.e.,

$$\begin{aligned} f(\mathbf{k},\mathbf{k}') &= \frac{\delta \Sigma(\mathbf{k},\mathscr{E}(\mathbf{k}))}{\delta n(\mathbf{k}')} \\ &= \left.\frac{\partial \Sigma(\mathbf{k},\omega)}{\partial \omega}\right|_{\omega=\mathscr{E}(\mathbf{k})} \frac{\delta \mathscr{E}(\mathbf{k})}{\delta n(\mathbf{k}')} + \left.\frac{\delta \Sigma(\mathbf{k},\omega)}{\delta n(\mathbf{k}')}\right|_{\omega=\mathscr{E}(\mathbf{k})} \end{aligned} \qquad [9.45]$$

Rearranging terms and using (9.41) this gives

$$f(\mathbf{k},\mathbf{k}') = Z \frac{\delta\Sigma(\mathbf{k},\omega)}{\delta n(\mathbf{k}')}\bigg|_{\omega=\mathscr{E}(\mathbf{k})} \qquad [9.46]$$

The change in the self-energy is due to the change in the Green's function [remember, $\delta n(\mathbf{k}')$ is an occupation number]. Using the identity

$$\frac{\delta\Sigma(\mathbf{k},\omega)}{\delta n(\mathbf{k}')} = \frac{i}{2\pi^4}\int \frac{\delta\Sigma(\mathbf{k},\omega)}{\delta G(\mathbf{q},\Omega)}\frac{\delta G(\mathbf{q},\Omega)}{\delta n(\mathbf{k}')}\, d\mathbf{q}\, d\Omega \qquad [9.47]$$

it is not difficult to show that

$$\frac{\delta\Sigma(\mathbf{k},\omega)}{\delta n(\mathbf{k}')} = iZ(2\pi)^4 \cdot \frac{\delta\Sigma(\mathbf{k},\omega)}{\delta G(\mathbf{k}', \mathscr{E}(\mathbf{k}'))} \qquad [9.48]$$

If we now use the first approximation to the self-energy

$$\Sigma(\mathbf{k},\omega) = \frac{i}{(2\pi)^4}\int W(\mathbf{q},\omega')G(\mathbf{k} - \mathbf{q},\omega - \omega')\, d\omega'\, d\mathbf{q} \qquad [9.49]$$

then

$$f(\mathbf{k},\mathbf{k}') = -Z^2 W(\mathbf{k} - \mathbf{k}',\mathscr{E}(\mathbf{k}) - \mathscr{E}(\mathbf{k}')) \qquad [9.50]$$

which, for energies close to the Fermi energy, simplifies to

$$f(\mathbf{k},\mathbf{k}') = -Z^2 W(\mathbf{k} - \mathbf{k}',0) \qquad [9.51]$$

Thus, the quasiparticles interact by way of the static screened interaction, which we would expect, but reduced by a factor Z^2 to allow for the new oscillator strength in the quasiparticle, which is not so obvious. From the basis of Landau theory we would expect that the quasiparticles would retain an interaction by way of the full screened interaction—the Z^2 is rather like the appearance of a "quasicharge."

Some results for Landau parameters calculated from the full screened Lindhard interaction are shown in Fig. 9.9. The full interaction does give reasonable results. The inclusion of a factor Z^2 (which in this

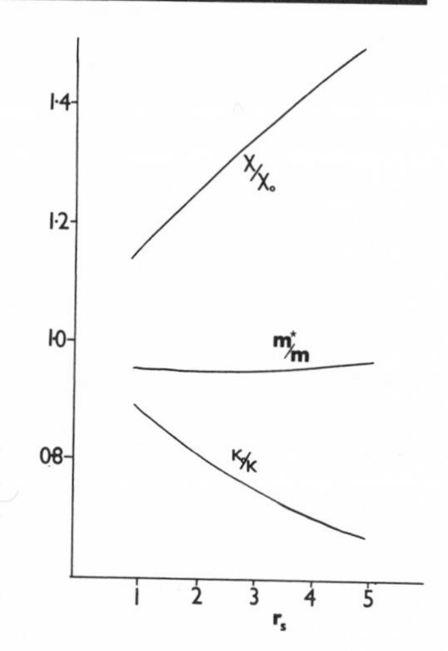

FIGURE 9.9. The spin susceptibility χ, compressibility K, and effective mass compared to the noninteracting value as calculated from the full-screened interaction [after L. Hedin and S. Lundqvist, *Solid State Phys.* **23**,1 (1969)], as a function of the electron–electron mean spacing r_5.

case would be $\approx$0.25) on any interaction-mediated parameter would, therefore, produce very poor results indeed.

9.5. INSULATING SYSTEMS

In the iteration series of Eq. (7.61), if the system is nonuniform, the problem of the calculation of the interaction is nontrivial. Taking the first iteration, one can easily calculate the polarization since this is simply

$$P(x,x',\omega) = \frac{i}{2\pi}\int G_0(x,x',\omega)G_0(x',x,\omega + \omega')e^{i\eta\omega'}\, d\omega' \qquad [9.52]$$

The Green's function may be represented in terms of the "best-fit" eigenfunctions and eigenvalues for the noninteracting system

$$G_0(x,x',\omega) = \sum_{\mathbf{k}} \frac{\phi_{\mathbf{k}}(x)\phi^*_{\mathbf{k}}(x')}{\omega - E(\mathbf{k}) + i\delta\,\mathrm{sgn}(E(\mathbf{k}) - \mu)} \qquad [9.53]$$

to give

$$P(x,x',\omega) = \sum_{\mathbf{k},\mathbf{j}} \frac{\phi_{\mathbf{k}}(x)\phi^*_{\mathbf{k}}(x')\phi_{\mathbf{j}}(x')\phi^*_{\mathbf{j}}(x)}{\omega + E(\mathbf{k}) - E(\mathbf{j}) + i\delta}\,(n(\mathbf{k}) - n(\mathbf{j})) \qquad [9.54]$$

The integrals over the allowed states are difficult but not impossible, even for quite complicated materials. The problem comes when one needs to calculate the inverse response $\epsilon^{-1}(x,x',\omega)$ from a knowledge of the dielectric response, i.e., from

$$\epsilon(x,x',\omega) = \delta(x - x') - \int v(x - x'')P(x'',x',\omega)\, dx'' \qquad [9.55]$$

since this is equivalent to an inversion of an infinite matrix. Most properties involve the interaction rather than the response function, but they may be divided into two categories:

(i) those which involve the details of the interaction (e.g., optical reflectivity, plasmons, phonons); and
(ii) those which involve integrations over the interaction, such as the self-energy.

For the first, one needs to do as complete a job as possible, but for the second it is often sufficient to use a simplified interaction. We will concentrate on the second type, since we are interested in self-energies at present.

In crystalline systems we can use translational symmetry, which requires

$$\epsilon^{-1}(\mathbf{q} + \mathbf{G}_1\, \mathbf{q} + \mathbf{G}_1) = [\epsilon(\mathbf{q} + \mathbf{G}_1, \mathbf{q} + \mathbf{G}_1)]^{-1} \qquad [9.58]$$

(**a** and **b** are lattice vectors) to transform the response function to a matrix in reciprocal space.

$$\epsilon(\mathbf{q} + \mathbf{G}_1, \mathbf{q} + \mathbf{G}_2, \omega) = \int e^{ix(\mathbf{q}+\mathbf{G}_1)}\epsilon(x,x',\omega)e^{+ix'(\mathbf{q}+\mathbf{G}_2)}\, dx\, dx' \qquad [9.57]$$

where $\mathbf{q}$ is a vector within the reduced Brillouin zone of the lattice and $\mathbf{G}_1$, $\mathbf{G}_2$ are reciprocal lattice vectors. If we now *neglect* the off-diagonal elements ($\mathbf{G}_1 \neq \mathbf{G}_2$), the matrix can, of course, be inverted easily to give

$$\epsilon^{-1}(\mathbf{q} + \mathbf{G}_1\, \mathbf{q} + \mathbf{G}_1)[\epsilon(\mathbf{q} + \mathbf{G}_1, \mathbf{q} + \mathbf{G}_1)]^{-1} \qquad [9.58]$$

This, surprisingly, is usually sufficient for the purpose of calculating self-energies in semiconductors, insulators, and metals even though in doing so one loses much of the structure of the response. What one calculates is the diagonal element given by

$$\epsilon(\mathbf{q},\mathbf{q},\omega) = 1 - \frac{4\pi e^2}{q^2} \sum_{\mathbf{k},\mathbf{j}} \frac{(n(\mathbf{k}) - n(\mathbf{j}))\,|M(\mathbf{k},\mathbf{j},\mathbf{q})|^2}{\omega + E(\mathbf{k}) - E(\mathbf{j}) + i\delta} \qquad [9.59]$$

where

$$M(\mathbf{k},\mathbf{j},\mathbf{q}) = \int \phi_{\mathbf{j}}^*(x) e^{i\mathbf{q}\cdot x} \phi_{\mathbf{k}}(x)\, dx \qquad [9.60]$$

In uniform systems, the matrix element is simply a delta function $\delta(\mathbf{k} + \mathbf{q} - \mathbf{j})$ and the energy is a single-valued function of the wave vector. In nonuniform systems the matrix element consists of a series of delta functions $\delta(\mathbf{k} + \mathbf{q} - \mathbf{j} - \mathbf{G})$, expressing the mixing of the plane waves to form Bloch functions. Alternatively, in terms of the crystal momentum $\mathbf{k}$, $\mathbf{j}$ (defined within the Brillouin zone), states of different bands are connected by scattering in the lattice through $\delta(\mathbf{k} + \mathbf{q} - \mathbf{j} - \mathbf{G})$. The matrix element becomes a connection between the crystal momenta $\mathbf{k}$ and $\mathbf{j}$ and different bands by the wave vector $\mathbf{q}$, i.e.

$$M(\mathbf{k},\mathbf{j},\mathbf{q}) \rightarrow \sum_{m,m'} M(m,m',\mathbf{k},\mathbf{k} + \mathbf{q})\delta(\mathbf{k} - \mathbf{j} + \mathbf{q})$$

Similarly the energy $E(\mathbf{k})$ becomes a function of the wave vector and band index m, $[E_m(\mathbf{k})]$.

Like the uniform system, the main contribution to the response comes from around the Fermi energy, which for insulators and semiconductors means the conduction and valence band edges. The electron–hole pairs described by the polarization bubble become excitations requiring a minimum energy—the band gap energy. At a small wavelength ($\mathbf{q} \approx 0$) the main energy denominator involves the band gap between valence and conduction bands $E_c(\mathbf{k}) - E_v(\mathbf{k})$. Figure 9.10 shows how one can consider the response developing from the metallic to the semiconductor case in a simple extended zone as the matrix elements of the crystal potential connect the plane-wave states. When one has a real three-dimensional semiconductor the regions indicated are folded back into the Brillouin zone to give the active regions of Fig. 9.11. From Fig. 9.10 it is also possible to see that if one has a wave vector larger than the region distorted by the crystal potential, then the difference between the metallic (i.e., Linhard) response and the semiconductor one will be small.

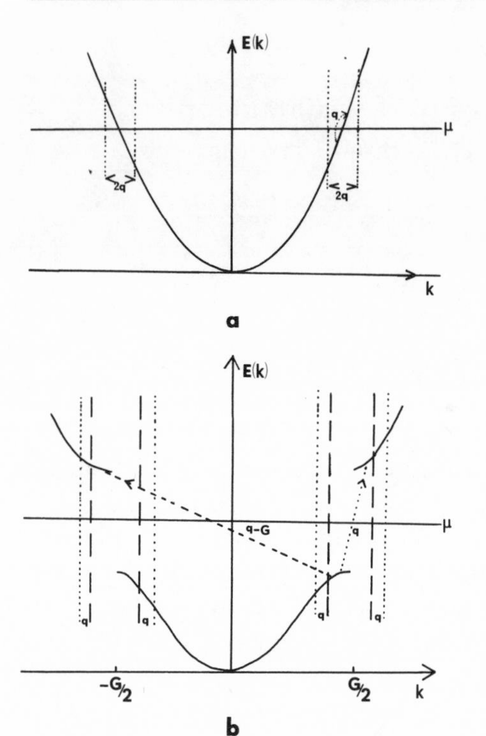

FIGURE 9.10. Development of the states contributing to the dielectric function of a simple insulator. (a) Simple metal, only the states within q of the Fermi wave vector contribute. (b) Mixing of states allows matrix elements between bands so that states connected by both q and $q \pm G$ are allowed. The vertical dashed lines show the region of significant matrix elements.

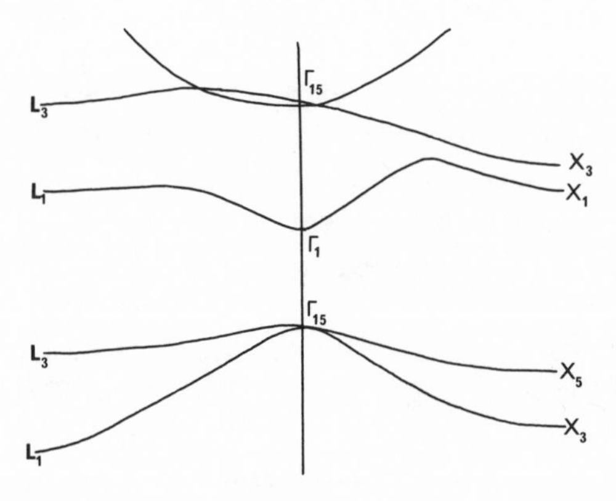

FIGURE 9.11. Active regions in the band structure of *ZnS*. This shows the region around the principal gap. The main structure in the response function comes from the energy bands $L_3 - \Gamma_{15} - X_s$ (valence band) to the conduction band $L_1 - \Gamma_1 - X_1$.

If one is considering the energy dependence, the major features will arise when the optical band gap $E_c(\mathbf{k}) - E_v(\mathbf{k} + \mathbf{q})$ is constant (E_0) for a range of crystal momentum values, for then one expects a resonance in the response, since

$$\epsilon(q,q,\omega) \approx (\text{consts}) - \frac{4\pi e^2}{q^2} \sum_{\mathbf{k}}{}' \frac{[n(\mathbf{k},m) - n(\mathbf{k}+\mathbf{q},m')]|M(m,m',\mathbf{k}+\mathbf{q},\mathbf{k})|^2}{\omega - E_0}, \qquad \omega \approx E_0$$

where $\Sigma'_{\mathbf{k}}$ contains only the region of interest and E_0 is the constant energy. The size of the resonance is obviously determined both by the number of states involved which are separated by the energy E_0 (the joint density of states is a common way of expressing this) and the matrix element. Figure 9.12a shows the response as a function of energy in the long-wavelength limit for silicon as measured by optical techniques. It is typical of most semiconductors in that there is a well-defined resonance at one energy with fairly minor structure around it. This energy corresponds to the vertical gap between conduction and valence bands, and it is a common feature of all the covalent semiconductors that this gap varies by only a small amount over the whole Fermi surface. In fact, if one thinks of the semiconductor as just a distorted Fermi sphere with a band gap at the Fermi surface (i.e., a three-dimensional version of Fig. 9.10), one has a remarkably constant gap in all directions. This is the basis of the *Penn model for semiconductor response,* in which the real band structure is replaced by a spherically symmetric two-band structure with a constant band gap. Figure 9.12b shows the calculated response as a function of wave vector for silicon using both the real wave functions and the simple Penn model. The difference is small and it is noticeable that, as one would expect, both show a very rapid return to the metallic response at finite momentum.

A complete calculation of the interaction effects on an insulating system has yet to be performed, but one can obtain most of the physical effects by looking at the contribution to the electron quasiparticle state from the self-energy. It is assumed that the wave functions of the best noninteracting Hamiltonian are a good approximation (this is better than one might at first suspect since most one-electron calculations include fitting parameters). In the effective single-particle Hamiltonian for the

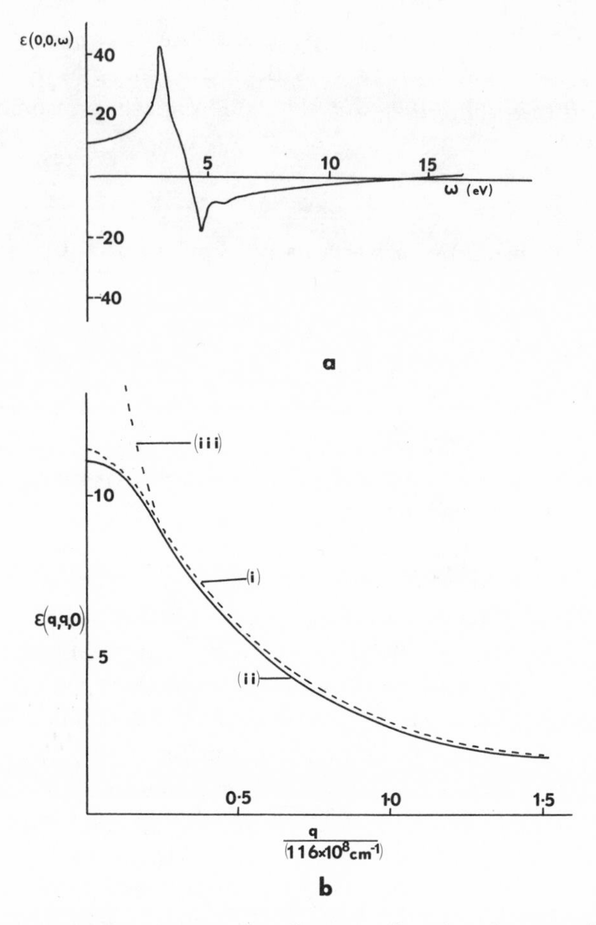

FIGURE 9.12. (a) Dielectric function for silicon as a function of energy. (b) Calculated static dielectric functions for silicon: (i) Penn model; (ii) Full calculation; (iii) Metallic response for same density.

eigenstates, one has the term at the quasiparticle peak

$$\int \Sigma(x,x',\mathcal{E}(\mathbf{k}))\phi_{\mathbf{k}}(x')\,dx'$$

The contribution to the energy from this term will be

$$\Delta\mathcal{E}(k) = \int \phi_{\mathbf{k}}^{*}(x)\Sigma(x,x',\mathcal{E}(\mathbf{k}))\phi_{\mathbf{k}}(x')\,dx\,dx' \qquad [9.62]$$

The self-energy, like the response, is a matrix in the reciprocal lattice vectors, so that in terms of the coefficients of the quasiparticle state $\phi_{\mathbf{k},m}(x)$ (m is a band index) defined by

$$\phi_{\mathbf{k},m}(x) = e^{i\mathbf{k}\cdot r} \sum_{\mathbf{G}} a_{\mathbf{G},m}(\mathbf{k}) e^{i\mathbf{G}\cdot\mathbf{r}} \times \text{(spin wave function)} \qquad [9.63]$$

The expression for the energy contribution is simply

$$\Delta\mathscr{E}_m(\mathbf{k}) = \sum_{\mathbf{G},\mathbf{G}'} a^{+}_{\mathbf{G},m}(\mathbf{k}) a_{\mathbf{G}'m}(\mathbf{k}) \Sigma(\mathbf{k} + \mathbf{G}, \mathbf{k} + \mathbf{G}'\ \mathscr{E}_m(k)) \qquad [9.64]$$

where, on neglecting the off-diagonal elements of the response function, the self-energy itself can be written

$$\Sigma(\mathbf{k} + \mathbf{G}, \mathbf{k} + \mathbf{G}', \mathscr{E}_m(\mathbf{k}))$$
$$= \frac{i}{(2\pi)^4} \sum_{m'} \int \frac{W(\mathbf{q},\omega') a_{\mathbf{G},m'}(\mathbf{k} - \mathbf{q}) a_{\mathbf{G}',m'}(\mathbf{k} - \mathbf{q})\, d\mathbf{q}\, d\omega'}{\mathscr{E}_m(\mathbf{k}) - \omega' - \mathscr{E}_{m'}(\mathbf{k} - \mathbf{q}) + i\delta\, \mathrm{sgn}(\mathscr{E}_{m'}(\mathbf{k} - \mathbf{q}) - \mu)} \qquad [9.65]$$

Apart from the weighting factors, due to the mixing of the plane waves into Bloch functions, the expressions are basically similar to those of Sec. 9.2. This difference does, however, make a significant change in the effect of the self-energy. One can see this best by considering the case where only one mixing term is important (i.e., the two-band model of Fig. 9.10). The wave function takes the form

$$\phi_{k,\pm}(\mathbf{r}) = \frac{1}{[1 + a^2_{\pm}(\mathbf{k})]^{1/2}} (e^{i\mathbf{k}\cdot\mathbf{r}} + a_{\pm}(\mathbf{k}) e^{i(\mathbf{k}-\mathbf{G})\cdot\mathbf{r}}) \qquad [9.66]$$

where the $\pm$ refer to the upper, lower bands. The mixing term is only important in the region of the band gap. Away from this

$$a_{\pm}(\mathbf{k}) \approx 0 \qquad [9.67]$$

and the self-energy reverts to the jellium case—effectively an added-on constant to the kinetic energy. At the band edge ($|k| = G/2$) we have

$a_{\pm}(G/2) = \pm 1$ and then

$$\phi_{\pm}(\mathbf{r}) = \frac{1}{\sqrt{2}}\left(e^{(i\mathbf{G}\cdot\mathbf{r}/2)} \pm e^{(-i\mathbf{G}\cdot\mathbf{r}/2)}\right) \qquad [9.68]$$

The self-energy contribution to the state becomes

$$\Delta E_{\pm}\left(\frac{G}{2}\right) = \Sigma\left(\frac{\mathbf{G}}{2}, \frac{\mathbf{G}}{2}, \mathscr{E}_{\pm}\left(\frac{\mathbf{G}}{2}\right)\right) \pm \Sigma\left(\frac{\mathbf{G}}{2}, -\frac{\mathbf{G}}{2}, \mathscr{E}_{\pm}\left(\frac{\mathbf{G}}{2}\right)\right) \qquad [9.69]$$

We first take the static limits of these self-energy terms. [This is a reasonable first approximation, since, for the coulomb and exchange terms in the self-energy, the energies involved in the integrals (i.e., plasmon) will be much larger than the band gap.] The contribution to the energies of the states on either side of the gap will be of the form

$$\Delta E_{\pm} \approx \Sigma\left(\frac{\mathbf{G}}{2}, \frac{\mathbf{G}}{2}, 0\right) \pm \Sigma\left(\frac{\mathbf{G}}{2}, -\frac{\mathbf{G}}{2}, 0\right) \qquad [9.70]$$

i.e., there is a contribution to the band gap from the self-energy of $2\Sigma(\mathbf{G}/2, -\mathbf{G}/2, 0)$. If we consider the integrals involved, we see that, for the coulumb hole term, the mixing factors on either side of the gap will produce a cancellation in $\Sigma(\mathbf{G}/2, -\mathbf{G}/2, 0)$ [i.e., $a_{\mathbf{G},n'}(\mathbf{k} - \mathbf{q})a_{\mathbf{G}'n'}(\mathbf{k} - \mathbf{q})$ is $+1$ on one side and -1 on the other], but the exchange contribution is finite because only the occupied states are sampled. Thus, the screened exchange term in semiconductors and insulators contributes directly to the band gap. Calculations performed for silicon and zinc sulfide suggest that this is a significant contribution (between $\frac{1}{4}$ and $\frac{1}{3}$ of the gap).

When one comes to consider the effect of the energy $\mathscr{E}_{\pm}(\mathbf{G}/2)$ on the states one need only consider the diagonal, screened exchange terms in the self-energy, since the coulomb hole terms are altered by only a very small amount, to obtain the major effect. In the screened exchange, the diagonal term is given by

$$\sum_{\mathrm{ex}}\left(\frac{\mathbf{G}}{2}, \frac{\mathbf{G}}{2}, \mathscr{E}_{\pm}\left(\frac{\mathbf{G}}{2}\right)\right) \approx \frac{-1}{(2\pi)^3}\int_{\mathrm{occ\ states}} W\left(\mathbf{q}, \mathscr{E}_{\pm}\left(\frac{\mathbf{G}}{2}\right) - \mathscr{E}_{-}\left(\frac{\mathbf{G}}{2} - \mathbf{q}\right)\right) d\mathbf{q}$$

[9.71]

Now for the state **G**/2 − **q** to be occupied, as we have indicated, it must be in the lower band. The difference, therefore, in the screened exchange for a state at the top of the occupied (i.e., valence) states and the bottom of the unoccupied (conduction) states will be due to the different energy range for the interaction; i.e., for conduction band state it goes from the value of the energy gap upwards while for the valence band it starts from zero. If one looks at the response function (Fig. 9.12) one sees that, since there is a resonance at the band gap energy, the difference in starting from zero or the energy gap is significant for longer wavelengths. Thus, there is a "dynamic" contribution to the band gap from the self-energy due to the difference in the exchange interaction for particles in the upper and lower bands. For normal (Si, Ge, etc.) semiconductors this is a small effect (~10% of the gap), but as one moves towards insulators, the estimates from such simple models are that this effect becomes very large as the band gap becomes comparable in size to the valence bandwidth. It has, in fact, always been difficult in single-particle calculations to obtain the correct band structures of insulators with sufficiently large band gaps. For semiconductors the calculations are fitted by the use of adjustable parameters so that the self-energy effects are masked. Figure 9.13 shows, however, the change one obtains if one takes into account the self-energy term in zinc sulfide. The fitted band gaps remain unaltered but the bands change so that the effective masses are altered. Of course, there are many other effects which come into such a calculation, but Fig. 9.13 can be taken as an indication of the magnitude of the self-energy variation.

We can go one step further. If the interaction through the self-energy affects the band gap, then changing that interaction should produce changes in the band gap. There is, in fact, some experimental evi-

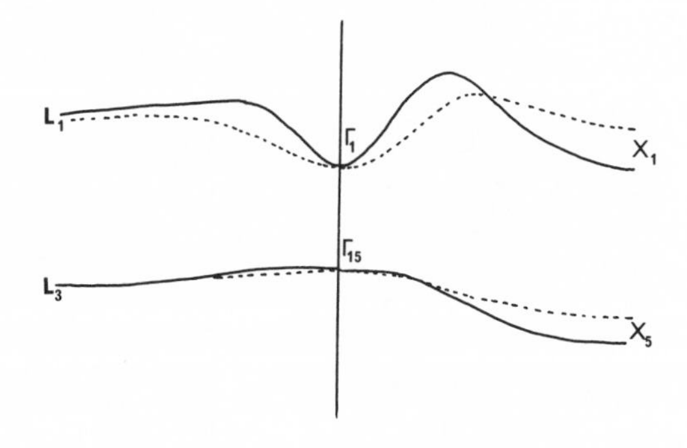

FIGURE 9.13. Effect of self-energy on the band structure of *ZnS* [after X. X. Bennett and J. C. Inkson, *J. Phys. C* **12**, 283 (1979)]. Solid line, "normal" band structure; dashed line, with self-energy correction (the gap has been adjusted to ensure agreement at Γ).

dence for this. In heavily doped semiconductors, one can consider, to a first approximation, that the electrons (or holes) populate significantly the conduction (or valence) bands up to some new Fermi level (Fig. 9.14). If one then optically excites an electron from the valence band to the conduction band, it should require a larger energy if the band gap remains constant (Fig. 9.14). This is called the "Burstein shift." Populating the valence or conduction bands, however, changes the interaction by effectively introducing a low-density "metallic" gas of carriers; i.e., at long wavelengths the interaction goes as

$$W(q \approx 0, \omega = 0) \approx \frac{4\pi e^2}{\epsilon(0)(q^2 + \lambda^2)} \qquad [9.72]$$

where λ^{-1} is a Thomas–Fermi screening length for the excess carriers. At higher energies (say, ≈energy gap) the relative change in the interaction is much smaller since the screening is strong anyway. Calculations have been performed of this effect and the change in the band gap energy is significant, in fact it is about the same order of magnitude as the increase in the energy due to the occupation of the band. For heavy-mass bands, where the Fermi energy changes only slowly with increasing occupancy and the screening effect is very good because of the large density of states at the Fermi level, the net effect is a decrease in the excitation energy as the doping increases. For light-mass bands the Fermi level rises rapidly and the screening effect is smaller so what one sees is a reduced increase in the excitation energy (Fig. 9.15). These effects are in qualitative agreement at least with the experimentally observed positive and negative Burstein shifts.

Thus, in contrast to metals, the self-energy, due to the interaction, can play an important role in the properties of the system. We have just

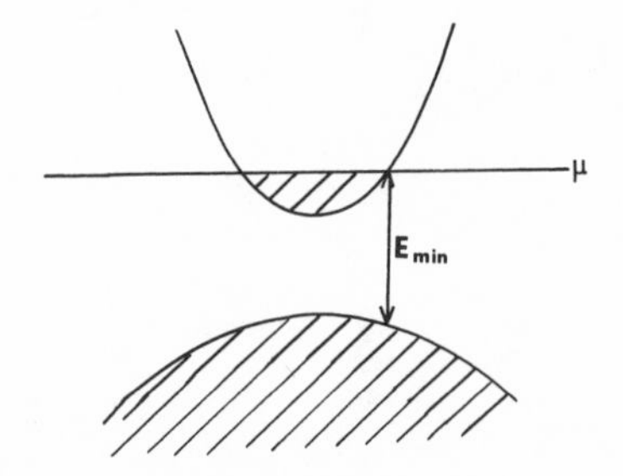

FIGURE 9.14. Effect of excess carriers on the optical threshold.

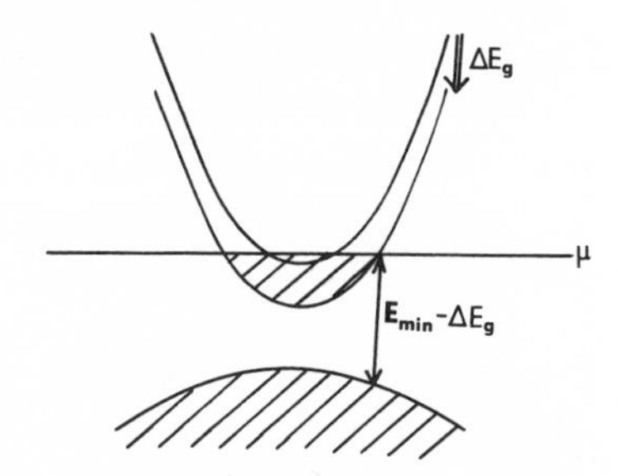

FIGURE 9.15. Effect of band gap narrowing (ΔE_g) on the optical threshold.

tried to indicate some of these, since it is true to say that this aspect of many-body theory is still being studied.

9.6. SURFACES

If one wishes to consider nonuniform systems it is instructive to look at a surface. As an electron leaves a metal, the large exchange and correlation contribution to the energy must disappear. What one would like to have to discuss such a system is not the energy change in a state but the effective spatial potential to put into the Schrödinger equation. As a replacement for the self-energy term, we could write,

$$V_k(x)\phi_k(x) = \int \Sigma(x,x',\mathcal{E}(k))\phi_k(x')\, dx' \qquad [9.73]$$

This would define the effective potential for the state $\phi_k(x)$. As long as quasiparticles exist, this is quite a sensible substitution.

The ingredients in such a calculation are the quasiparticle states and the interaction. If we consider electrons which have an energy appreciably higher than the Fermi energy in the metal then the $\phi_k(x)$ become simple plane waves. This substitution gives

$$V_k(x) = e^{-ikx}\Sigma(x,-k,\mathcal{E}(\mathbf{k})) \qquad [9.74]$$

and using the plane-wave states in the Green's function

$$\Sigma(x,-k,\mathcal{E}(\mathbf{k})) = \frac{\imath e^{\imath kx}}{(2\pi)^4}\int \frac{e^{i\delta\omega'}W(x,q,\omega')\, d\omega'\, dq}{\mathcal{E}(\mathbf{k}) - \mathcal{E}(\mathbf{k}-\mathbf{q}) - \omega' + i\delta\, \mathrm{sgn}(\mathcal{E}(\mathbf{k}-\mathbf{q}) - \mu)} \qquad [9.75]$$

or

$$V_k(x) = \frac{i}{(2\pi)^4} \int \frac{e^{i\delta\omega'} W(x,\mathbf{q},\omega')\, d\mathbf{q}\, d\omega'}{\mathscr{E}(\mathbf{k}) - \mathscr{E}(\mathbf{k}-\mathbf{q}) - \omega' - i\delta\, \mathrm{sgn}(\mathscr{E}(\mathbf{k}-\mathbf{q}) - \mu)} \qquad [9.76]$$

where $W(x,q,\omega)$ is the single-sided Fourier transform of the interaction $W(x,x',\omega)$. Since we required the energy of the particle considered to be quite high, the exchange terms are negligible [$\mathscr{E}(k - q) < \mu$ requires a very large momentum transfer] and so the term we consider is the coulomb hole contribution.

The problem with such expressions, as we have stressed, is the interaction. In this case the problem of the metal–vacuum system response function needs to be solved and inverted and it is not sufficient to ignore the off-diagonal elements, since it is these which express the presence of the surface. Instead, to extract the physics, one can perform a semiclassical type of matching problem. If one specifies the response function in the metal [$\epsilon_M(\mathbf{q},\omega)$] and the vacuum ($\epsilon_V \equiv 1$, of course), then one can calculate the total response by requiring the potential and its derivative to be continuous at the surface (on a microscopic level it is **E** not **D** which is important). This enables one to derive a complete model interaction (model since this type of calculation can only handle the simplest response functions). The interaction in a surface problem always has poles corresponding to both bulk plasmons and new collective oscillation mode, the surface plasmon, which is localized at the surface.

The actual calculation of the interaction, the self-energy and hence the effective potential, is beyond the scope of this section. In just the same way as the coulomb hole term in the jellium solid resulted in an effective induced potential for the self-energy [i.e., Eq. (9.25)]

$$\sum_{\text{C.H.}} \approx \tfrac{1}{2} V_{\text{IND}} \qquad [9.77]$$

in the present case one obtains the same simplification and so, using Eq. (9.76), one obtains (removing the spin variable)

$$V_k(r) = -\frac{e^2}{4r}, \qquad r \to \infty \qquad [9.78]$$

The limit outside the surface of the coulomb hole is the classical image potential. On a strictly semiclassical level, this is not surprising. Figure 9.16 shows how one can consider the development of the coulomb hole as the particle moves into the solid as a modification of the surface screening charge. Figure 9.17 shows a calculation of the effective potential where the contributions from the surface plasmon and bulk plasmon poles have been separated out. The physical points to note are firstly, the dominance of the surface plasmon contribution outside the metal. This is to be expected—the image potential is, after all, a surface effect. The second is the net cancellation inside the metal of changes in the bulk and surface plasmon contributions. One can consider this as a result of the balance between surface and bulk collective modes; since they arise from

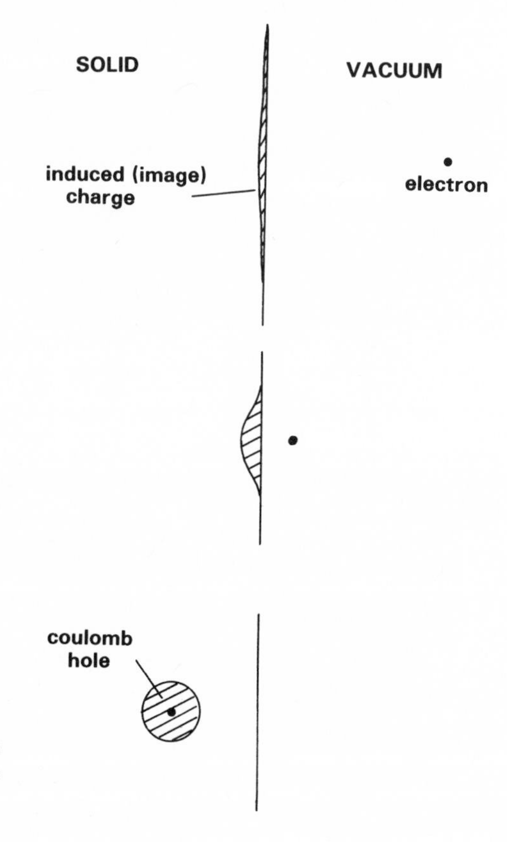

FIGURE 9.16. Transformation of the image charge into the coulomb hole as an electron enters a metal.

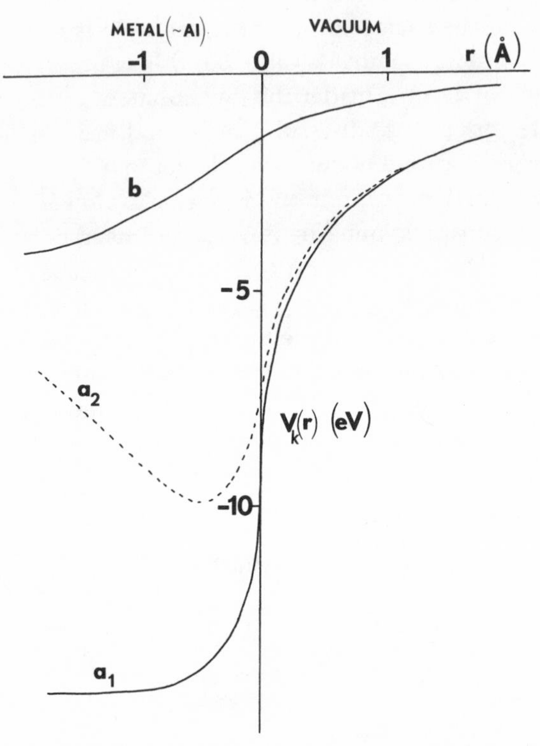

FIGURE 9.17. Optical potential [$V_k(r)$] for an electron approaching a metal surface. (a) Real part showing separation into total (a_1) and surface (a_2) contributions. (b) Imaginary part. [After J. C. Inkson, *J. Phys. F* **3**, 2143 (1973).]

oscillations of the same carriers, as one becomes more important the other decreases.

The possibility that the particle can excite real plasmon excitations reflected in the imaginary part of the self-energy produces in this case an imaginary part to the potential $V_k(x)$; i.e., the potential becomes a complex "optical" potential similar to the scattering potential used in nuclear physics. The imaginary part describes plasmon emission processes for both bulk and surface modes. Figure 9.17 shows a typical potential, though one must be careful, since the imaginary part to the potential is very dependent upon the state considered (unlike the real part).

These potentials and the application of self-energy techniques to surfaces are, it is fair to say, at a quite primitive level. Like the insulator case, however, it does show that the calculation of the jellium is rather a special case in the sense of producing a result which does not have a significant effect upon the one-electron states.

BIBLIOGRAPHY

Dielectric Response

LINDHARD, J., *Dan. Math. Phys. Medd* **28,** 8, 1954.
EHRENREICH, H., and COHEN, M. H., *Phys. Rev.* **115,** 786, 1959.

Self-Energy Calculations

LUNDQVIST, B. I., *Phys. Kondens. Mater.,* **6,** 206, 1967; and **7,** 117, 1968.
HEDIN, L., *Phys. Rev.,* **139,** A796, 1965.

Landau Quasiparticles

PINES, D., and NOZIÈRES, P., *Theory of Quantum Liquids,* Benjamin, Menlo Park: CA, 1966.
HEDIN, L., and LINDQVIST, S., *Solid State Phys.,* **23,** 1, 1969.

Insulating Systems

GUINEA, F., and TREJEDOR, C., *J. Phys. C.,* **13,** 5515, 1980.
KANE, E. O., *Phys. Rev. B* **4,** 1917, 1971; and *B* **5,** 1493, 1972.
INKSON, J. C., and BENNETT, M., *J. Phys. C.,* **11,** 2017, 1978; and **10,** 987, 1978.
INKSON, J. C., *J. Phys. C,* **9,** 1177, 1976.

Surfaces

HARRIS, J., and JONES, R., *J. Phys. C,* **7,** 3751, 1974.
INKSON, J. C., *J. Phys. F,* **3,** 2143, 1973.

PROBLEMS

1. The Hamiltonian of a one-dimensional insulating system is given by

$$H = -\frac{\hbar^2}{2m}\nabla^2 + V_0 \cos Gx$$

Show that the Lehman representation has two peaks even for the non-interacting system. [The more appropriate Lehman representation is in terms of the single-particle states ϕ_n (i.e., $\langle\phi_n|A(x,x',\omega)|\phi_{n'}\rangle$).]

2. If single-particle excitations (i.e., electron–hole) dominate the scattering below the plasmon energy, show that for insulating systems with a band gap E_g the quasiparticle energies are still real for energies within $\pm E_g$ of the valence and conduction band edges.

3. A metal whose response is determined by the Thomas-Fermi equation

$$(\nabla^2 - \lambda^2)V(\mathbf{r}) = 0$$

occupies the $z < 0$ infinite half space. Show that the general form of the induced potential in the metal is given by

$$V_M(\mathbf{r}) = \int e^{(m2+\lambda 2)1/2z} J_0(m\rho)g(m)\, dm$$

where m, ρ are the wave vector and coordinate parallel to the surface (in cylindrical polars) and $g(m)$ is a general function. Show also that the potential outside the metal produced by a charge Q at the point a must be of the form

$$V_V(r) = \int e^{-mz} J_0(m\rho)f(m)\, dm + Q\int e^{-m|z-a|} J_0(m\rho)\, dm$$

Use these two equations to calculate the induced potential due to the charge and hence show that the image potential for the charge is finite at the metal surface, and also that at large distances the asymptotic form of the image potential is given by

$$V_{\text{IM}}(z) = -\frac{e^2}{4(z + \lambda^{-1})}$$

4. Calculate the effective response to a charge in a two-component system in which the negative z half space is occupied by a medium with a response function $\epsilon(\omega)$ and the other by a vacuum [$\epsilon(\omega) = 1$]. Hence show that this system has collective oscillations at frequencies given by

$$\epsilon(\omega) = 0 \quad \text{and} \quad \epsilon(\omega) + 1 = 0$$

5. Calculate the self-energy at large distances for such a response and show that it is equal to the image potential and independent of state

concerned; that is,

$$\Sigma(k,\mathrm{E}(k)) = -\frac{e^2}{4z}\frac{\epsilon(0) - 1}{\epsilon(0)}$$

CHAPTER TEN

THERMAL EFFECTS ON THE GREEN'S FUNCTION

Up to now we have avoided any discussion of thermal effects upon the interacting system apart from drawing attention, in Chap. 5, to the similarity between the existence of inexact states in interacting quantum mechanics and the equipartition theorem in statistical mechanics. In fact, quantum statistical mechanics is based upon the same framework as classical statistical mechanics. For this reason we will not try to develop a detailed theory here but simply cover the ground sufficiently to derive the required properties.

10.1. THE DENSITY MATRIX

Suppose we have a state $|n\}$ (n is a general label) and an observable described by the operator $\hat{O}$. In terms of a complete set of eigenstates $|i\rangle$ spanning the space of $|n\}$, we can write

$$\begin{aligned}\langle\hat{O}\rangle_n &= \{n|\hat{O}|n\} \\ &= \sum_{i,j}\{n|i\rangle\langle i|\hat{O}|j\rangle\langle j|n\} \\ &= \sum_{i,j}\langle j|n\}\{n|i\rangle\langle i|\hat{O}|j\rangle\end{aligned} \qquad [10.1]$$

Therefore

$$\langle\hat{O}\rangle_n = \sum_{i,j}\rho_{ji}O_{ij}$$

where

$$\rho_{ji} = \langle j|n\rangle\langle n|i\rangle$$
$$= \langle j|\hat{\rho}_n|i\rangle \qquad [10.2]$$

and $\hat{\rho}_n$ is the density matrix operator for the state $|n\rangle$

$$\hat{\rho}_n = |n\rangle\langle n| \qquad [10.3]$$

Using (10.3) we can write Eq. (10.1) in a representation independent form as

$$\langle\hat{O}\rangle_n = \mathrm{Tr}(\hat{\rho}_n\hat{O}) \qquad [10.4]$$

Although this definition is relatively trivial in noninteracting systems, in large interacting systems it is very useful since it may be more easily approximated. Suppose now we move to a statistical system, that is we consider a system which has a probability P_n of being in the state $|n\rangle$ and consider an ensemble X of such systems. In the ensemble, there will be P_nX in the state $|n\rangle$. The average of the observable corresponding to the operator $\hat{O}$ will now be

$$\langle\hat{O}\rangle = \sum_n P_n\langle O\rangle_n$$
$$= \sum_n P_n\,\mathrm{Tr}(\hat{\rho}_n\hat{O}) \qquad [10.5]$$

If we wish to express the resulting statistical average in the same way as Eq. (10.4), we can define a statistical density matrix as

$$\hat{\rho} = \sum_n P_n|n\rangle\langle n| \qquad [10.6]$$

whence

$$\langle\hat{O}\rangle = \mathrm{Tr}(\hat{\rho}\hat{O}) \qquad [10.7]$$

This has the property

$$\mathrm{Tr}(\hat{\rho}) = \sum_n P_n = 1 \qquad [10.8]$$

which is to be expected. It also follows from Eq. (10.6) that

$$(\hat{\rho})^2 = \sum_n P_n |n\}\{n| \qquad [10.9]$$

In an exact state ($P_n = 1$ for one $|n\}$ only) we also have

$$(\hat{\rho}_n)^2 = \hat{\rho}_n \qquad [10.10]$$

10.2. STATISTICAL MECHANICS

We will only consider the grand canonical ensemble, where both particles and energy may be transferred between the systems comprising the ensemble. This is the most general situation and the derivation is no more difficult than in the canonical and microcanonical ensembles.

Consider an ensemble of X systems each of which may exist in a state $|n\}$ with a probability P_n. The number of systems in the state $|n\}$ will then be $P_n X$ ($\equiv X_n$). The quantum number n may represent a range of energies or particle numbers and need not, of course, be the ground state of the system for those particular quantities.

For the ensemble as a whole, therefore, the probability of any configuration (Ω) is given by

$$\Omega = \frac{X!}{X_1!X_2!X_3!\cdots} \qquad [10.11]$$

Just as in classical statistical mechanics, we now use Stirling's approximation

$$\ln X! = X \ln X - X \qquad [10.12]$$

to give

$$\Omega = \exp\left[-\sum_n (X_n \ln X_n) - X \ln X\right] \qquad [10.13]$$

In terms of probabilities this simplifies further to

$$\Omega = \exp\left[-\sum_n XP_n \ln P_n\right] \qquad [10.14]$$

We may now write this in terms of the density matrix

$$\Omega = \exp[-X\,\mathrm{Tr}(\hat{\rho}\ln\hat{\rho})] \qquad [10.15]$$

The probability can thus be written as

$$\Omega = \exp\left(+\frac{X}{k_B}S\right) \qquad [10.16]$$

where S (which will turn out to be the entropy) is given by

$$S = k_B\,\mathrm{Tr}(\hat{\rho}\ln\hat{\rho}) \qquad [10.17]$$

and k_B is the Boltzman constant.

Maximizing the probability of the configuration Ω corresponds to maximizing S. We follow the standard treatment of Lagrange multipliers, but the formulation is slightly different from the classical case. For the total ensemble we require

(i) Conservation of energy:

$$\delta\langle\hat{H}\rangle = \delta(\mathrm{Tr}(\hat{\rho}\hat{H})) = 0 \qquad [10.18]$$

(ii) Conservation of number of particles:

$$\delta\langle\hat{N}\rangle = \delta(\mathrm{Tr}(\hat{\rho}\hat{N})) = 0 \qquad [10.19]$$

(iii) Conservation of probability:

$$\delta(\mathrm{Tr}(\hat{\rho})) = 0 \qquad [10.20]$$

These provide three Lagrange multipliers; β for (i), $\mu\beta$ for (ii), and λ for (iii), so we have

$$\frac{\delta}{\delta\hat{\rho}}[-k_B\,\mathrm{Tr}(\hat{\rho}\ln\hat{\rho}) - \beta k_B\mu\,\mathrm{Tr}(\hat{\rho}\hat{N}) - k_B\beta\,\mathrm{Tr}(\hat{\rho}\hat{H}) + \lambda\,\mathrm{Tr}(\hat{\rho})] = 0 \qquad [10.21]$$

or

$$k_B \ln \hat{\rho} - (\lambda - k_B) + k_B\beta\mu\hat{N} - k_B\beta\hat{H} = 0 \qquad [10.22]$$

If we define a function Z by

$$k_B \ln Z = \lambda - k_B$$

we can solve Eq. (10.22) to give

$$\hat{\rho} = \frac{1}{Z} e^{-\beta(\hat{H} - \mu\hat{N})} \qquad [10.23]$$

and

$$Z = \mathrm{Tr}(e^{-\beta(H - \mu N)}) \qquad [10.24]$$

Z is immediately recognizable as the partition function for the *grand canonical ensemble* while we can identify S as the entropy by the relation [from (10.22)]

$$\begin{aligned} S &= -k_B \, \mathrm{Tr}(\hat{\rho}(\mu\beta\hat{N} - \beta\hat{H} - \ln Z)) \\ &= -k_B\beta\mu\langle\hat{N}\rangle + k_B\beta\langle\hat{H}\rangle + k_B \ln Z \end{aligned} \qquad [10.25]$$

With Z as the partition function we define

$$k_B \ln Z = -\frac{F}{T} \qquad [10.26]$$

and expression (10.25) becomes the well-known relation for the free energy (dropping the expectation value brackets in the classical limit)

$$F = E - TS - \mu N \qquad [10.27]$$

Thus we recover the classical thermodynamic functions.

The expectation value of any operator is now given by Eqs. (10.7) and (10.23) as

$$\langle\hat{O}\rangle = \mathrm{Tr}\left(\frac{e^{-\beta(\hat{H} - \mu\hat{N})}}{Z} \hat{O}\right) \qquad [10.28]$$

As an example, let us take the number operator for the state l.

$$\hat{n}_l = \hat{a}_l^+ \hat{a}_l \tag{10.29}$$

The expectation value is given by

$$\begin{aligned}\langle \hat{n}_l \rangle &= \frac{1}{Z} \operatorname{Tr}(e^{-\beta(\hat{H}-\mu\hat{N})} \hat{a}_l^+ \hat{a}_l) \\ &= \frac{1}{Z} \operatorname{Tr}(e^{-\beta(\hat{H}-\mu\hat{N})} \hat{a}_l^+ e^{+\beta(\hat{H}-\mu\hat{N})} e^{-\beta(\hat{H}-\mu\hat{N})} \hat{a}_l) \\ &= \frac{1}{Z} \operatorname{Tr}(e^{-\beta(E_l-\mu)} \hat{a}_l^+ e^{-\beta(\hat{H}-\mu\hat{N})} \hat{a}_l) \\ &= \frac{1}{Z} e^{-\beta(E_l-\mu)} \operatorname{Tr}(\hat{a}_l^+ e^{-\beta(\hat{H}-\mu\hat{N})} \hat{a}_l)\end{aligned} \tag{10.30}$$

using the identity

$$e^{-\beta(E_l-\mu)} \hat{a}_l^+ = e^{-\beta(\hat{H}-\mu\hat{N})} a_l^+ e^{\beta(\hat{H}-\mu\hat{N})} \tag{10.31}$$

Now we use the cyclic property of the trace and the commutator (anti-commutator) relationships appropriate to bose (fermion) particles.

$$\begin{aligned}\langle \hat{n}_l \rangle &= \frac{e^{-\beta(E_l-\mu)}}{Z} \operatorname{Tr}(e^{-\beta(\hat{H}-\mu\hat{N})} \hat{a}_l \hat{a}_l^+) \\ &= e^{-\beta(E_l-\mu)}[1 \pm \langle n_l \rangle]\end{aligned} \tag{10.32}$$

or

$$\langle \hat{n}_l \rangle = [e^{\beta(E_l-\mu)} \mp 1]^{-1} \tag{10.33}$$

the standard Bose-Einstein (Fermi-Dirac) distribution for bosons (fermions).

Because the operator $\hat{H} - \mu\hat{N}$ appears so often in the grand canonical ensemble it is usually written as $\hat{K}$—the *grand canonical Hamiltonian*—and then

$$\langle \hat{O} \rangle = \operatorname{Tr}\left[\frac{e^{-\beta\hat{K}}}{Z} \hat{O}\right] \tag{10.34}$$

These expectation value relationships describe a wealth of equilibrium properties, but without the basic knowledge of the interacting system, relationships such as (10.34) are useless. In order to accommodate the microscopic properties we need, as before, to calculate a property akin to the Green's function.

10.3. THE "THERMAL" HEISENBERG REPRESENTATION

The exponentials that appear in statistical mechanics (i.e., $e^{-\beta\hat{K}}$) bear a very close resemblance to the operator appearing in the normal Heisenberg–Schrödinger transformation [Eq. (4.9)]. The statistical average of Eq. (10.28) is also similar to the expectation value expression in the Green's function formulation:

$$\langle \hat{O} \rangle = \pm i \lim_{\substack{t' \to t-\delta \\ x' \to x}} \{\mathrm{Tr}[\hat{O}(x)G(x,t,x',t')]\} \qquad [10.35]$$

In generalizing the previous Green's function definitions to finite temperature, it is as well to take full advantage of both of these facts. In analogy to the Heisenberg transformation

$$\hat{O}_H(x,t) = e^{(i\hat{H}t)/\hbar}\hat{O}_S(x)e^{(-i\hat{H}t)/\hbar} \qquad [10.36]$$

we define a "thermal" Heisenberg transformation

$$\hat{O}_K(x,\tau) = e^{(\hat{K}\tau)/\hbar}\hat{O}_S(x)e^{(-\hat{K}\tau)/\hbar} \qquad [10.37]$$

where we have used the grand canonical Hamiltonian K ($= \hat{H} - \mu\hat{N}$ as before) and changed to a new variable τ. (This is sometimes considered as an "imaginary" time $\tau = it$). The resulting transformation is not Hermitian, since

$$\hat{O}_K^+(x,\tau) = e^{-\hat{K}\tau/\hbar}\hat{O}_S^+(x)e^{(+\hat{K}\tau)/\hbar} \neq \hat{O}_K(x,\tau) \qquad [10.38]$$

Just as in the Heisenberg notation, we may now, by differentiating $O_K(x,\tau)$, derive an "equation of motion"

$$\hbar \frac{\partial}{\partial\tau} \hat{O}_K(x,\tau) = [\hat{K},\hat{O}_K(x,\tau)] \qquad [10.39]$$

Extending the analogy even further, we may define an interaction representation so that

$$\hat{O}_I(x, \tau) = e^{(+\hat{K}_0\tau)/\hbar}\hat{O}_S(x)e^{-\hat{K}_0\tau/\hbar} \quad [10.40]$$

where

$$\hat{K} = \hat{K}_0 + \hat{H}_{\text{INT}} \quad [10.41]$$

and $\hat{K}_0$ is considered soluble. The corresponding equation of motion is

$$\hbar \frac{\partial}{\partial \tau} \hat{O}_I(x,\tau) = [\hat{K}_0, \hat{O}_I(x,\tau)] \quad [10.42]$$

Following the derivation in Sec. 4.3, we can write

$$\hat{O}_K(x,\tau) = \left[T_\tau \exp\left(\frac{1}{\hbar} \int_0^\tau \hat{H}_I \, d\tau\right)\right] \hat{O}_I(x,\tau) \left[T_\tau \exp\left(-\frac{1}{\hbar} \int_0^\tau \hat{H}_I \, d\tau\right)\right] \quad [10.43]$$

in the normal case, where $\hat{H}_1$ is τ dependent and T_τ is a "τ" ordering operator.

The Green's function has been defined as a time-ordered expectation value of the pair of field operators $\hat{\psi}_H(x,t)\hat{\psi}_H^+(x',t')$. It seems sensible, therefore, to extend this to the thermal situation, replacing the expectation value by a thermal average and the time ordering by a τ ordering so that we define a function $\mathcal{G}(x,\tau,x',\tau')$ as

$$\begin{aligned} \mathcal{G}(x,\tau,x',\tau') &= -\text{Tr}\{\hat{\rho} T_\tau[\hat{\psi}_K(x,\tau)\hat{\psi}_K^+(x',\tau')]\} \\ &= -\langle T_\tau[\hat{\psi}_K(x,\tau)\hat{\psi}_K^+(x',\tau')]\rangle \end{aligned} \quad [10.44]$$

where T_τ orders the operators according to their τ values and includes, as before, a change of sign for interchange of fermion operators. Consider now Eq. (10.7).

$$\langle \hat{O} \rangle = \text{Tr}[\hat{\rho}\hat{O}(x)] \quad [10.45]$$

In terms of Eq. (10.44), in the thermal Heisenberg representation, this becomes

$$\langle \hat{O} \rangle = \mp \lim_{\substack{x \to x' \\ \tau \to \tau' + \delta}} \text{Tr}[\hat{O}(x)\mathcal{G}(x,\tau,x',\tau')] \quad [10.46]$$

where the proof follows exactly as in the case of the expectation value at zero temperature. This is the property we require for our Green's function; in a thermal system it is statistical averages which are of most importance.

10.4. EVALUATION OF THE PERTURBATION EXPANSION

The equation of motion for the Green's function may be derived from Eq. (10.39). For our "standard" interacting system we have [cf. Eq. (6.54)]

$$\hat{K} = \int dx\, \hat{\psi}_K^+(x,\tau)(H_0(x) - \mu)\hat{\psi}_K(x,\tau) + \int dx\, dx'\, \hat{\psi}_K^+(x,\tau)\hat{\psi}_K^+(x',\tau')v(x-x')\,\delta(\tau-\tau')\,\hat{\psi}_K(x',\tau')\hat{\psi}_K(x,\tau) \quad [10.47]$$

The equation of motion for the field operator is then

$$\left(\hbar\frac{\partial}{\partial\tau} + H_0(x) - \mu\right)\hat{\psi}_K(x,\tau) + \int dx''\, \hat{\psi}_K^+(x'',\tau)\hat{\psi}_K(x'',\tau)v(x-x'')\hat{\psi}_K(x,\tau) = 0 \quad [10.48]$$

where again, formally, the only change is the replacement of "it" by τ. Similarly the equation of motion for the Green's function becomes

$$\left(\hbar\frac{\partial}{\partial\tau} + H_0(x) - \mu\right)\mathcal{G}(x,\tau,x',\tau') + \int dx'\, v(x-x')\,\mathrm{Tr}\{\hat{\rho}T_\tau[\hat{\psi}_K^+(x'',\tau)\hat{\psi}_K(x'',\tau)\hat{\psi}_K(x,\tau)\hat{\psi}_K^+(x',\tau')]\} = -\hbar\delta(x-x')\,\delta(\tau-\tau') \quad [10.49]$$

Interpreting the four-field operator trace as a two-particle Green's function, we see that we are retracing exactly the same steps as previously with the zero-temperature formalism. To determine the perturbation expansion we introduce an external perturbation $\phi(x,\tau)$ so that

$$K = K_0 + \phi(x,\tau) \quad [10.50]$$

This enables us to work in the interaction representation and define

$$\hat{S} = T_\tau \exp\left(\frac{1}{\hbar}\int_0^{\beta\hbar} \hat{\phi}(\tau)\, d\tau\right) \qquad [10.51]$$

which enables us to write

$$\mathcal{G}(x_1\tau_1 x_2\tau_2) = \frac{\mathrm{Tr}\{\hat{\rho}T_\tau[S\hat{\psi}_{k0}(x_1,\tau_1)\hat{\psi}_{k0}(x_2,\tau_2)]\}}{\mathrm{Tr}\{\hat{S}\}} \qquad [10.52]$$

where all operators are expressed in the unperturbed system. In the limit of zero temperature and $\tau \rightarrow it$, this is the same equation as (7.6). One can consider the difference in the integral limits on $\hat{S}$ as simply developing the system from zero temperature, when the perturbation is switched on rather than $t = -\infty$. All one is required to do now is repeat mechanically the development and definitions of Secs. 7.1 to 7.4. *Nothing* changes in this development except that the internal temperature variables are integrated from $0 \rightarrow \beta$ rather than $-\infty \rightarrow +\infty$. It is in this way that the temperature appears in the perturbation expansion. We again have the set of functions: polarization, $P(x_1,\tau_1,x_2,\tau_2)$; noninteracting Green's function, $\mathcal{G}_0(x_1,\tau_1,x_2,\tau_2)$; self-energy, $\Sigma(x_1,\tau_1,x_2,\tau_2)$, etc., which may be evaluated at various levels of approximation by expanding the vertex function $\Gamma(x_1,\tau_1,x_2,\tau_2,x_3,\tau_3)$ in terms of the screened interaction $W(x_1,\tau_1,x_2,\tau_2)$.

One must remember in this context, however, that τ is *not* a real physical variable in the sense time is; it is a formal device for introducing thermal properties. Because of this the physical interpretations we had in the real-time, zero-temperature perturbation series do not carry over—it is the Green's function and particularly its use in Eq. (10.46) that is the important physical quantity now.

10.5. PERIODICITY OF $\mathcal{G}$ AND THE EXTENSION TO ENERGY DEPENDENCY

Just as τ relates to an imaginary time, so can we find a variable corresponding to the energy ω. In order to do this we derive a very important property of the thermal Green's function—its periodicity in the τ

variables. We have

$$\mathcal{G}(x_1,\tau_1,x_2,\tau_2) = -\mathrm{Tr}\left\{\frac{e^{-\beta\hat{K}}}{Z}\, T_\tau[\hat{\psi}_K(x_1,\tau_1)\hat{\psi}_K^+(x_2,\tau_2)]\right\} \qquad [10.53]$$

Suppose $\tau_2 > \tau_1$ then

$$\mathcal{G}(x_1,\tau_1,x_2,\tau_2) = \mp\mathrm{Tr}\left\{\frac{e^{-\beta\hat{K}}}{Z}\,\hat{\psi}_K^+(x_2,\tau_2)\hat{\psi}_K(x_1,\tau_1)\right\} \qquad [10.54]$$

Using the cyclic property of the trace, we have

$$\begin{aligned}\mathcal{G}(x_1,\tau_1,x_2,\tau_2) &= \mp\mathrm{Tr}\left\{\hat{\psi}_K(x_1,\tau_1)\frac{e^{-\beta\hat{K}}}{Z}\,\hat{\psi}_K^+(x_2,\tau_2)\right\} \\ &= \mp\mathrm{Tr}\left\{\frac{e^{-\beta\hat{K}}}{Z}\, e^{+\beta\hat{K}}\hat{\psi}_K(x_1,\tau_1)e^{-\beta\hat{K}}\hat{\psi}_K^+(x_2,\tau_2)\right\} \qquad [10.55] \\ &= \mp\mathrm{Tr}\left\{\frac{e^{-\beta\hat{K}}}{Z}\,\hat{\psi}_K(x_1,\tau_1+\beta\hbar)\hat{\psi}_K^+(x_2,\tau_2)\right\}\end{aligned}$$

Thus the time variable τ_1 is undefined to within a factor of $n\beta\hbar$. It follows that we can, therefore, always limit τ_1 to the region $0 \rightarrow \beta\hbar$ without loss of generality. If we consider $\tau_1 > \tau_2$ we can derive a similar result for τ_2, so that τ_2 need also only be defined in the region $0 \rightarrow \beta\hbar$. It follows from Eq. (10.55) that

$$\mp\mathrm{Tr}\left\{\frac{e^{-\beta\hat{K}}}{Z}\,\psi_K(x_1,\tau_1+\beta\hbar)\psi_K(x_2,\tau_2)\right\} = \pm\mathcal{G}(x_1,\tau_1+\beta\hbar,x_2,\tau_2) \qquad [10.56]$$

or

$$\mathcal{G}(x_1,\tau_1,x_2,\tau_2) = \pm\mathcal{G}(x_1,\tau_1+\beta\hbar,x_2,\tau_2) \qquad [10.57a]$$

similarly

$$\mathcal{G}(x_1,\tau_1,x_2,\tau_2) = \pm\mathcal{G}(x_1,\tau_1,x_2,\tau_2+\beta\hbar) \qquad [10.57b]$$

So the Green's function is periodic (antiperiodic) in the τ variables for Bose (Fermi) systems.

For most practical systems the Hamiltonian is independent of τ. This means that for all of the functions appearing in the perturbation analysis

$$F(x_1,\tau_1,x_2,\tau_2) \Rightarrow F(x_1,x_2,\tau_1 - \tau_2)$$

and so

$$\begin{aligned} F(x_1,x_2,\tau_1 - \tau_2) &= \pm F(x_1,x_2,\tau_1 - \tau_2 + \beta\hbar) \\ &= F(x_1,x_2,\tau_1 - \tau_2 + 2\beta\hbar) \end{aligned} \tag{10.58}$$

Because of this periodicity, the Fourier transform is defined by the pair of equations

$$\mathcal{G}(x_1,x_2,\tau) = \frac{1}{\beta}\sum_n e^{(-i\omega_n\tau)/\hbar}\mathcal{G}(x_1,x_2,\omega_n) \tag{10.59}$$

$$\mathcal{G}(x_1,x_2,\omega_n) = \frac{1}{\hbar}\int_{-\beta\hbar}^{+\beta\hbar} d\tau\, e^{(i\omega_n\tau)/\hbar}\mathcal{G}(x_1,x_2,\tau) \tag{10.60}$$

where we define

$$\omega_n = \frac{n\pi}{\beta} \tag{10.61}$$

From Eq. (10.60) we have, using the periodicity

$$\mathcal{G}(x_1,x_2,\omega_n) = \frac{1}{2\hbar}(1 \pm e^{i\omega_n\beta})\int_0^{\beta\hbar} d\tau\, e^{(i\omega_n\tau)/\hbar}\mathcal{G}(x_1,x_2,\tau) \tag{10.62}$$

From the definition of ω_n we see that

$$\mathcal{G}(x_1,x_2,\omega_n) = 0$$

when n is odd for bosons and even for fermions. Thus the statistics of the particles are built automatically in by using the even values for bosons and the odd values for fermions. In a uniform system we can, of course, make a further simplification and use a momentum representation

$$\mathcal{G}(x_1,x_2,\omega_n) = \mathcal{G}(x_1 - x_2,\omega_n) \Rightarrow \mathcal{G}(q,\omega_n) \tag{10.63}$$

This enables us to use the simpler form of the perturbation series so that for instance

$$\mathcal{G}(q,\omega_n) = \mathcal{G}_0(q,\omega_n) + \mathcal{G}_0(q,\omega_n)\Sigma(q,\omega_n)\mathcal{G}(q,\omega_n) \tag{10.64}$$

or

$$\mathcal{G}^{-1}(q,\omega_n) = \mathcal{G}_0^{-1}(q,\omega_n) - \Sigma(q,\omega_n) \tag{10.65}$$

The important function is, of course, $\mathcal{G}(x_1,x_2,-\eta)$, where η is a positive infinitesimal, which is to be used in Eq. (10.46) to calculate the statistical limits. From (10.59), we have

$$\mathcal{G}(x_1,x_2,-\eta) = \frac{1}{\beta}\sum_n e^{(i\omega_n\eta)/\hbar}\mathcal{G}(x_1,x_2,\omega_n) \tag{10.66}$$

To illustrate the properties inherent in these equations, consider the non-interacting Green's function. We have a set of eigenstates $\phi_p(x)$ so that the field operators are given by

$$\hat{\psi}(x) = \sum_p \phi_p(x)\hat{a}_p \tag{10.67a}$$

$$\hat{\psi}^+(x) = \sum_p \phi_p^*(x)\hat{a}_p^+ \tag{10.67b}$$

In the "thermal Heisenberg" representation

$$\hat{\psi}_K(x,\tau) = \sum_p \phi_p(x)e^{[-(E_p-\mu)\tau]/\hbar}\hat{a}_p \tag{10.68a}$$

$$\hat{\psi}_K^+(x,\tau) = \sum_p \phi_p^+(x)e^{[(E_p-\mu)\tau]/\hbar}\hat{a}_p^+ \tag{10.68b}$$

where we have used the equation of motion

$$\begin{aligned}\hbar\frac{\partial}{\partial\tau}\hat{a}_p(\tau) &= [\hat{K},\hat{a}_p(\tau)] \\ &= -(E_p-\mu)\hat{a}_p(\tau)\end{aligned} \tag{10.69}$$

Substituting into the expression for the Green's function we have ($\tau_1 > \tau_2$)

$$\mathcal{G}_0(x_1,\tau_1,x_2,\tau_2) = -\sum_p \phi_p(x_1)\phi_p^*(x_2)e^{[-(E_p-\mu)(\tau_1-\tau_2)]/\hbar}\ \mathrm{Tr}\left[\frac{e^{-\beta\hat{K}}}{Z}\hat{a}_p\hat{a}_p^+\right] \quad [10.70]$$

From Eq. (10.33) we see that this is

$$\mathcal{G}_0(x_1,\tau_1,x_2,\tau_2) = -\sum_p \phi_p(x_1)\phi_p(x_2)e^{[-(E_p-\mu)(\tau_1-\tau_2)]/\hbar}[1 \pm \langle\hat{n}_p\rangle^{\mp}] \quad [10.71]$$

with $+$, $-$ according to whether we have fermions or bosons. The term $\langle\hat{n}_p\rangle^{\mp}$ is the occupation number

$$\langle\hat{n}_p\rangle^{\mp} = (e^{\beta(E_p-\mu)} \mp 1)^{-1} \quad [10.72]$$

for bosons and fermions. Using Eq. (10.60) we see that

$$\begin{aligned}\mathcal{G}_0(x_1,x_2,\omega_n) &= -\sum_p \phi_p(x_1)\phi_p^+(x_2)(1 \pm \langle n_p\rangle^{\mp})\frac{1}{\hbar}\int_0^{\beta\hbar} d\tau\ e^{-(E_p-\mu-i\omega_n)\tau/\hbar} \\ &= \sum_p \frac{\phi_p(x_1)\phi_p^*(x_2)(1 \pm \langle n_p\rangle^{\mp})}{(i\omega_n - \epsilon_p - \mu)}(e^{[i\omega_n-(E_p-\mu)]\beta/\hbar} - 1)\end{aligned}$$

[10.73]

For n even or odd the expressions $1 \pm \langle n_p\rangle^{\mp}$ and $e^{[i\omega_n-(E_p-\mu]\beta/\hbar} - 1$ cancel so that *for either case*

$$\mathcal{G}_0(x_1,x_2,\omega_n) = +\sum_p \frac{\phi_p(x_1)\phi_p(x_2)}{i\omega_n - (E_p - \mu)} \quad [10.74]$$

The statistics being maintained by n being even or odd. This is a simple extension of the normal Green's function to a *complex discrete variable* ω_n. The expectation values can now be obtained from Eqs. (10.46) and (10.66).

$$\langle O\rangle = -\sum_p \int \phi_p^*(x_1)\hat{O}(x_1)\phi_p(x_1)\,dx_1 \sum_n \frac{e^{(i\eta\omega_n)/t}}{i\omega_n - (E_p - \mu)} \quad [10.75]$$

The final sum is over even or odd values of n according to whether the system is a fermion or boson system. This sum crops up frequently and is given by the relationship

$$\sum_{\substack{n \\ \text{even} \\ \text{odd}}} \frac{e^{(i\eta\omega_n)/\hbar}}{i\omega_n - (E_p - \mu)} = \mp(e^{\beta(E_p-\mu)} \mp 1) \qquad [10.76]$$

Thus

$$\langle O \rangle = \pm \sum_p \frac{\langle O_p \rangle}{(e^{\beta(E_p-\mu)} \mp 1)} \qquad [10.77]$$

the standard result expected from classical statistical physics. In the case of a uniform system

$$\phi_p(x) = \frac{1}{\sqrt{V}} e^{ip\cdot x} \qquad [10.78]$$

so that

$$\mathcal{G}_0(p,\omega_n) = \frac{1}{i\omega_n - (E_p - \mu)} \qquad [10.79]$$

From Eq. (10.65) we have

$$\mathcal{G}(p,\omega_n) = \frac{1}{i\omega_n - (E_p - \Sigma(p, \omega_n) - \mu)} \qquad [10.80]$$

so that again the self-energy can be interpreted as a contribution to the energy of the state. There is no concept of lifetime, however, since we are in a purely statistical (i.e., equilibrium) system, but the complex nature of the self-energy will change the statistical properties of the system through the relationship (10.46).

10.6. REAL-TIME THERMAL GREEN'S FUNCTIONS

The thermal Green's function suffers from the disadvantage that many processes, such as the interaction of a system with an external probe, are often time dependent. For this purpose it is necessary to work with a time-dependent, but finite temperature, Green's function. The obvious definition is to take the *grand canonical ensemble* average of the single-particle Green's functions we have used at zero temperature. That is

$$G_\beta(x_1,t_1,x_2,t_2) = -\, i\, \mathrm{Tr}[\hat{\rho} T[\hat{\psi}(x_1,t_1)\hat{\psi}^+(x_2,t_2)]] \qquad [10.81]$$

It is best to study this via the Lehman representation. If we explicitly write out the trace in terms of the complete set of many body states, we have

$$G_\beta(x_1,t_1,x_2,t_2) = -i \sum_n \frac{e^{-\beta K_n}}{Z} \langle n | T[\hat{\psi}(x_1,t_1)\hat{\psi}^+(x_2,t_2)] | n \rangle \qquad [10.82]$$

Separating the field operators and Fourier transforming to the energy-dependent Green's function, as in Sec. 6.3, we have

$$G_\beta(x_1,x_2,\omega) = \sum_{m,n} \frac{e^{-\beta K_n}}{Z} \frac{\langle n|\hat{\psi}(x_1)|m\rangle\langle m|\hat{\psi}^+(x_2)|n\rangle}{\omega - (K_n - K_m) - i\delta} \pm \frac{e^{-\beta K_m}}{\overline{Z}} \frac{\langle n|\hat{\psi}(x_1)|m\rangle\langle m|\hat{\psi}^+(x_2|n\rangle}{\omega - (K_n - K_m) - i\delta} \qquad [10.83]$$

The infinitesimals give the time-ordering properties and arise as in Sec. 6.3. We see that the poles of the Green's function again give the excitation energy of the system in the sense that

$$K_n - K_m = E_n - E_m - \mu(N_n - N_m) \qquad [10.84]$$

But now the residues are weighted by the thermal factors $e^{-\beta K_n/Z}$. The advanced (G_β^{A}) and retarded (G_β^{R}) Green's functions, similarly defined as in the single-particle case, are given by

$$G_\beta^{\mathrm{A}}(x_1,x_2,\omega) = \sum_{n,m} \frac{e^{-\beta K_n}}{Z} \frac{\langle n|\hat{\psi}(x_1)|m\rangle\langle m|\hat{\psi}^+(x_2)|n\rangle}{\omega - (K_n - K_m) - i\delta} (1 \mp e^{-\beta(K_m - K_n)}) \qquad [10.85a]$$

$$G_\beta^{\mathrm{R}}(\kappa_1,\kappa_2,\omega) = \sum_{n,m} \frac{e^{-\beta K_n}}{Z} \frac{\langle n|\hat{\psi}(x_1)|m\rangle\langle m|\hat{\psi}^+(x_2)|n\rangle}{\omega - (K_n - K_m) + i\delta} (1 \mp e^{-\beta(K_m - K_n)}) \qquad [10.85b]$$

where the $\pm i\delta$ ensures the time dependence of the Green's function:

$$G_\beta^A(x_1,x_2,t_1 - t_2) = 0, \qquad t_1, > t_2 \tag{10.86a}$$
$$G_\beta^R(x_1,x_2,t_1 - t_2) = 0, \qquad t_1, < t_2 \tag{10.86b}$$

We can perform a similar expansion on the thermal Green's function using Eqs. (10.44) and (10.60) to obtain

$$\mathcal{G}(x_1,x_2,\omega_n) = \sum_{m,n} \frac{e^{-\beta K_n}}{Z} \frac{\langle n|\hat{\psi}(x_1)|m\rangle\langle m|\hat{\psi}^+(x_2)|n\rangle}{i\omega_n - (K_n - K_m)} (1 \mp e^{-\beta(K_m - K_n)}) \tag{10.87}$$

This may be written as a Lehman expansion in the form

$$\mathcal{G}(x_1,x_2,\omega_n) = \frac{1}{2\pi} \int_{-\infty}^{+\infty} \frac{A(x_1,x_2,\omega')\, d\omega'}{i\omega_n - \omega'} \tag{10.88}$$

where

$$A(x_1,x_2,\omega') = \sum_{n,m} \frac{e^{-\beta K_n}}{Z} \langle n|\hat{\psi}(x_1)|m\rangle\langle m|\hat{\psi}^+(x_2)|n\rangle \times (1 \mp e^{\beta\omega'})\, \delta(\omega' - (K_n - K_m)) \tag{10.89}$$

In terms of the *same* function $A(x_1,x_2,\omega')$ we may write Eqs. (10.85) as

$$G_\beta^{\substack{A\\R}}(x_1,x_2,\omega) = \frac{1}{2\pi} \int_{-\infty}^{+\infty} \frac{A(x_1,x_2,\omega')}{\omega - \omega' \mp i\delta} \tag{10.90}$$

Then in terms of the same function $[A(x_1,x_2,\omega')]$ we may also write the time-ordered Green's function as

$$G_\beta(x_1,x_2,\omega) = \frac{1}{2\pi} \int \frac{A(x_1,x_2,\omega')\, d\omega'}{(\omega - \omega' - i\delta)(1 \mp e^{-\beta\omega'})} \pm \frac{1}{2\pi} \int \frac{A(x_1,x_2,\omega')}{(\omega - \omega' + i\delta)(1 \mp e^{-\beta\omega'})} \tag{10.91}$$

Thus, *all* of the various Green's functions, *real time* (advanced, retarded, time ordered) and *thermal*, are related by the same function

$A(x_1,x_2,\omega')$. If we knew $A(x_1,x_2,\omega')$ all the properties of the system could be derived. In practice, what we have is a perturbation expansion which may be used to calculate the thermal Green's function to any level of sophistication required. We then know (approximately) the function $A(x_1,x_2,\omega')$ at the infinite set of points $i\omega_n$. Analytic continuation can then be used to derive the complete function $A(x_1,x_2,\omega')$. This is not strictly a unique process, since we only know the functions $A(x_1,x_2,\omega')$ at a set of points, but the process is saved by the fact that all Green's functions have the same limiting form at high frequency ($\sim 1/\omega$). This makes the analytical continuation unique so that the knowledge of the thermal Green's function implies an equivalent knowledge of all of the other Green's functions at the same level of accuracy.

Just as the thermal Green's function has the ability to calculate thermal averages of observables and connect to the equilibrium thermodynamics, the real-time thermal Green's functions have their uses as well. These are basically the extension of the dynamic properties, such as those described in Chap. 9, to the case of finite temperature. In studying the zero-temperature formulation, we saw that useful results were obtained in the first iteration that used the noninteracting Green's function. The thermal noninteracting Green's function is given by [Eq. (10.74)]

$$\mathcal{G}_0(x_1,x_2,\omega_n) = \sum_p \frac{\phi_p(x_1)\phi_p^*(x_2)}{i\omega_n - (\epsilon_p - \mu)} \qquad [10.92]$$

From Eq. (10.90) we see that

$$A(x_1,x_2,\omega') = \sum_p 2\pi\delta(\omega' - (\epsilon_p - \mu))\phi_p(x_1)\phi_p^*(x_2) \qquad [10.93]$$

Thus we have that

$$G_\beta(x_1,x_2,\omega) = \sum_p \phi_p(x_1)\phi_p^*(x_2)\left\{\frac{(1 \mp e^{-\beta(\epsilon_p-\mu)})^{-1}}{\omega - (\epsilon_p - \mu) + i\delta} \pm \frac{(1 \mp e^{\beta(\epsilon_p-\mu)})^{-1}}{\omega - (\epsilon_p - \mu) - i\delta}\right\} \qquad [10.94]$$

In the limit of zero temperature this reduces to the familiar Green's function

$$G(x_1,x_2,\omega) = \sum_p \phi_p(x_1)\phi_p^*(x_2) \left\{ \frac{\theta(\epsilon_p - \mu)}{\omega - (\epsilon_p - \mu) + i\delta} \pm \frac{\theta(\mu - \epsilon_p)}{\omega - (\epsilon_p - \mu) - i\delta} \right\} \qquad [10.95]$$

The θ functions denote the changeover of particle- and hole-type excitations at the Fermi energy. The finite-temperature Green's function broadens this out since now it is possible to have hole and particle excitations on either side of the Fermi energy (or chemical potential). These occupation factors for the holes and particles carry through the various relationships in the iteration series modifying the results. It must be emphasized, however, that in most cases it is the simple independent-particle aspects, i.e., the occupation, which is changed, while the self-energies, polarization, etc., are all dominated by their zero-temperature value. This is to be expected since the characteristic energies are so very high compared to the typical thermal energy. It is only in the case of very low energy excitations such as phonons or near-phase transitions that thermal effects dominate the interacting system behavior.

BIBLIOGRAPHY

LANDAU, L., and LIFSHITZ, E. M., *Statistical Physics,* Pergamon, Elmsford: NY, 1980.

ABRIKOSOV, A. A., GORKOV, L. P., and DZYALOSHINKI, I. E., *Quantum Field Theoretical Methods in Statistical Physics,* Dover, New York, 1975.

KADANOFF, L. P., and BAYM, G., *Quantum Statistical Mechanics,* Benjamin, Menlo Park: CA, 1976.

PROBLEMS

1. Prove Eq. (10.42) and show also that

$$\hbar \frac{\partial}{\partial \tau} \hat{O}^+(x,\tau) = -[\hat{K},\hat{O}(x,\tau)]$$

2. Prove Eq. (10.46).

3. With the Hamiltonian of Eq. (10.47) show that the thermal Green's function does in fact satisfy Eq. (10.49).

4. Repeat Prob. 3 with the Hamiltonian of Eq. (10.50) and the Green's function of (10.52).

5. Derive Eq. (10.83).

6. Derive the first-order polarization propagator for a system $P_0(\mathbf{x}_1,\mathbf{x}_2,\omega)$ at a finite temperature and show that it corresponds to taking into account the thermal redistribution of the carriers.

CHAPTER ELEVEN

BOSON PARTICLES

Up to now, the application of the Green's function formalism has been almost exclusively aimed at the electron gas, a system of interacting fermions. Bosons differ in their basic statistical properties, so it is not surprising that when one comes to treat them in detail, there is a quite significant change in the structure and behavior of the perturbation series.

The property of bosons that creates the difficulty is the possibility of *multiple occupancy of levels*. In Fermi systems it is only possible for a level to be occupied at most by one particle—there is no such restriction in boson systems. At this point, it is usual to make a distinction between two types of bosons, those which are conserved (such as ^{4}He) and those which are not (plasmon, photon, phonon, etc.). In the case where bosons are conserved, at zero temperature we would expect all particles to be in the lowest possible energy level that would become macroscopically occupied. This means that if we apply the creation or annihilation operators directly to this particular state, we obtain a result which is large and not immediately amenable to perturbative analysis. Basically the lowest (or condensate) state is special and must be treated as such.

For bosons which are not conserved, this phenomena of multiple occupancy of one particular state is not so common but can occur (an example is given in Sec. 12.4), so for either system we may have to consider a "special" state whose occupancy may be much greater than any other.

Firstly, however, we will consider the two most common bosons in solids, the plasmon and phonon. These are well behaved and give good examples of boson systems coupled to a fermion system (the electron gas).

11.1. COLLECTIVE EXCITATIONS IN SOLIDS

In previous chapters we considered the ionic part of the solid as an inert background that simply serves to balance the electronic charge. The ions, however, interact between themselves and with the electrons. Because of the constraints on the system, this interaction shows itself as collective oscillations, the phonons. These were discussed briefly in Chap. 3 in the context of the quantization of a field; now let us consider them in more detail.

Consider a set of ions interacting by way of an (as yet) unspecified two-body interaction $U(\mathbf{R}_i - \mathbf{R}_j)$, where $\mathbf{R}_i$ and $\mathbf{R}_j$ are the ionic coordinates. We can expand this interaction around the assumed equilibrium positions $\mathbf{R}_{i0}$, $\mathbf{R}_{j0}$ to give

$$U(\mathbf{R}_i - \mathbf{R}_j) = U(\mathbf{R}_{i0} - \mathbf{R}_{j0}) + \frac{1}{2}\frac{\partial^2 U(\mathbf{R}_i - \mathbf{R}_j)}{\partial\mathbf{R}_i\partial\mathbf{R}_j}\Bigg|_{\substack{\mathbf{R}_{i0}\\ \mathbf{R}_{j0}}} \delta\mathbf{R}_i\delta\mathbf{R}_j \qquad [11.1]$$

where the first-order term in the displacements is zero because the lattice is assumed stable. Leaving aside the constant terms, we can write the Hamiltonian for the ions as

$$H = \sum_i \frac{|\mathbf{P}_i|^2}{2M_i} + \frac{1}{4}\sum_{i,j}{}' \frac{\partial^2 U(\mathbf{R}_i - \mathbf{R}_j)}{\partial\mathbf{R}_i\partial\mathbf{R}_j}\Bigg|_{\substack{\mathbf{R}_{i0}\\ \mathbf{R}_{j0}}} \delta\mathbf{R}_i\delta\mathbf{R}_j \qquad [11.2]$$

where M_i, $\mathbf{p}_i$ are the mass and momentum of the ion. We can now use the standard second quantization technique of Chap. 3. The normal modes are given by the transformations

$$\delta\mathbf{R}_i = \frac{1}{(NM)^{1/2}}\sum_{\mathbf{k},\lambda} q_{\mathbf{k},\lambda}\mathscr{E}_{\mathbf{k}\lambda}e^{i\mathbf{k}\cdot\mathbf{R}_{i0}} \qquad [11.2a]$$

$$\mathbf{p}_i = \left(\frac{M}{N}\right)^{1/2}\sum_{\mathbf{k},\lambda} p_{\mathbf{k}\lambda}\mathscr{E}_{\mathbf{k}\lambda}e^{i\mathbf{k}\cdot\mathbf{R}_{i0}} \qquad [11.2b]$$

where $\mathscr{E}_{\mathbf{k},\lambda}$ is the polarization vector (one longitudinal mode, two transverse). This gives the Hamiltonian in the standard form

$$\hat{H} = \sum_{\mathbf{k},\lambda} p^+_{\mathbf{k},\lambda}p_{\mathbf{k},\lambda} + \left(\frac{\omega_{\mathbf{k}\lambda}}{\hbar}\right)^2 q^+_{\mathbf{k},\lambda}q_{\mathbf{k},\lambda} \qquad [11.3]$$

where $\omega_{\mathbf{k}\lambda}/\hbar$ is the frequency of the mode $\mathbf{k}$, λ. This transformation, of course, depends upon the details of the potential and a classical solution to the oscillation problem—not a trivial problem, but for the purposes of this section, we assume this solved for any given $U(\mathbf{R}_i - \mathbf{R}_j)$.

Creation and annihilation operators can now be defined (Sec. 3.2) so that

$$\hat{H} = \sum_{\mathbf{k},\lambda} \omega_{\mathbf{k},\lambda}\left(a^{+}_{\mathbf{k}\lambda} a_{\mathbf{k}\lambda} + \frac{1}{2}\right) \tag{11.4}$$

and the momentum operator is

$$\hat{\mathbf{p}} = \sum_{\mathbf{k},\lambda} \hbar \mathbf{k} \hat{a}^{+}_{\mathbf{k}\lambda} \hat{a}_{\mathbf{k},\lambda} \tag{11.5}$$

This defines a set of elementary excitations (the phonons) for any given interionic potential. The actual form of that potential will depend upon how the electrons respond to the movement of the ions. If the electrons do not react at all, the interionic potential will be coulombic, the electrons simply forming a background neutralizing charge. This situation is the reciprocal of the one we have considered in the jellium solid, where the electron-electron interaction is coulombic. The coulombic interaction is, therefore, an important case. It also gives rise to a set of collective modes. Consider, for example, the Hamiltonian for the electron system.

$$H = \sum_{n=1}^{N} \frac{|\mathbf{p}_n|^2}{2m} + \frac{1}{2} \sum_{n,m}{}' \frac{e^2}{|\mathbf{r}_n - \mathbf{r}_m|} - H_b \tag{11.6}$$

where H_b is the interaction with the background neutralizing charge density.

If we use the density operator

$$\rho(\mathbf{r}) = \sum_{n=1}^{N} \delta(\mathbf{r} - \mathbf{r}_n) \tag{11.7}$$

a set of normal modes can be written as

$$\begin{aligned} \rho_{\mathbf{k}} &= \int \rho(\mathbf{r}) e^{-i\mathbf{k}\cdot\mathbf{r}}\, d\mathbf{r} \\ &= \sum_n e^{-i\mathbf{k}\cdot\mathbf{r}_n} \end{aligned} \tag{11.8}$$

If we assume that the background charge density (N) is a *uniform positive charge*, the Hamiltonian becomes

$$H = \sum_{n=1}^{N} \frac{|\mathbf{p}_n|^2}{2m} + \sum_{k} \frac{2\pi e^2}{k^2} (\rho_{\mathbf{k}}^{+} \rho_{\mathbf{k}} - N) \qquad [11.9]$$

The terms $\rho_{\mathbf{k}}^{+} \rho_{\mathbf{k}} - N$ indicates the physical fact that we are looking at charge fluctuations around the equilibrium value. To quantize, one requires a coordinate σ_k conjugate to ρ_k such that

$$[\hat{\rho}_{\mathbf{k}}, \hat{\sigma}_{\mathbf{k}'}] = \frac{\hbar}{i} \delta_{\mathbf{k},\mathbf{k}'} \qquad [11.10]$$

This coordinate is

$$\hat{\sigma}_{\mathbf{k}} = -\frac{\hbar}{kN} \sum_{n=1}^{N} e^{-i\mathbf{k}\cdot\mathbf{r}_n} \frac{\partial}{\partial \mathbf{r}_n} \qquad [11.11]$$

which, as one might expect, is essentially the Fourier transform of the momentum operator

$$\hat{\mathbf{p}} = \sum_{n} \frac{\hbar}{i} \frac{\partial}{\partial \mathbf{r}_n} \qquad [11.12]$$

On substituting one has

$$\hat{H} = \sum_{\mathbf{k}} \frac{Nk^2}{2m} \hat{\sigma}_{\mathbf{k}}^{+} \hat{\sigma}_{\mathbf{k}} + \left(\frac{2\pi e^2}{k^2}\right) \hat{\rho}_k^{+} \hat{\rho}_k - \sum_{\mathbf{k}} \frac{2\pi e^2}{k^2} N + \sum_{\mathbf{k},n} \hbar^2 \mathbf{k} \cdot \frac{\partial}{\partial \mathbf{r}_n} \qquad [11.13]$$

One can again transform to creation and annihilation operators to obtain the Hamiltonian

$$H = \sum_{\mathbf{k}} \omega_p \left(\hat{b}_{\mathbf{k}}^{+} \hat{b}_{\mathbf{k}} + \frac{1}{2}\right) - \sum_{\mathbf{k}} \frac{2\pi e^2}{k^2} N + \sum_{\mathbf{k},n} \hbar^2 \mathbf{k} \cdot \frac{\partial}{\partial \mathbf{r}_n} \qquad [11.14]$$

where $\omega_p = \hbar(4\pi e^2 N/m)^{1/2}$. The first term is a standard elementary excitation Hamiltonian giving the quantized collective oscillations of the electron "gas." These have an energy ω_p and are the plasmons which were discussed briefly in Sec. 9.1. The transformation is not, however, complete. The remaining two terms consist of the background energy, which is simply a constant and, more importantly, a term consisting of mixed coordinates $\mathbf{k}$ and $\mathbf{r}_n$. This latter term indicates that one cannot completely transform the interacting electron gas into a set of noninteracting collective oscillations; there remains a residual particle aspect. Only at very long wavelengths ($k \approx 0$) is the transformation a good one. The effect of the extra terms can be incorporated into the plasmon system by perturbation theory, where it results in a change in the plasmon energy and a finite lifetime for the excitations. Thus, we write

$$\hat{H} = \sum_{\mathbf{k}} \omega_p(\mathbf{k})\left(\hat{b}_{\mathbf{k}}^{+}\hat{b}_{\mathbf{k}} + \frac{1}{2}\right) + (\text{corrections}) \qquad [11.15]$$

where $\omega_p(\mathbf{k})$ is complex.

In general then both the electron and the ion systems can be described by a collective set of elementary excitations. For the phonons, however, we must include the electron response if a realistic description is to be obtained of the excitation spectrum. The coulomb interaction is a good approximation for the electrons but it bears little relation to the real ion–ion interaction.

11.2. ELECTRON–PHONON SYSTEM

The Hamiltonian for the electron–ion system may be written as

$$H = \mathrm{H}_{\mathrm{el}} + H_{\mathrm{ion}} + H_{\text{el-ion}} \qquad [11.16]$$

The ion system can be treated in terms of the previous section with the ion–ion interaction given by the bare coulomb interaction:

$$U(\mathbf{R}_i - \mathbf{R}_j) = \frac{Z^2e^2}{|\mathbf{R}_i - \mathbf{R}_j|} = Z^2 v(\mathbf{R}_i - \mathbf{R}_j) \qquad [11.17]$$

(Z = ion charge). This will give phonons akin to the plasmon modes and the associated creation and annihilation operators.

The electron–ion term is

$$H_{\text{el-ion}} = \sum_{i,\alpha} \frac{Ze^2}{|\mathbf{R}_\alpha - \mathbf{r}_i|} = Z \sum_{i,\alpha} v(\mathbf{r}_i - \mathbf{R}_\alpha) \qquad [11.18]$$

If we expand this in terms of the deviations of the ions from their equilibrium positions, as in Eq. (11.1), we obtain

$$H_{\text{el-ion}} = \sum_{i,\alpha} \frac{Ze^2}{|\mathbf{R}_{\alpha 0} - \mathbf{r}_i|} + \sum_{i,\alpha} Z \frac{\partial v(\mathbf{r}_i - \mathbf{R}_\alpha)}{\partial \mathbf{R}_\alpha}\bigg|_{\mathbf{R}_{\alpha 0}} \delta \mathbf{R}_\alpha \qquad [11.19]$$

The first term represents the stable lattice potential and can be incorporated into the single-particle electron potential, where it will give rise to Bloch states. The second term is the electron–ion interaction required.

Substituting from Eq.(11.2a) and writing the normal-mode displacement vector in terms of the creation and annihilation operator [cf. Eqs. (3.13)]

$$\hat{H}_{\text{INT}} = \sum_{\substack{\mathbf{k},\lambda \\ \mathbf{p}}} \frac{V^{\mathbf{p}}_{\mathbf{k},\lambda}}{(2w_{\mathbf{k},\lambda})^{1/2}} \hat{c}^+_{\mathbf{p}+\mathbf{k},\sigma} \hat{c}_{\mathbf{p},\sigma} (\hat{a}^+_{-\mathbf{k}} + \hat{a}_{\mathbf{k},\lambda}) \qquad [11.20]$$

where $\hat{c}^+$, $\hat{c}$ are the electron creation and annihilation operators and $V^{\mathbf{p}}_{\mathbf{k},\lambda}$ is the scattering matrix element

$$V^{\mathbf{p}}_{\mathbf{k},\lambda} = -\frac{1}{(NM_i)^{1/2}} \left\langle \mathbf{p} + \mathbf{k} \left| \sum_{\mathbf{R}_{\alpha 0}} Z \mathscr{E}_{\mathbf{k}\lambda} \frac{\partial v(\mathbf{r} - \mathbf{R}_\alpha)}{\partial \mathbf{R}_\alpha}\bigg|_{\mathbf{R}_{\alpha 0}} e^{i\mathbf{k}\cdot\mathbf{R}_{\alpha 0}} \right| \mathbf{p} \right\rangle \qquad [11.21]$$

Ignoring umklapp terms, i.e., for a uniform system

$$V^{\mathbf{p}}_{\mathbf{k},\lambda} = \left(\frac{N}{M}\right)^{1/2} Zv(k) = \left(\frac{N}{M_i}\right)^{1/2} \frac{\pi e^2}{k^2} \qquad [11.22]$$

In terms of field operators the electron–phonon interaction becomes

$$H_{\text{INT}} = \int \hat{\psi}^+_\sigma(r) \gamma(\mathbf{r}) \hat{\phi}(\mathbf{r}) \hat{\psi}_\sigma(\mathbf{r})\, d\mathbf{r} \qquad [11.23]$$

where

$$\hat{\phi}(\mathbf{r}) = \frac{1}{\sqrt{V}} \sum_{\mathbf{k},\lambda} (e^{i\mathbf{k}\cdot\mathbf{r}}\hat{a}_{\mathbf{k},\lambda} + e^{-i\mathbf{k}\cdot\mathbf{r}}\hat{a}^{+}_{-\mathbf{k},\lambda}) \qquad [11.24]$$

is the field operator for the phonon system: this form is chosen specifically to make the interaction Hamiltonian as simple as possible but does not affect the considerations of Sec. 6.3. The term $\gamma(\mathbf{r})$ contains the details of the interaction and is often taken to be a constant so that

$$\hat{H}_{\mathrm{INT}} = \gamma \int \hat{\psi}_{\sigma}(\mathbf{r})\hat{\phi}(\mathbf{r})\hat{\psi}_{\sigma}(\mathbf{r})\, d\mathbf{r} \qquad [11.25]$$

The Green's function for the phonon is defined as (cf. Chap. 6)

$$D(\mathbf{r}_1,t_1,\mathbf{r}_2,t_2) = \frac{1}{i} \langle 0 | T[\hat{\phi}_H(\mathbf{r}_1,t_1)\hat{\phi}_H(\mathbf{r}_2,t_2)] | 0 \rangle \qquad [11.26]$$

Remembering that the field operators for the electron and phonons commute, the equation of motion for the electron field operators can now be written from the Heisenberg equation of motion as

$$\left[i\hbar \frac{\partial}{\partial t_1} - H_e(\mathbf{r}_1,t_1) \right] \hat{\psi}_{\sigma}(\mathbf{r}_1,t_1) - \gamma\hat{\phi}(\mathbf{r}_1,t_1)\hat{\psi}_{\sigma}(\mathbf{r}_1,t_1) = 0 \qquad [11.27]$$

For the phonon field operator we have

$$\left[i\hbar \frac{\partial}{\partial t_1} - H_{\mathrm{ph}}(\mathbf{r}_1,t_1) \right] \hat{\phi}(\mathbf{r}_1,t_1) - \gamma\hat{\psi}^{+}_{\sigma}(\mathbf{r}_1,t_1)\hat{\psi}_{\sigma}(\mathbf{r}_1,t_1) = 0 \qquad [11.28]$$

These form a pair of coupled field operator equations. Writing Eq. (11.28) as

$$D_0^{-1}(\mathbf{r}_1,t_1,\ \mathbf{r}_2,t_2)\hat{\phi}(\mathbf{r}_2,t_2) - \gamma\hat{\psi}^{+}_{\sigma}(r_1t_1)\hat{\psi}_{\sigma}(r_1,t_1) = 0 \qquad [11.29]$$

where D_0 is the uncoupled (or bare) phonon propagator we have

$$\hat{\phi}(\mathbf{r}_1,t_1) = \gamma \int D_0(\mathbf{r}_1,t_1,\mathbf{r}_2,t_2)\hat{\psi}^{+}_{\sigma}(\mathbf{r}_2,t_2)\hat{\psi}_{\sigma}(\mathbf{r}_2,t_2)d\mathbf{r}_2\, dt_2 \qquad [11.30]$$

On substituting into Eq.(11.27) we have

$$\left[i\hbar \frac{\partial}{\partial t_1} - H_e(\mathbf{r}_1,t_1) \right] \hat{\psi}_\sigma(\mathbf{r}_1,t_1)$$

$$- \gamma^2 \int D_0(r_1,t_1,r_2,t_2) \hat{\psi}_\sigma^+(r_2,t_2) \hat{\psi}_\sigma(\mathbf{r}_2,t_2) \hat{\psi}_\sigma(\mathbf{r}_1,t_1)\, dr_2\, dt_2 = 0 \quad [11.31]$$

The electron–phonon coupling term now has exactly the same form as the electron–electron coulomb interaction term implicit in $H_e(\mathbf{r}_1,t_1)$ [Eq. (6.55)]. We can, therefore, write a Green's function equation for the electron system that is *identical* to that used previously [Eq. (7.1)] except for the replacement of the coulomb interaction by an effective interaction $U(\mathbf{r}_1,t_1,\mathbf{r}_2,t_2)$, given by

$$U(\mathbf{r}_1,t_1,\mathbf{r}_2,t_2) = v(\mathbf{r}_1,t_1,\mathbf{r}_2,t_2) + \gamma^2 D_0(\mathbf{r}_1,t_1,\mathbf{r}_2,t_2) \quad [11.32]$$

The interaction between electrons can now take place either through the coulomb field or by way of the exchange of a phonon, one electron emitting, the other absorbing, as shown in Fig. 11.1. The coupling parameter (γ) can be associated with each vertex as shown.

Using the interaction of Eq. (11.32), we can go further. We can now repeat *all* of the previous formal development, replacing v by U wherever it occurs. Thus all of Secs. 7.1 to 7.5 and the diagrammatic analysis of Chap. 8 carry over unchanged, except for a redefinition of the screened and bare interactions.

One can, however, perform a very useful resummation of the interaction, which includes a great deal of physics. The screened interaction is diagrammatically given by

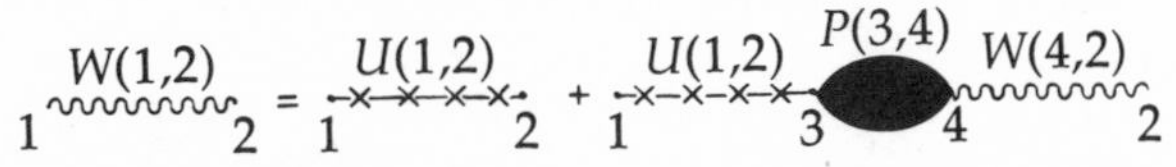

W(1,2) (1 — 2) = U(1,2) (1 — 2) + U(1,2) (1 — 3) P(3,4) (3 — 4) W(4,2) (4 — 2) [11.33]

If we expand this equation (Fig. 11.2), we can identify the following terms.

1. A phonon line enters and leaves, between which are a series of

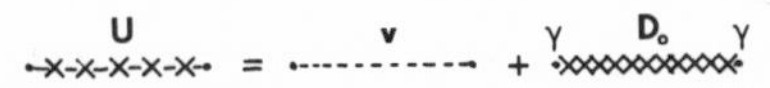

FIGURE 11.1. Electron–electron interaction by way of exchange of photons and the coulomb field.

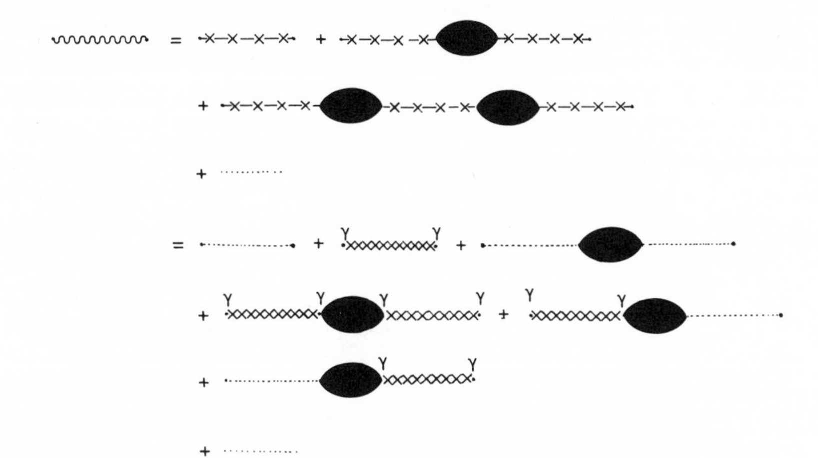

FIGURE 11.2. Expansion of the screened interaction in terms of the coulomb and phonon exchange parts.

polarization inserts connected by the coulomb interaction only; for example

These may be included automatically in the series expansion by defining a screened or "renormalized" coupling constant

$$\begin{aligned} \gamma' &= \gamma + \gamma \mathsf{vP} + \gamma \mathsf{vPvP} + \cdots \\ &= \gamma(1 - \mathsf{vP})^{-1} \end{aligned} \qquad [11.34]$$

That is

$$\gamma'(1,2) = \gamma\epsilon^{-1}(1,2)$$

where $\epsilon^{-1}(1,2)$ is the coulomb response function of Eq. (7.49). The example term becomes part of

2. Using the renormalized coupling constant, those parts of the interaction which begin and end with a phonon line and are separated

by mixed phonon and coulomb parts, for example

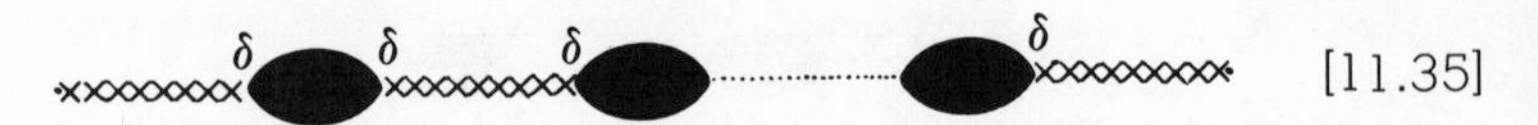

[11.35]

may be included by defining a screened phonon propagator $D(\mathbf{r}_1,t_1,\mathbf{r}_2,t_2)$ defined through the equation

$D(1,2)$ (1, 2) = $D_o(1,2)$ (1, 2) + $D_o(1,3)\gamma$ (1, 3) $P(3,4)$ $\gamma'(4,5)$ $D(5,2)$ (4,5, 2) [11.36]

3. In terms of the renormalized coupling constant and renormalized phonon propagator, any term of the form

may be written as

γ' γ'

Taking all of these together, we find that the interaction may be written in two ways—either as Eq. (11.33) or, in the renormalized form, as

$$W(1,2) = W_e(1,2) + W_{\text{ph}}(1,2) \qquad [11.37]$$

$W_e(1,2)$ is the "normal" electron screened interaction

$$W_e(1,2) = v(1,2) + \int v(1,3)P(3,4)W_e(4,2)\, d[3]\, d[4] \qquad [11.38]$$

and

$$W_{\text{ph}}(1,2) = \int \gamma'(1,3)D(3,4)\gamma'(4,2)\, d[3]\, d[4] \qquad [11.39]$$

is the renormalized phonon exchange. In both of these equations the polarization propagator is calculated with the *total* interaction rather than either of the two contributions so the separation is not quite complete.

The new phonon propagator is related to the bare phonon propagator through Eq.(11.36), which may be rewritten as

$$\int D_0^{-1}(1,3)D(3,2)\,d[3] - \gamma^2\int P(1,3)\epsilon^{-1}(3,4)D(4,2)\,d[3]\,d[4] = \delta(1-2) \qquad [11.40a]$$

or

$$\left(i\hbar\,\frac{\partial}{\partial t_1} - H_{\text{ph}}(1)\right)G(1,2) - \gamma^2\int P(1,3)\epsilon^{-1}(3,4)D(4,2)\,d[3]\,d[4] = \delta(1-2) \qquad [11.40b]$$

The coupling to the electron system corresponds to an addition to the phonon Hamiltonian. Since the bare phonon propagator is the collective oscillation for a *coulomb* potential, we have [returning to Eqs. (11.21)–(11.23)] as the definition of the coupling constant, and considering a uniform material

$$\begin{aligned}\gamma^2P(\mathbf{q},\omega)\epsilon^{-1}(\mathbf{q},\omega) &= \left(\frac{N}{M_i}\right)\frac{v^2(\mathbf{q})P(\mathbf{q},\omega)Z^2}{\epsilon(\mathbf{q},\omega)} \\ &= \frac{N}{M_i}Z^2\frac{v(\mathbf{q})}{\epsilon(\mathbf{q},\omega)} - Z^2\left(\frac{N}{M_i}\right)v(\mathbf{q})\end{aligned} \qquad [11.41]$$

The second term cancels the bare coulomb interaction in H_{ph} and we are left with a *screened* coulomb interaction. Thus, the phonon Green's function $D(1,2)$ corresponds to phonons in a lattice where the interionic potential is *screened.* This is what we would expect physically, of course. It also means that the coupling constant

$$\gamma^1(1,2) = \gamma\epsilon^{-1}(1,2)$$

or

$$\gamma'(\mathbf{q}) = \left(\frac{N}{M_i}\right)^{1/2}\frac{v(\mathbf{q})}{\epsilon(\mathbf{q},w)} \qquad [11.42]$$

That is, the electrons couple to the renormalized phonons by way of the much weaker screened coulomb interaction.

We have two ways of treating the problem:

(i) bare phonons, whose frequencies are determined by the coulomb interaction (essentially the ionic–plasmon system), coupling to the electron system by way of a strong coupling parameter, or

(ii) renormalized phonons, whose frequencies are determined by the screened coulomb interaction, weakly coupled to the electron system by a screened coupling parameter.

The renormalized phonon is not of course a "true" phonon in that it corresponds to the collective motion of both ions and electrons, but it is the *physical* excitation which is seen (rather than the "theoretical" bare-ion phonon).

The two-phonon propagators have the same form:

$$D_0(q,\omega) = \left(\frac{1}{\omega - \omega_{\text{ph}} - i\delta} + \frac{1}{\omega + \omega_{\text{ph}} - i\delta}\right) \quad [11.43a]$$

$$D(q,\omega) = \left(\frac{1}{\omega - \omega_{\text{ph}}(q) - i\delta} + \frac{1}{\omega + \omega_{\text{ph}}(q) - i\delta}\right) \quad [11.43b]$$

differing formally only in the energy of the excitation. The difference between the electron and phonon system is that the renormalization of the electron system due to the phonons is generally small, while the change in the phonon energy due to the electrons is large. In a uniform material it is given by $\omega_{\text{ph}}(q) = \omega_{\text{ph}}/[\epsilon(q,0)]^{1/2}$, where ω_{ph} is the bare phonon frequency.

11.3. PLASMONS AND THE TOTAL INTERACTION

Consider the interaction of an added electron with an N-particle system. Neglecting the phonons, we may write

$$H = H_N + H_1 + H_{\text{INT}} \quad [11.44]$$

The Hamiltonian for the N-particle system can be written in terms of the plasmon field operators defined by analogy to the phonon operators as

$$\hat{\Phi}(\mathbf{r}) = \frac{1}{\sqrt{V}} \sum_{k} (\hat{b}_{\mathbf{k}} e^{i\mathbf{k}\cdot\mathbf{r}} + \hat{b}^{+}_{-\mathbf{k}} e^{-i\mathbf{k}\cdot\mathbf{r}}) \qquad [11.45]$$

so that the equation of motion for the plasmon Green's function becomes

$$\left(i\hbar \frac{\partial}{\partial t_1} - H_N(\mathbf{r}_1, t_1) \right) R\,(\mathbf{r}_1, \mathbf{t}_1, \mathbf{r}_2, \mathbf{t}_2,) = \hbar\delta(\mathbf{r}_1 - \mathbf{r}_2)\,\delta(t_1 - t_2) \qquad [11.46]$$

The single-particle Green's function satisfies the equation

$$\left(i\hbar \frac{\partial}{\partial t_1} - H_1(\mathbf{r}_1, t_1) \right) G_{\sigma 1,\sigma 2}(\mathbf{r}_1, t_1, \mathbf{r}_2, t_2) = \hbar\delta(\mathbf{r}_1 - \mathbf{r}_2)\,\delta(t_1 - t_2)\delta_{\sigma 1,\sigma 2} \qquad [11.47]$$

and the interaction term is

$$H_{\mathrm{INT}} = \sum_{i=1}^{N} \frac{e^2}{|\mathbf{r} - \mathbf{r}_n|} \qquad [11.48]$$

We can rewrite this in terms of the creation and annihilation operators for the electron $(\hat{c}^{+}_{\mathbf{k}\sigma}, \hat{c}_{\mathbf{k}\sigma})$ and plasmon system $(\hat{b}^{+}_{\mathbf{k}}, \hat{b}_{\mathbf{k}})$ as

$$\hat{H}_{\mathrm{INT}} = \sum_{\mathbf{k}} \frac{v(\mathbf{k})}{[2\omega_p(\mathbf{k})]^{1/2}} \hat{c}^{+}_{\mathbf{p}+\mathbf{k},\sigma} \hat{c}_{\mathbf{p},\sigma} (\hat{b}_{\mathbf{k}} + \hat{b}^{+}_{-k}) \qquad [11.49]$$

The analogy to the phonon system is inescapable. We can write the coupled equations

$$\left[i\hbar \frac{\partial}{\partial t_1} - H_N(\mathbf{r}_1, t_1) \right] \hat{\Phi}(\mathbf{r}_1, t_1) - \gamma_p \hat{\psi}_{\sigma 1}(\mathbf{r}_1, t_1) \hat{\psi}_{\sigma 1}(\mathbf{r}_1, t_1) = 0 \qquad [11.50]$$

and

$$\left[i\hbar \frac{\partial}{\partial t_1} - H_1(\mathbf{r}_1, t_1) \right] \hat{\psi}_{\sigma 1}(\mathbf{r}_1, t_1) - \gamma_p \hat{\Phi}(\mathbf{r}_1, t_1) \hat{\psi}_{\sigma 1}(\mathbf{r}_1, t_1) = 0 \qquad [11.51]$$

for the electron plasmon system, where γ_p is the electron–plasmon coupling. Rather than go through the whole derivation again we see that on substitution of Eq. (11.50) into Eq. (11.51), we have

$$\left[i\hbar\frac{\partial}{\partial t_1} - H_1(\mathbf{r}_1,t_1)\right]\hat{\psi}_{\sigma 1}(\mathbf{r}_1,t_1)$$
$$- \gamma_p^2 \int R(\mathbf{r}_1,t_1,\mathbf{r}_2,t_2)\hat{\psi}^+_{\sigma 1}(\mathbf{r}_2,t_2)\hat{\psi}_{\sigma 1}(\mathbf{r}_2,t_2)\hat{\psi}_{\sigma 1}(\mathbf{r}_1,t_1)\, d\mathbf{r}_2\, dt_2 = 0 \quad [11.52]$$

The second term replaces the electron–electron interaction of Eq. (6.56) by *plasmon* exchange.

Thus, the total interaction of the added electron (or hole) can be written as a sum of the bare coulomb, plasmon exchange, and phonon exchange as shown schematically in Fig. 11.3. We can now understand the concept of screening in an alternative way to the response function of Chap. 7. The bare coulomb interaction is basically repulsive; the plasmon exchange produces an attractive interaction which cancels, to a large extent, the repulsion. For instance, if we neglect the phonon interaction, we can use the response function of Eq. (9.13) to explicitly show the plasmon Green's function

$$\begin{aligned} W(q,\omega) &= \frac{v(q)}{\epsilon(q,\omega)} \\ &= v(q) + v(q)\frac{\omega_p^2}{\omega^2 - \omega_p^2(q)} \end{aligned} \quad [11.53]$$

The second term is the plasmon exchange term and is attractive for $\omega < w_p(q)$. In fact for $q \to 0$ we have

$$W(q,\omega) = v(q)\left(1 + \frac{\omega_p^2}{\omega^2 - \omega_p^2}\right), \qquad q \to 0 \quad [11.54]$$

i.e., perfect cancellation at zero frequency. As ω approaches the plasmon energy the net interaction appears highly attractive. This is not as critical

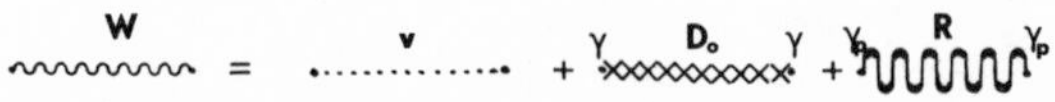

FIGURE 11.3. Electron-Electron interaction W by way of exchange of phonons, plasmons, and coulomb field.

as it may seem, however, since the electron–electron interaction involves momentum *and* energy transfers. For electrons in the solid this always produces a net repulsive interaction. The inclusion of the phonon exchange results in a quite different situation; we have now

$$W(q,\omega) = v(q) + v(q)\frac{\omega_p}{\omega^2 - \omega_p^2(q)} + \gamma_{\text{ph}}'^2 \frac{\omega_{\text{ph}}(q)}{\omega^2 - \omega_{\text{ph}}^2(q)} \qquad [11.55]$$

At zero frequency and wave vector this gives

$$W(0,0) = -\gamma_{\text{ph}}'^2 \qquad [11.56]$$

i.e., *overscreening*. For most systems this has little effect, but because the phonons have small energy and a slow increase in energy with frequency (i.e., $\omega_{\text{ph}}(q) \propto q$), it is possible to obtain a net attractive electron–electron interaction for electrons near the Fermi surface. We shall see in Chap. 13 that this produces a pairing of the electrons into bosons and the formation of a Bose condensate at low temperatures.

11.4. BOSON SYSTEMS WITH A CONDENSATE

Consider a boson system. To be specific we will take a Hamiltonian

$$\begin{aligned}\hat{H} = &\int \hat{\psi}^+(x_1,t_1)(H_0(x_1) + \phi(x_1,t_1))\hat{\psi}(x_1,t_1)\,dx_1 \\ &+ \int \hat{\psi}^+(x_1,t_1)\hat{\psi}^+(x_2,t_1)U(x_1, x_2)\hat{\psi}(x_2,t_1)\hat{\psi}(x_1,t_1)dx_1\,dx_2 \qquad [11.57]\end{aligned}$$

where H_0 is the one-particle Hamiltonian, $\phi(x_1,t_1)$ is an external potential, and $U(x_1,x_2)$ is a general two-body interaction. Within the Heisenberg representation, we may calculate the equation of motion for the field operators and essentially retrace the steps of Sec. 7.1 to obtain

$$\begin{aligned}&\left[i\hbar\frac{\partial}{\partial t_1} - H_0(x_1) - \phi(x_1,t_1) - V_H(x_1,t_1)\right] G(x_1,t_1, x_2,t_2) \\ &\quad - \int \Sigma(x_1,t_1, x_3,t_3)G(x_3,t_3,x_2,t_2)\,dx_3\,dt_3 = \hbar\delta(x_1 - x_2)\,\delta(t_1 - t_2) \qquad [11.58]\end{aligned}$$

The Hartree potential is

$$V_H(x_1,t_1) = \frac{1}{i}\int U(x_1,x_3)G(x_3,t_3,\ x_3,t_3^+)\ dx_3\ dt_3 \qquad [11.59]$$

and the self-energy

$$\Sigma(x_1,t_1,x_2,t_2) = -\,i\int U(x_1,x_4)\ \delta(t_1 - t_4)G(x_1,t_1,x_3,t_3) \times \frac{\delta G^{-1}(x_3,t_3,x_2,t_2)}{\delta\phi(x_4,t_4)}\ dx_3\ dt_3\ dx_4\ dt_4 \qquad [11.60]$$

The only difference being the replacement of the coulomb potential by a more general two-body interaction.

As they stand, however, Eqs. (11.58)–(11.60) tell us nothing about the physics of the boson system unless it is a normal system, i.e., one without a condensate. For those systems, one may continue the iteration process as in the fermion case. To progress further in the condensate case, we need to introduce some simplifications, in particular we need to bring in explicitly the properties of the macroscopically occupied state.

Consider the field operators of the system; we extract that part of the field operator which corresponds to the condensate and write

$$\hat{\psi}(x,t) = \hat{\xi}(x,t) + \hat{\chi}(x,t) \qquad [11.61a]$$

$$\hat{\psi}^+(x,t) = \hat{\xi}^+(x,t) + \hat{\chi}^+(x,t) \qquad [11.61b]$$

The single-particle Green's function can now be split

$$\begin{aligned} G(x_1,t_1,\ x_2,t_2) &= \frac{1}{i}\,\langle N|\,T[(\hat{\xi}(x_1,t_1) + \hat{\chi}(x_1,t_1))(\hat{\xi}^+(x_2,t_2) + \hat{\chi}^+(x_2,t_2))\,|N\rangle \\ &= \frac{1}{i}\,\langle N|\,T[\hat{\xi}(x_1,t_1)\hat{\xi}^+(x_2,t_2)]\,|N\rangle \\ &\quad + \frac{1}{i}\,\langle N|\,T[\hat{\chi}(x_1,t_1)\hat{\chi}^+(x_2,t_2)]\,|N\rangle \\ &= G_c(x_1,t_1,\ x_2,t_2) + G'(x_1,t_1,\ x_2,t_2) \end{aligned} \qquad [11.62]$$

where G_c refers to the condensate part of the system and G' to the remainder. The cross product terms in the equation vanish identically, since the states $\hat{\xi}|N\rangle$ and $\hat{\chi}|N\rangle$ must be orthogonal since $\hat{\xi}$ and $\hat{\chi}$ refer to different parts of the system.

Consider first the macroscopically occupied state. We can introduce a complete set of states $|M,j\rangle$ (where M is the number of particles and j the eigenvalue index) as in Sec. 6.2 to give

$$\begin{aligned} G_c(x_1,t_1,x_2,t_2) &= \frac{1}{i}\sum_{M,j} \langle N|\hat{\xi}(x_1,t_1)|M,j\rangle\langle j,M|\hat{\xi}^+(x_2,t_2)|N\rangle \\ &= \frac{1}{i}\sum_{M,j} \langle N|\hat{\xi}(x_1)|M,j\rangle\langle j,M|\hat{\xi}^+(x_2)|N\rangle\, e^{[i(E_j\hat{M}-E_0\hat{N})(t_1-t_2)]/\hbar}, \\ & t_1 > t_2 \end{aligned} \tag{11.63}$$

Since ξ applies to the lowest energy state only, we have

$$\hat{\xi}(x_2) = \xi_0(x_2)\hat{a}_0 \tag{11.64}$$

where $\xi_0(x_2)$, a_0 are the eigenfunction and annihilation operator corresponding to the condensate state. This gives

$$\hat{\xi}^+(x_2)|N\rangle = (n + \tfrac{1}{2})^{1/2}\xi_0(x_2)|N + 1\rangle \tag{11.65}$$

where $|N + 1\rangle$ is the ground state of the $N + 1$ particle system (i.e., $M = N + 1$, $j = 0$) and there are n ($\gg 1$) particles in the condensed state. This reduces Eq. (11.63) to

$$G_0(x_1,t_1,x_2,t_2) = \frac{1}{i}(\sqrt{n}\,\xi_0(x_1)e^{(-i\mu t_1)/\hbar})(\sqrt{n}\,\xi_0(x_2)e^{(i\mu t_2)/\hbar}) \tag{11.66}$$

where μ is the chemical potential ($E^0_{N+1} - E^0_N$ or $E^0_N - E^0_{N-1}$), and we have neglected all of the microscopically occupied states ($|N \pm 1, j \neq 0\rangle$) compared to the condensate. Physically the analysis corresponds to the (obvious) assumption that if we have a state containing a macroscopic number of particles, adding or subtracting one particle by applying the operators $\hat{\xi}^+$ or $\hat{\xi}$, respectively, will not change that state except to increase (or decrease) its energy by the chemical potential. It also splits the interacting condensate Green's function into two *independent* parts associated with the addition or subtraction of a particle from the condensate; so we write

$$G_c(x_1,t_1,x_2,t_2) = \frac{1}{i}\,\xi(x_1,t_1)\xi^*(x_2,t_2) \qquad [11.67]$$

where

$$\xi(x_1,t_1) = \sqrt{n}\,\xi_0(x_1)e^{(-i\mu t_1)/\hbar} \qquad [11.68]$$

We now take out the condensate Green's function from Eq. (11.58) to leave

$$\left[i\hbar\frac{\partial}{\partial t_1} - H_0(x_1) - \mu - V_H(x_1,t_1)\right] G'(x_1,t_1,x_2,t_2)$$
$$- \int \Sigma(x_1,t_1,x_3,t_3)G'(x_3,t_3,x_2,t_2)\,dx_3\,dt_3 = \hbar\delta(x_1 - x_2)\,\delta(t_1 - t_2) \qquad [11.69]$$

In G' the field operators $\hat{\chi}$, $\hat{\chi}^+$ refer to the system outside the condensate so there are no occupied states for the associated noninteracting Green's function. This means that integrals involving $G_0'(x_1,t_1,x_2,t_2)$ $(t_1 < t_2)$ automatically vanish. It also means that Eq. (11.69) is now a normal Green's function equation except for the changes in the self-energy and Hartree potential. In this sense the condensate acts rather like an external potential, modifying the propagation of the particles outside the condensate.

Equation (11.69) can now be formally solved by iteration. Since the vertex function involves the *full* Green's function, we first derive the self-energy term as in the *bare* coulomb expansion of Sec. 7.1 and then substitute using Eq. (11.62) and then Eq. (11.66) to obtain the actual contribution. The results are best illustrated graphically. The second-order self-energy diagram may be written as

$\Sigma_2(1,2)$ = 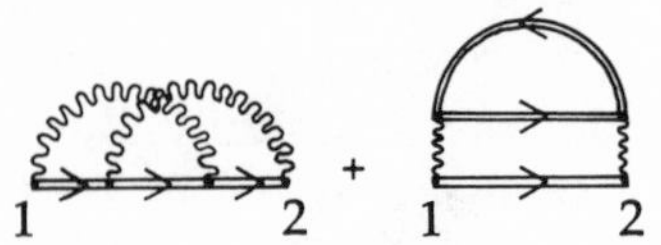

[11.70]

where

$G(1,2) \rightarrow \;\equiv$ 

$U(1,2) \rightarrow \;\equiv$ 

We now split G into its constituents

$$G(1,2) = \frac{1}{i}\,\xi(1)\cdot\xi^*(2) + G'(1,2) \qquad [11.71]$$

which we write as

$G(1,2) =$ 1 ⇢ ⇢ 2 + 1 ⟶ 2

Substituting for the Green's function we obtain all possible combinations and so, for instance

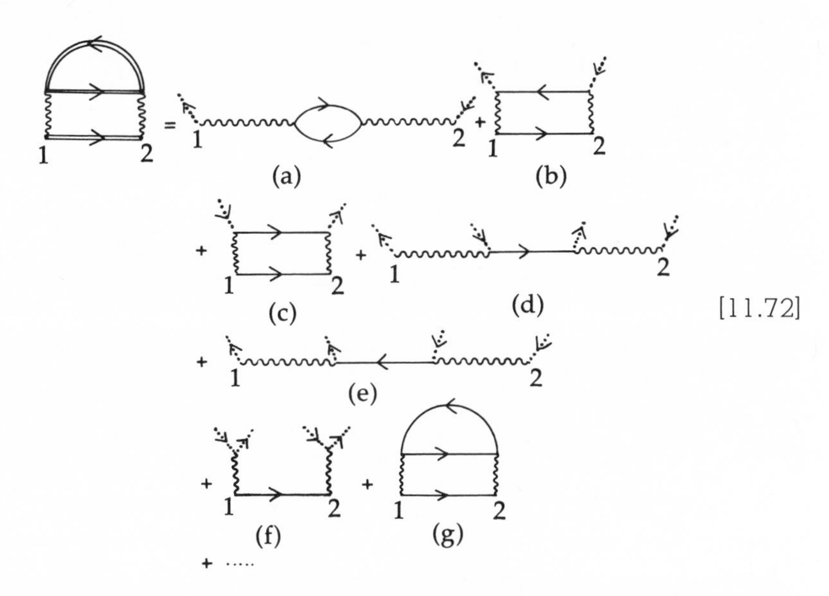

In terms of the zero-order Green's function, diagram (a) of Eq. (11.72) vanishes but, more importantly, (d), (e), and (f) are now *separable* in the sense of a Dyson equation for G'. Including (d) in the Dyson equation (writing G_0' as - - - > - - -)

⟶ = - - - > - - - + - - - > - - - ● ⟶ [11.73]

would imply double counting, since the terms arising from (d) are already included in the first-order self-energy term

which arises from splitting the Green's function in

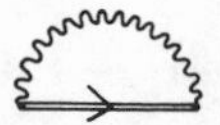

Even more difficult is the problem presented by diagram (e) of Eq. (11.72). This is clearly separable but the Green's function is reversed. In terms of "irreducible" self-energies for use in a new Dyson equation it is of the form

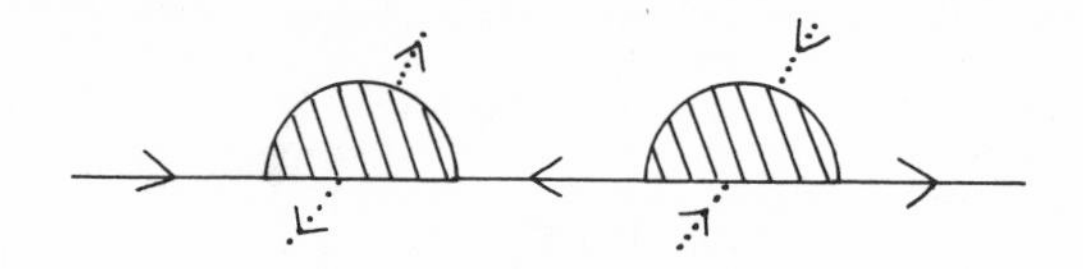

The condensate lines ensure *overall* conservation of particles in the total diagram but they are not conserved in the sense of the noncondensate Green's function—the condensate is acting as an effective source and sink for particles. We thus have to separate the diagrams of (11.72) and all subsequent orders of the self-energy into three types of irreducible self-energies, identified by the behavior of the condensate lines

$\Sigma_{11} =$ [11.74a]

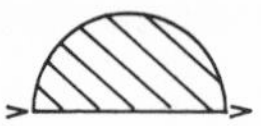

$\Sigma_{12} =$ [11.74b]

$\Sigma_{21} =$ [11.74c]

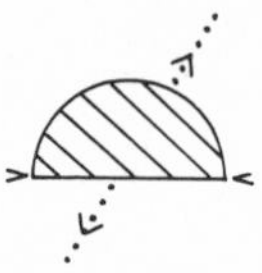

The Dyson equation can be written down as usual as

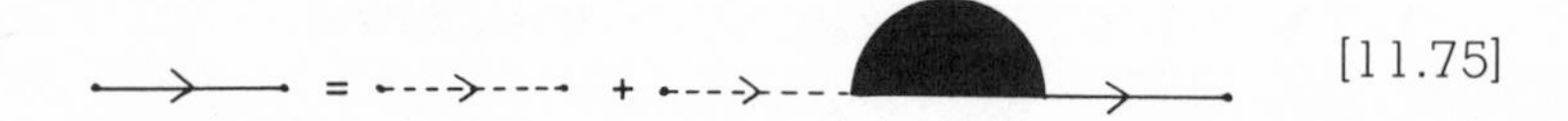

[11.75]

But separating Σ into its constituent parts, we can identify two possible terms

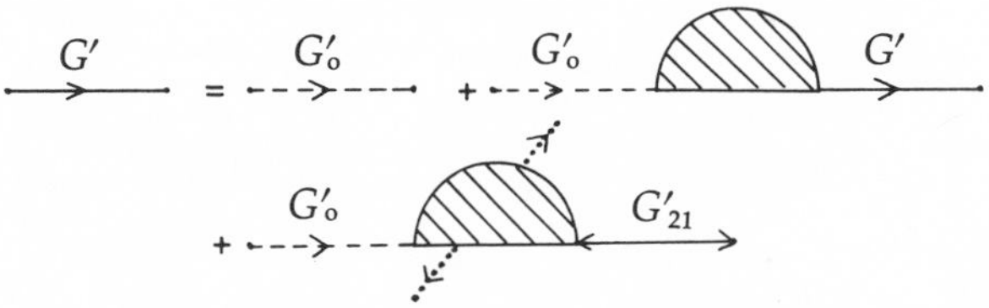

G' = G'_0 + G'_0 G'

+ G'_0 G'_{21} [11.76]

where the anomalous Green's function $G'_{21}(x_1,t_1,x_2,t_2)$ corresponds to all possible combinations of diagrams resulting in a net *gain* from the condensate of two particles, i.e.

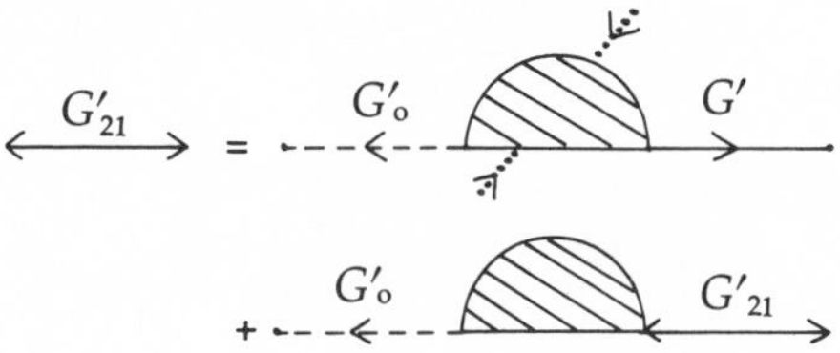

G'_{21} = G'_0 G'

+ G'_0 G'_{21} [11.77]

The interchange of indices give us two more equations

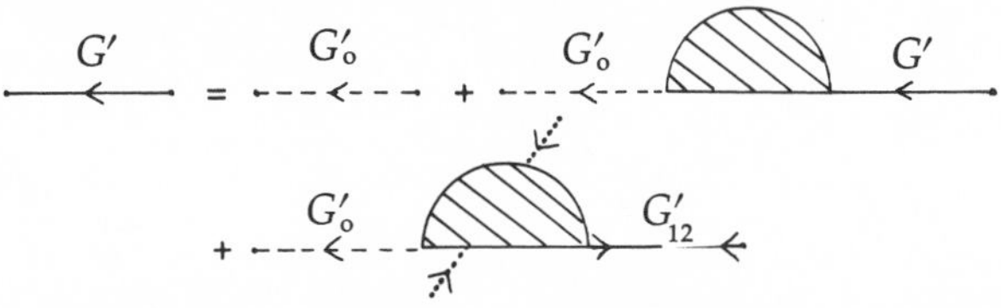

G' = G'_0 + G'_0 G'

+ G'_0 G'_{12} [11.78]

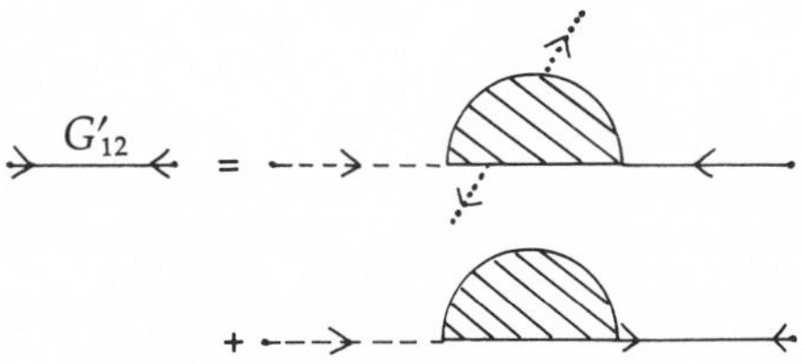

G'_{12} = [11.79]

+

in terms of anomalous Green's functions $G'_{12}(x_1,t_1,x_2,t_2)$, which corresponds to a net *loss* of two particles to the condensate. The anomalous Green's functions both disappear in the noninteracting limit, since all self-energies are zero. Algebraically these four equations can be coupled into one matrix equation by defining a matrix self-energy and Green's function:

$$\mathbf{G}'(x_1,t_1,x_2,t_2) = \begin{bmatrix} G'(x_1,t_1,x_2,t_2) & G'_{12}(x_1,t_1,x_2,t_2) \\ G'_{21}(x_1,t_1,x_2,t_2) & G'(x_2,t_2,x_1,t_1) \end{bmatrix} \qquad [11.80a]$$

$$\Sigma(x_1,t_1,x_2,t_2) = \begin{bmatrix} \Sigma_{11}(x_1,t_1,x_2,t_2) & \Sigma_{12}(x_1,t_1,x_2,t_2) \\ \Sigma_{21}(x_1,t_1,x_2,t_2) & \Sigma_{11}(x_2,t_2,x_1,t_1) \end{bmatrix} \qquad [11.80b]$$

so that we return formally to the Dyson equation

$$\mathbf{G}'(1,2) = \mathbf{G}'_0(1,2) + \int \mathbf{G}'_0(1,3)\Sigma\,(3,4)\mathbf{G}(4,3)\; d[3]\; d[4] \qquad [11.81]$$

The anomalous Green's functions may be written in terms of the field operators. They are given by the relationships

$$G'_{12}(x_1,t_1,x_2,t_2) = \langle N - 2|\,T[\hat{\chi}(x_1t_1)\hat{\chi}(x_2,t_2)]\,|N\rangle \qquad [11.82a]$$

and

$$G'_{21}(x_1,t_1,x_2,t_2) = \langle N + 2|\,T[\hat{\chi}^+(x_1,t_1)\hat{\chi}^+(x_2,t_2)]\,|N\rangle \qquad [11.82b]$$

where the coupling is between ground states of $N \pm 2$ particles. In a uniform system all of the equations simplify and we can transform to energy–momentum space. The condensate is the zero-momentum state and we have

$$\mathbf{G}(p,\omega) = \begin{bmatrix} G'(p,\omega) & G'_{12}(p,\omega) \\ G'_{21}(p,\omega) & G'(-p,-\omega) \end{bmatrix} \qquad [11.83a]$$

$$\Sigma(p,\omega) = \begin{bmatrix} \Sigma_{11}(p,\omega) & \Sigma_{12}(p,\omega) \\ \Sigma_{21}(p,\omega) & \Sigma_{11}(-p_1,-\omega) \end{bmatrix} \qquad [11.83b]$$

and so

$$\mathbf{G}'(p,\omega) = \mathbf{G}_0'(p,\omega) + \mathbf{G}_0'(p,\omega)\Sigma(p,\omega)\mathbf{G}'(p,\omega) \qquad [11.84]$$

We can use the zero-order Green's function

$$G_0'(p,\omega) = \frac{1}{\omega - (\epsilon(p) - \mu)} \qquad [11.85]$$

to give a solution for the anomalous and normal Green's functions in terms of the self-energies. The results are

$$G'(p,\omega) = \frac{w + \epsilon(p) + \Sigma_{11}(p,\omega)}{S(p,\omega)} \qquad [11.86]$$

$$G'_{\alpha\beta}(p,\omega) = -\frac{\Sigma_{\alpha\beta}(p,\omega)}{S(p,\omega)} \qquad [11.87]$$

where

$$S(p,\omega) = \left[\omega - \frac{(\Sigma_{11}(p,\omega) - \Sigma_{11}(-p,-\omega))}{2}\right]^2 - \left[\epsilon(p) + \frac{(\Sigma_{11}(p,\omega) + \Sigma_{11}(-p,\omega))}{2} - \mu\right]^2 + \Sigma_{12}(p,\omega)\,\Sigma_{21}(p,\omega) \qquad [11.88]$$

The minimum energy pole of the Green's function ($\omega = 0, p = 0$) should give the chemical potential. From Eq. (11.88) this is

$$S(p,\omega) = 0 \qquad (p \to 0, \omega \to 0) \qquad [11.89]$$

or

$$\mu = \Sigma_{11}(0,0) - [\Sigma_{12}(0,0)\,\Sigma_{21}(0,0)]^{1/2} \qquad [11.90]$$

It is easy to show from general arguments that $\Sigma_{12}(p,\omega) = \Sigma_{21}(p,\omega)$ so that

$$\mu = \Sigma_{11}(0,0) - \Sigma_{21}(0,0) \qquad [11.91]$$

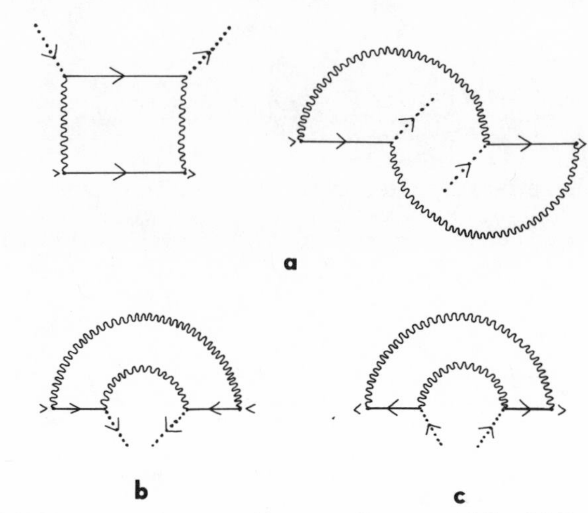

FIGURE 11.4. Second-order self-energy diagrams: (a) Σ_{11}. (b) Σ_{21}. (c) Σ_{12}.

The calculation of the self-energy terms is a complicated process beyond first order and we will not try to describe them here. Figure 11.4 shows the results for second order [many diagrams such as (a) and (g) of Eq. (11.72) are removed because the lowest-order Green's function $G_0'(x_1,t_1,x_2,t_2)$ is zero for $t_1 < t_2$].

The inclusion of a condensate complicates but does not destroy the basic concepts of the iteration procedure. The diagrammatic techniques, in particular, become much more essential in keeping track of the possible terms in the iterative expansion because of the appearance of reducible diagrams in forming any Dyson equation. Physically the formation of the condensate changes the system because it produces an effectively macroscopic potential, against which the noncondensate particles will scatter without changing the condensate to any great extent. This "rigidity" of the condensate is an important property. All of the equations can be extended to incorporate temperature effects as in Chap. 10. At high temperatures the condensate will disappear, as we shall see in Sec. 12.4 and Chap. 13, and a phase transition occurs.

BIBLIOGRAPHY

FEENBERG, E., *Theory of Quantum Fluids,* Academic Press, New York, 1969.

PINES, D., *Elementary Excitations in Solids,* Benjamin, Menlo Park: CA, 1964.

PINES, D., *The Many-Body Problem*, Benjamin, Menlo Park: CA, 1962.
SCHRIEFFER, J. R., *Superconductivity*, Benjamin, Menlo Park: CA, 1964.
ZIMAN, J. M., *Electrons and Phonons*, Cambridge University Press, New York, 1959.
BELIAEV, S. T., *Sov. Phys. JETP*, **7**, 289, 1958.

PROBLEMS

1. Derive Eq. (11.4) and obtain the plasmon creation and annihilation operators in terms of the $\hat{p}_k$ and $\hat{q}_k$.

2. Obtain the equation of motion for the field operators for electrons and phonons [(11.27), (11.28)] from the Hamiltonian of Eqs. (11.16) with the interaction term as given in Eq. (11.25).

3. Show that with a screened coulomb interaction

$$W(q) = \frac{v(q)}{\epsilon(q,w)}$$

the phonon energy $\omega_{\text{ph}}(q)$ is given by

$$\omega_{\text{ph}}(q) = \frac{\omega_{\text{ph}}}{[\epsilon(q,\omega)]^{1/2}}$$

where ω_{ph} is the ionic plasmon energy.

4. Split the Green's function into the condensate and noncondensate parts in the second-order self-energy diagram given by

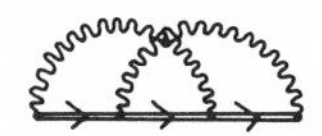

and hence produce the set of diagrams corresponding to Eq. (11.72).

5. Show that the nonzero second-order diagrams are given by Fig. 11.4.

CHAPTER TWELVE

SPECIAL METHODS

The Green's function self-energy formulation can describe the physics of interacting particles in a very wide variety of systems. Because of its generality, however, it is not always the most suitable method for any particular problem. The iteration series must be developed in many different directions in order to include the correct physical properties, and it is not always obvious which direction (or, alternatively, which subset of diagrams) will actually describe the system being studied. For this reason, a number of alternative formulations and special models have been developed that concentrate upon a particular physical property of the system to give either a useful alternate viewpoint or bring out the physics involved more clearly.

12.1. THE DENSITY FUNCTIONAL METHOD (NEARLY UNIFORM ELECTRON GASES)

In the derivation of the iteration procedure, the application of an external field and the variation of that field played a vital role. The density functional method developed by Hohenberg, Kohn, and Sham uses a similar technique to derive the ground-stage properties of the system in terms of the electron density, which is considered as a variable.

Given a Hamiltonian in the general form

$$H_\phi = H_0 + H_{\mathrm{INT}} + \phi(\mathbf{r}) \qquad [12.1]$$

where H_0 contains all the single-particle potentials, H_{INT} is the electron coulomb interaction and $\phi(\mathbf{r})$ the external potential. The equation

$$\hat{H}_\phi | N,\phi\rangle = E_\phi | N,\phi\rangle \qquad [12.2]$$

gives the ground state for a particular $\phi(\mathbf{r})$, and changing $\phi(\mathbf{r})$ will change the ground state.

The electron density

$$n(r) = \langle N,\phi | \hat{\psi}^{+}(\mathbf{r})\hat{\psi}(\mathbf{r}) | N,\phi \rangle \qquad [12.3]$$

is obviously, then, a functional of the external potential ($n[\phi]$); i.e., the density depends upon some function of the external potential. Conversely, if we consider the possible external potentials which may be responsible for the *same* density $n(\mathbf{r})$, it is easy to show that the external potential is a unique functional of the density. This follows from the fact that the ground-state energy is a minimum, i.e.,

$$E' = \langle N,\phi' | \hat{H}_{\phi'} | N,\phi' \rangle$$

Therefore

$$\begin{aligned} E' &< \langle N,\phi | \hat{H}_{\phi'} | N,\phi \rangle \\ E' &< \langle N,\phi | \hat{H}_{\phi} + \int(\phi(r') - \phi(r))\hat{\psi}^{+}(\mathbf{r})\hat{\psi}(\mathbf{r})\, d\mathbf{r}\; | N,\phi \rangle \\ E' &< E + \langle N,\phi | \int(\phi'(\mathbf{r}) - \phi(\mathbf{r}))\hat{\psi}^{+}(\mathbf{r})\hat{\psi}(\mathbf{r})\, d\mathbf{r}\; | N,\phi \rangle \end{aligned} \qquad [12.4]$$

Equally, we can show that

$$E < E' + \langle N,\phi' | \int(\phi(\mathbf{r}) - \phi'(\mathbf{r}))\hat{\psi}^{+}(\mathbf{r})\hat{\psi}(\mathbf{r})\, dr\; | N,\phi' \rangle \qquad [12.5]$$

But

$$\langle N,\phi | \int(\phi(\mathbf{r}) - \phi'(\mathbf{r}))\hat{n}(\mathbf{r})\, dr\; | N,\phi \rangle = \int(\phi(\mathbf{r}) - \phi'(\mathbf{r}))n(\mathbf{r})\, d\mathbf{r} \qquad [12.6]$$

which, by the assumption of identical densities for the different external potentials, gives

$$\langle N,\phi | \int(\phi(\mathbf{r}) - \phi'(\mathbf{r}))\hat{n}(\mathbf{r})\, d\mathbf{r}\; | N,\phi \rangle = \langle N,\phi' | \int(\phi(\mathbf{r}) - \phi'(\mathbf{r}))\hat{\mathbf{r}}(\mathbf{r})\, dr\; | N,\phi' \qquad [12.7]$$

The inequalities (12.4) and (12.5) are incompatible and so the density of the system $n(\mathbf{r})$ defines the unique $\phi(\mathbf{r})$ which brings it about and hence a unique Hamiltonian H. Thus, schematically, we have the sequence

$$n(\mathbf{r}) \rightarrow \phi(\mathbf{r}) \rightarrow \hat{H}[n] \rightarrow | N,\phi \rangle \rightarrow | N,n(\mathbf{r}) \rangle \qquad [12.8]$$

The combination of $|N,n(\mathbf{r})\rangle$ and $\hat{H}$ gives the total energy as a functional of the density

$$E[n] = \langle N,n(\mathbf{r})|\hat{H}|N,n(\mathbf{r})\rangle \tag{12.9}$$

This, by similar arguments as before, is a minimum for the correct density, which immediately suggests a variational calculation of the ground-state energy. Normally one separates out the single-particle terms so that

$$E[n] = \int \phi(\mathbf{r})n(\mathbf{r})\, d\mathbf{r} + \frac{1}{2}\int e^2 \frac{n(\mathbf{r})n(\mathbf{r}')}{|\mathbf{r}-\mathbf{r}'|}\, d\mathbf{r}'\, d\mathbf{r} + \int U(\mathbf{r})n(\mathbf{r})\, d\mathbf{r} + G[n] \tag{12.10}$$

The second term is the classical coulomb energy, $U(\mathbf{r})$ is the single-particle potential from ions, etc., and $G[n]$ contains the kinetic, exchange, and correlation contributions to the total energy. This, again, is a unique functional of the electron density.

The variational principle can be expressed as the pair of equations

$$\frac{\delta E[n]}{\delta n(\mathbf{r})} = 0 \tag{12.11a}$$

$$\int \delta n(\mathbf{r})\, d\mathbf{r} = 0 \tag{12.11b}$$

The application of Eq. (12.11) is, however, difficult except in very special cases because it leaves the most difficult part of the calculation, the identification of the functional, undefined. In the case of a slowly varying density, however, one can make some progress.

Let us consider what a functional really means; it says that if we have a system with a varying electron density, then G is a function of that *total* distribution. Suppose, for instance, we were to write the functional $G[n]$ in the form

$$G[n] = \int g_{\mathbf{r}}[n]\, d\mathbf{r} \tag{12.12}$$

where $g_{\mathbf{r}}[n]$ would be an energy density functional. Changing the electron density by a small amount at $\mathbf{r}'$ would then change the energy density functional not only at $\mathbf{r}'$ but also *over the whole system*. Thus the functional is nonlocal and depends upon the *whole* electron density.

Physically we might expect that the result of small changes in the electron density distribution at large distances from the point of interest would not really affect the density functional. For sufficiently slowly varying densities, the functional $g_r[n]$ might then be expanded in a density gradient series so that

$$g_r[n] = g_0(n(\mathbf{r})) + g_1(n(\mathbf{r}))\nabla n(\mathbf{r}) + \cdots \qquad [12.13]$$

where the g_0, g_1, etc., are local *functions* (not functionals) of the electron density and the information about the changes in the electron density are incorporated in a Taylor expansion of the density functional.

Using the first approximation to the functional, i.e., a "local density approximation," one can write, separating out the noninteracting kinetic energy part of the functional ($T[n]$)

$$G[n] = T[n] + E_{xc}[n] \qquad [12.14]$$

with

$$E_{xc}[n] = \int n(\mathbf{r})\epsilon_{xc}(n(\mathbf{r}))\, d\mathbf{r} \qquad [12.15]$$

$\epsilon_{xc}(n(\mathbf{r}))$ is the exchange and correlation energy per electron for a *uniform* gas of interacting electrons of density equal to the local density $n(\mathbf{r})$. Equation (12.10) now becomes

$$E = \int (\phi(\mathbf{r}) + U(\mathbf{r}))n(\mathbf{r})\, d\mathbf{r} + \frac{1}{2}\int e^2 \frac{n(\mathbf{r})n(\mathbf{r}')}{|\mathbf{r} - \mathbf{r}'|}\, d\mathbf{r}\, d\mathbf{r}' + T[n] + \int n(\mathbf{r})\epsilon_{xc}(n(\mathbf{r}))\, d\mathbf{r} \qquad [12.16]$$

If we now vary the energy so as to obtain the energy minimum, that is, require that

$$\frac{\delta E}{\delta n(\mathbf{r})} = 0 \qquad [12.17]$$

Subject, of course, to keeping the total number of particles constant

$$\int n(\mathbf{r})\, d\mathbf{r} = N \qquad [12.18]$$

one obtains

$$\int \delta n(\mathbf{r}) \left[\phi(\mathbf{r}) + U(\mathbf{r}) + \int e^2 \frac{n(\mathbf{r}')\, d\mathbf{r}'}{|\mathbf{r} - \mathbf{r}'|} + \frac{\delta T[n]}{\delta n(\mathbf{r})} + \frac{d(n(\mathbf{r}))\epsilon_{xc}(n(\mathbf{r}))}{dn(\mathbf{r})} \right] = 0$$

[12.19]

Comparing this to the case of a *noninteracting* electron gas moving in an effective potential

$$V(\mathbf{r}) = U(\mathbf{r}) + \phi(\mathbf{r}) + \int e^2 \frac{n(\mathbf{r}')\, d\mathbf{r}}{|\mathbf{r} - \mathbf{r}'|} + \frac{d[n(\mathbf{r})\epsilon_{xc}(n(\mathbf{r}))]}{d(n(\mathbf{r}))} \qquad [12.20]$$

one sees that the solution to Eq. (12.20) is equivalent to the set of one-electron Schrödinger equations

$$\left(-\frac{\hbar^2 \nabla^2}{2m} + V(\mathbf{r}) \right) \psi_i(\mathbf{r}) = E_i \psi_i(\mathbf{r}) \qquad [12.21]$$

for the lowest eigenstages $i = 1 \rightarrow N$ and

$$n(\mathbf{r}) = \sum_{i=1}^{N} |\psi_i(\mathbf{r})|^2 \qquad [12.22]$$

to ensure self-consistency.

These three equations [(12.20)–(12.22)] produce a set of single-particle states in the same sense as the Hartree and Hartree–Fock equations. The states $\psi_i(\mathbf{r})$ only have meaning if they are *occupied,* since one has only considered the ground state of the system. The advantage is, of course, that a set of equations like (12.20)–(12.22) are much easier to handle than the Green's function equation of motion and they do produce a set of *effective single-particle states.* The potential function of Eq. (12.20) can be split into the usual components: $U(\mathbf{r})$ and $\phi(\mathbf{r})$ are the one-electron potentials, the next term is easily identified as the Hartree potential (including the self-interaction term), and the final part is an effective single-particle exchange and correlation potential.

If we calculate this potential using the Hartree–Fock exchange of

Sec. 1.4 we find that

$$V_{ex}(\mathbf{r}) = \frac{d(n(\mathbf{r}))\epsilon_{xc}(n(\mathbf{r}))}{d(n(\mathbf{r}))} \qquad [12.23]$$
$$= -\frac{1}{\pi}(3\pi^2 n(\mathbf{r}))^{1/3}$$

which is, in fact, identical to the Slater averaged exchange [Eq. (1.35)] except for a numerical factor $\frac{2}{3}$. Thus the density functional method reproduces and so, to a large extent, justifies the ansatz used over many years for the calculation of single-particle eigenstates and eigenvalues in solids. This is the real importance of this version of many-body theory. It has been found to be remarkably accurate in this lowest-order approximation and has even been used with success in many areas where the basic assumption of a slowly varying density no longer applies.

One must be careful, however, to remember that the local density functional theory applies to the ground state. It does not easily generalize to give excited-state properties. Of course it is very tempting to use Eq. (12.21) as a normal Schrödinger equation and interpret the extra eigenfunctions and eigenvalues in terms of higher single-particle states, just as one does in Hartree–Fock; as then, however, one has the problems of relaxation and change in the effective exchange and correlation potentials to contend with. If one uses such an approximation in the calculations of insulator single-particle energies, for instance, it is found that the band gaps around the Fermi energy are consistently too small. This reflects the difference in the ground-state exchange and correlation potential and the value for even the first excited state with an electron in the conduction band. Within quite broad limits it still remains, however, the most successful of techniques for actual numerical calculations.

12.2 HIGHLY LOCALIZED SYSTEMS (ANDERSON–HUBBARD MODELS)

Consider the system of N one-electron states $\Phi_l(\mathbf{r})$ localized in space as shown schematically in Fig. 12.1a. If we now commence to occupy these states then, we can put one electron on each site until we have

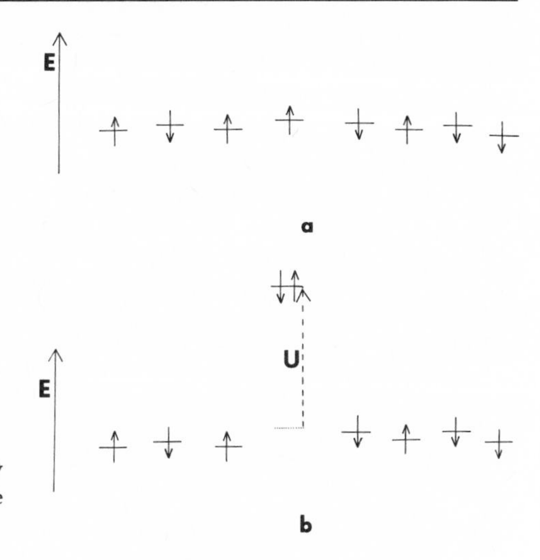

FIGURE 12.1. (a) Electron occupancy of localized states. (b) Effect of double occupancy on $(N + 1)$th state.

filled them all, forming a narrow "valence" band. As indicated in the diagram, the spin may be quite random from site to site. If we now try to put the $(N + 1)$th electron into the system it must go onto an already occupied site. This means that it will have to be of opposite spin to the existing particle and also that there will be an extra coulombic repulsion to overcome between the electrons, given by

$$U = \int |\Phi_l(\mathbf{r})|^2 \frac{e^2}{|\mathbf{r} - \mathbf{r}'|} |\Phi_l(\mathbf{r}')|^2 \, d\mathbf{r} \, d\mathbf{r}' \qquad [12.24]$$

as indicated in Fig. 12.1b. Thus, the energy spectrum of this system has a gap in it of value U between the N- and $(N + 1)$-particle states. This gives an insulating system where the gap arises from the coulomb interaction.

Suppose one contrasts this with the situation where there is a significant overlap between the states. The single-particle eigenstages become extended over the whole system and, in a periodic crystal, become Bloch states

$$\psi_k(\mathbf{r}) = \frac{1}{\sqrt{N}} \sum_l e^{i\mathbf{k}\cdot\mathbf{l}} \phi(\mathbf{r} - \mathbf{l}) \qquad [12.25]$$

If we now consider filling these eigenstates, as before, the coulomb repulsion for opposite spin electrons in the same Bloch state is given by

$$U_B = \frac{1}{N^2} \sum_{\mathbf{l},\mathbf{j}} \int |\phi(\mathbf{r} - \mathbf{l})|^2 \frac{e^2}{|\mathbf{r} - \mathbf{r}'|} |\phi(\mathbf{r}' - \mathbf{j})|^2 \, d\mathbf{r} \, d\mathbf{r}' \quad [12.26]$$

(neglecting screening which, in any case, will reduce the magnitude). For electrons on the same site we have the integral equal to U while for off-sites interactions we may approximate it by

$$\int |\phi(\mathbf{r} - \mathbf{l})|^2 \frac{e^2}{|\mathbf{r} - \mathbf{r}'|} |\phi(\mathbf{r} - \mathbf{j})|^2 \, d\mathbf{r} \, d\mathbf{r}' \approx \frac{e^2}{|\mathbf{l} - \mathbf{j}|} \quad [12.27]$$

Thus the total energy penalty is

$$U_B \approx \frac{U}{N} + \frac{1}{N^2} \sum_{\substack{l,j \\ l \neq j}} \frac{e^2}{|\mathbf{l} - \mathbf{j}|} \quad [12.28]$$

It is easy to show that for N large both of these terms vanish so that one returns to the effectively single-particle spectrum we are familiar with. (In practice, because of screening, the second term would be negligible in any case.)

For a localized system, then, one can devise the following model Hamiltonian to incorporate the physical effect of coulomb repulsion

$$H = \hat{H}_0 + \hat{H}_{\text{CORR}} \quad [12.29]$$

where

$$\hat{H}_0 = \sum_{l,\sigma} \epsilon_l \hat{a}^+_{l\sigma} \hat{a}_{l\sigma} \quad [12.30]$$

is the one-electron part and ϵ_l contains all of the single-particle effects.

$$\begin{aligned} \hat{H}_{\text{CORR}} &= \sum_l U_l \hat{a}^+_{l\sigma} \hat{a}_{l\sigma} \hat{a}^+_{l,-\sigma} \hat{a}_{l,-\sigma} \\ &= \sum_l U_l \hat{n}_{l\sigma} \hat{n}_{l,-\sigma} \end{aligned} \quad [12.31]$$

is the effect of the coulomb repulsion term when the electrons are on the same site. If each site has a number of degenerate levels, the model easily extends to include a coulomb repulsion on the same site between these degenerate levels.

As it stands this is a rather simple model but there are a large number of systems which contain localized and delocalized states in the same energy region. Consider the situation shown in Fig. 12.2a. We have a free electron band filled up to some Fermi level and we introduce into this system an atom with a highly degenerate but localized state—the obvious example is an atom with a partially filled d or f shell. As the electrons drop into the localized state we have to take into account the effects of the U due to the coulomb repulsion of the electrons localized on the atom. Thus, as each degenerate level in the state becomes filled, the energy required to introduce the *next* electron rises until putting the final electron into the localized state would raise the energy above the Fermi level (Fig. 12.2b). This means that the occupation of the degenerate

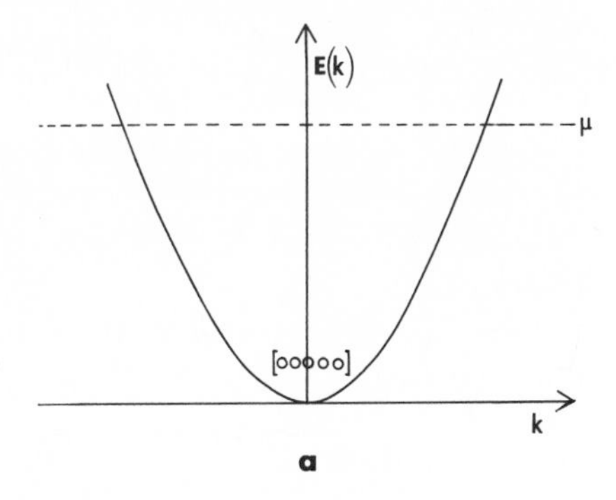

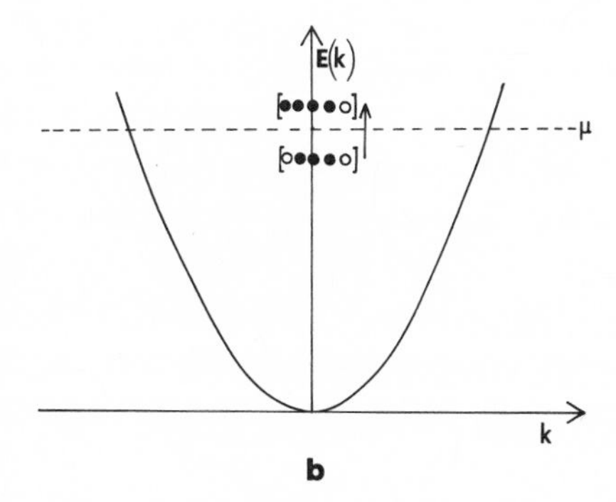

FIGURE 12.2. (a) Coexisting localized and delocalized states. (b) Equilibrium occupancy is achieved when the next electron to be added to the localized state would move the whole localized system above the Fermi level.

set of states on the atom will depend upon the value of the coulomb repulsion term as well as the normal single-particle properties.

A typical Hamiltonian for a mixed system of levels would be of the form

$$\hat{H} = \hat{H}_{0,f} + \hat{H}_{0,l} + H_{l,\mathrm{CORR}} + \hat{H}_{l,f} \qquad [12.32]$$

where $\hat{H}_{0,f}$ is the Hamiltonian for the delocalized states

$$\hat{H}_{0,f} = \sum_{k,0} \epsilon(k)\hat{a}^{+}_{k,\sigma}\hat{a}_{k,\sigma} \qquad [12.33]$$

$\hat{H}_{0,l}$ is the single-particle part of the localized Hamiltonian

$$\hat{H}_{0,l} = \sum_{l,\sigma} \epsilon_l \hat{a}^{+}_{l,\sigma}\hat{a}_{l,\sigma} \qquad [12.34]$$

$\hat{H}_{l,\mathrm{CORR}}$ is the coulomb term

$$\hat{H}_{l,\mathrm{CORR}} = \sum_{l,\sigma} U\hat{a}^{+}_{l,\sigma}\hat{a}_{l,\sigma}\hat{a}^{+}_{l,-\sigma}\hat{a}_{l,-\sigma} \qquad [12.35]$$

while $H_{l,f}$ would be the coupling between the localized and free electrons

$$\hat{H}_{l,f} = \sum_{\substack{k,l\\ \sigma}} V_{k,l}(\hat{a}^{+}_{k,\sigma}\hat{a}_{l,\sigma} + \hat{a}^{+}_{l,\sigma}\hat{a}_{k,\sigma}) \qquad [12.36]$$

Obviously this could easily be extended to the case of a set of degenerate levels on each site. This is one form of the Anderson Hamiltonian, which has been used with great success in treating many systems. In its basic form it does not appear to bear much resemblance to the self-energy equations we have been working with, though it is a special case of the general equation of motion. It does, however, in its very simplicity allow for the understanding of some very basic physical properties.

Consider briefly the case of a single transition metal impurity in a semiconductor. It is known from experiment that such an impurity can exist in a number of charge states (exactly how many depends upon the impurity and semiconductor concerned). In a free atom the equilibrium number of electrons in the *d* shell is determined by the coulomb repulsion. Although the *d* shell is incomplete and lower in energy than the *s* shell, the transfer of the electron from the s shell results in the d shell rising above the now-empty s shell (Fig. 12.3) so that the situation is stable (the U for this system is several electron volts). If the atom is placed into a semiconductor, the s level hybridizes, i.e., mixes with the semiconductor electron states and we are left with a coupled system described by a Hamiltonian similar to Eq. (12.32). Without the correlation term, one result of this Hamiltonian would be to produce a single impurity level in the gap due to the hybridization term in the Hamiltonian. With the correlation term, however, a number of states become possible. In general, the impurity state will be described by an operator of the form

$$b^+ \approx \alpha a^+_{l\sigma} + \beta a^+_{k\sigma} \qquad [12.37]$$

With a large U value, the contribution to the state which is d-like ($\approx\alpha^2$) results in a raising of the d level by a *proportion* ($\approx\alpha^2$) of the full U value. As the Fermi level is raised through the energy gap of the semiconductor, one then sees a "telescoping" of the possible free electron d levels into the energy gap regions, resulting in a number of possible charged states for the impurity (Fig. 12.4a). In practice the calculation is somewhat complicated, but Fig. 12.4b shows typical results of Haldane and Anderson on this system.

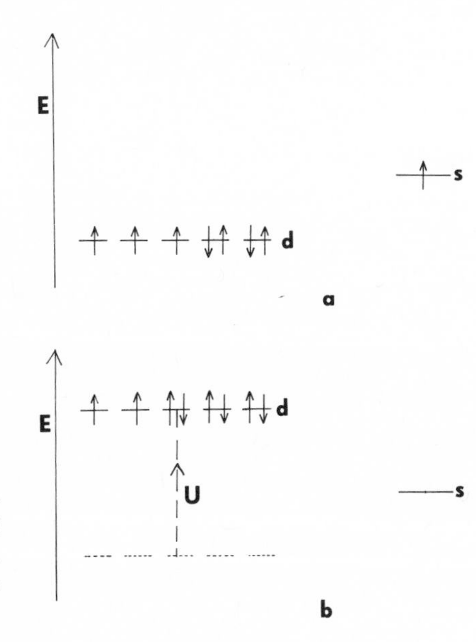

FIGURE 12.3. (a) Schematic representation of the equilibrium configuration of the s and d electrons in a transition metal. (b) Effect of transferring one electron from the s to d states.

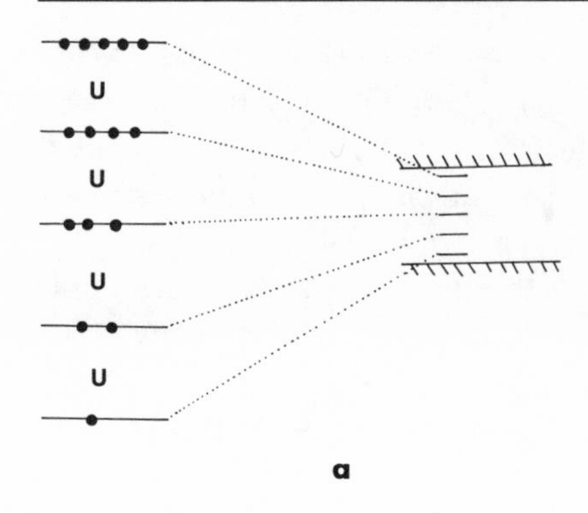

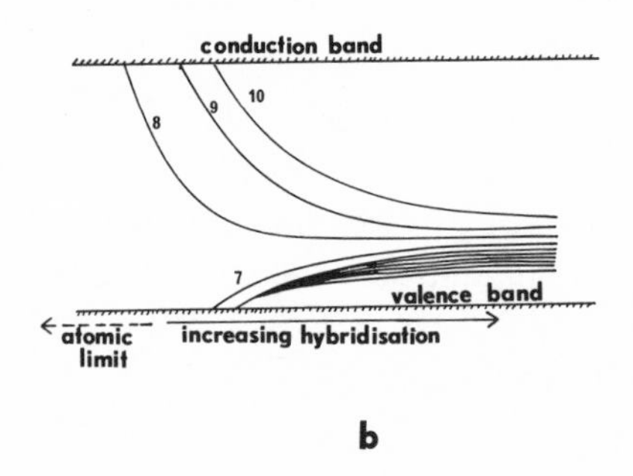

FIGURE 12.4. (a) Telescoping of the d levels into the semiconductor band gap. (b) Results of the Anderson model calculation [after D. Haldane and P. W. Anderson, *Phys. Rev. B* **13**, 2553 (1976)].

Characteristically these model calculations do not produce results which can be compared directly with experiment, but they do serve to illustrate the basic physics of what is going on in a situation where the full equations of motion would not be applicable. That is the importance of a model Hamiltonian of the type of Eq. (12.32). It provides a soluble approximation to a set of systems, which contains the essential physics.

12.3. CANONICAL TRANSFORMATIONS

There are many situations where one is presented with a Hamiltonian of the form

$$\hat{H} = \hat{H}_0 + \hat{H}_{\mathrm{INT}} \qquad [12.38]$$

where $\hat{H}_0$ is a soluble Hamiltonian and may be written in the standard form

$$H_0 = \sum_n \epsilon_n \hat{a}_n^+ \hat{a}_n \qquad [12.39]$$

The operators $\hat{a}_n^+$, $\hat{a}_n$ will obey the commutator or anticommutator relationships according to whether they are fermion or boson operators, i.e.

$$\hat{a}_{n'}\hat{a}_n^+ \pm \hat{a}_n^+ \hat{a}_{n'} = \delta_{nn'} \qquad [12.40]$$

The interaction part of the Hamiltonian (normally a perturbation, but not necessarily) is usually expressed in terms of the set of operators $\hat{a}_n^+$, $\hat{a}_n$. The solution to the problem then requires the evaluation of a new set of operators such that $\hat{H}$ may be expressed in the standard form

$$\hat{H} = \sum_j E_j \hat{b}_j^+ \hat{b}_j \qquad [12.41]$$

which is simply another way of saying that there will be a set of eigenstates and eigenvalues associated with the complete Hamiltonian. The $\hat{b}$, $\hat{b}^+$ operators will also need to obey commutator or anticommutator relationships like Eq. (12.40). The canonical transformation method simply amounts to expressing the operators $\hat{b}_j^+$, $\hat{b}_j$ in terms of the operators $\hat{a}_n^+$, $\hat{a}_n$ so that Eq. (12.39) is transformed into Eq. (12.41), subject to the conditions that the new operators still satisfy the commutator or anticommutator relationship. The technique is best illustrated with a simple example.

Consider the Hamiltonian for a boson system

$$\hat{H} = \sum_k \hbar\omega_k\left(\hat{a}_k^+ \hat{a}_k + \frac{1}{2}\right) + \frac{\alpha_k}{2}(\hat{a}_k^+ + \hat{a}_k) \qquad [12.42]$$

If we make the substitution

$$\hat{b}_k = \hat{a}_k + s_k \qquad [12.43a]$$

$$\hat{b}_k^+ = \hat{a}_k^+ + t_k \qquad [12.43b]$$

where s_k, t_k are real constants, the transformation is canonical if

$$[\hat{b}_k, \hat{b}_{k'}^+] = [\hat{a}_k, \hat{a}_{k'}^+] = \delta_{kk'} \qquad [12.44]$$

which gives

$$s_k = t_k \qquad [12.45]$$

Substituting into Eq. (12.42), we have

$$\begin{aligned} H &= \sum_k \hbar\omega_k\left((\hat{b}_k^+ - s_k)(\hat{b}_k - s_k) + \frac{1}{2}\right) + \alpha_k(\hat{b}_k^+ + \hat{b}_k) - 2\alpha_k s_k \\ &= \sum_k \hbar\omega_k\left(\hat{b}_k^+\hat{b}_k + \frac{1}{2}\right) - \hbar\omega_k s_k(\hat{b}_k^+ + \hat{b}_k) + \alpha_k(\hat{b}_k^+ + \hat{b}_k) \\ &\quad + (\hbar\omega_k s_k^2 - 2\alpha_k s_k) \end{aligned} \qquad [12.46]$$

The first term is the one we want. The second term cancels with the interaction term provided

$$s_k = \frac{\alpha_k}{\hbar\omega_k} \qquad [12.47]$$

while the fourth term is simply a constant and so may be ignored as far as the dynamics of the system is concerned. This gives

$$H = \sum_k \hbar\omega_k\left(\hat{b}_k^+\hat{b}_k + \frac{1}{2}\right) \qquad [12.48]$$

and

$$\hat{b}_k = \hat{a}_k + \frac{\alpha_k}{\hbar\omega_k} \qquad [12.49a]$$

$$\hat{b}_k^+ = \hat{a}_k^+ + \frac{\alpha_k}{\hbar\omega_k} \qquad [12.49b]$$

defines the canonical transformation. The vacuum state is not normally required, but since it changes as well, it may be derived in terms of the possible states of the unperturbed system by using the relationships

$$\hat{b}_k|0\} = \hat{a}_k|0\rangle = 0 \qquad [12.50]$$

If we write

$$|0\} = \sum_{k,n} \gamma_{k,n} |k,n\rangle \qquad [12.51]$$

then from (12.49a) and (12.50) we have for each value of k

$$\sum_{n} \gamma_{k,n} \left(\hat{a}_k + \frac{\alpha_k}{\hbar \omega_k} \right) |k,n\rangle = 0 \qquad [12.52]$$

or

$$\sum_{n} \gamma_{k,n} \left(\sqrt{n} |k,n-1\rangle + \frac{\alpha_k}{\hbar \omega_k} |k,n\rangle \right) = 0 \qquad [12.53]$$

Equating coefficients of the independent states $|k,n\rangle$ then gives us the coefficients $\gamma_{k,n}$.

Equation (12.42) represents a particularly simple Hamiltonian, but the method follows in the same way for any Hamiltonian, though the transformation attempted would depend upon the form of the perturbation. Section 13.2 deals with the use of a particular canonical transformation applied to *superconductivity*.

12.4. MEAN-FIELD THEORY

The use of the self-energy techniques is essentially perturbative. It is of vital importance that one picks, as a zero-order state, one which contains all of the properties of the actual physical state. Many systems, however, undergo phase transitions which are characterized by the appearance of a new ordering in the system, for example solid to liquid or paramagnet to ferromagnet. It is very difficult to treat these sorts of systems in a general manner, but fortunately there are a large group that may be treated, in an approximate manner at least, by a technique known as *mean-field theory*.

In many cases the appearance of a new property or "order parameter" of the system is generated by a driving force which itself is proportional to that order parameter (or some average of it), to a first approximation at least. This feedback mechanism often induces a bistable situation where one has either a zero- or large-order parameter and is a common feature of phase transitions. The best-known example is that of classical ferromagnetism, and since the technique we wish to describe is not restricted to quantum systems, it provides a very useful illustrative example.

Suppose we have a system of classical magnetic dipoles of magnetic moment μ_0. Then according to statistical mechanics, each dipole has an expectation value for the magnetic moment along the direction of a field $\mathcal{H}$ given by

$$\langle \mu(\mathcal{H}) \rangle = \frac{\iint \mu_0 \cos \phi e^{\beta\mu\mathcal{H}\cos\phi} \, d\,(\cos \phi)\, d\phi}{\iint e^{\beta\mu\mathcal{H}\cos\phi} \, d\,(\cos \phi)\, d\phi} \qquad [12.54]$$

This gives the standard result

$$\langle \mu(\mathcal{H}) \rangle = \mu_0 \left(\coth \beta\mu_0\mathcal{H} - \frac{1}{\beta\mu_0\mathcal{H}} \right) \qquad [12.55]$$

In terms of a collection of such dipoles (as in a solid), this means that the total magnetic dipole moment of the solid will be $N\langle \mu(\mathcal{H}) \rangle$. As long as the magnetic field $\mathcal{H}$ is considered to be the external field, Eq. (12.56) is well behaved [$\langle \mu(\mathcal{H}) \rangle$ going to zero as $\mathcal{H}$ is removed] and contains the physics of the paramagnetic solid. This is too simple, however, since acting upon each dipole will be an internal field $\mathcal{H}_I$ due to the nonzero expectation values of all of the other dipoles. That is

$$\mathcal{H} = \mathcal{H}_0 + \mathcal{H}_I \qquad [12.56]$$

where

$$\mathcal{H}_I = N\alpha \langle \mu(\mathcal{H}) \rangle \qquad [12.57]$$

and α is some function (coupling parameter) which incorporates the actual physics of the internal field, which in fact derives from the spin interaction between atoms. Since the elementary dipoles cannot distinguish between the external and internal fields, Eq. (12.57) becomes

$$\langle \mu(\mathcal{H}) \rangle = \mu_0 \left\{ \coth[\beta\mu_0(\mathcal{H}_0 + \alpha N\langle \mu(\mathcal{H}) \rangle)] + \frac{1}{\beta\mu_0(\mathcal{H}_0 + \alpha N\langle \mu(\mathcal{H}) \rangle)} \right\}$$

[12.58]

If the external field is reduced to zero, there will still be the possibility of a residual magnetic moment, given by the solution of

$$\langle \mu \rangle_0 = \mu_0 \left[\coth(\beta\mu_0\alpha N\langle \mu \rangle_0) + \frac{1}{\beta\mu_0\alpha N\langle \mu \rangle_0} \right] \qquad [12.59]$$

Figure 12.5 illustrates the graphical solution to this equation and shows that depending upon the values of α, β, etc., there is a possibility of a nonzero value for $\langle \mu \rangle_0$ characteristic of a ferromagnet. It is also clear that there is a critical temperature below which the ordering takes place and above which the only solution is $\langle \mu \rangle_0$ equal to zero, i.e., we have a phase transition.

This example serves to illustrate the general theme of what is commonly called mean-field theory:

1. There exists an order parameter ($\langle \mu \rangle$) whose value depends upon a driving field ($\mathcal{H}$). It is important to include the possibility that the order parameter is nonzero in all of the equations.
2. The driving field is, in turn, a function (usually linear) of the order parameter [Eq. (12.57)].

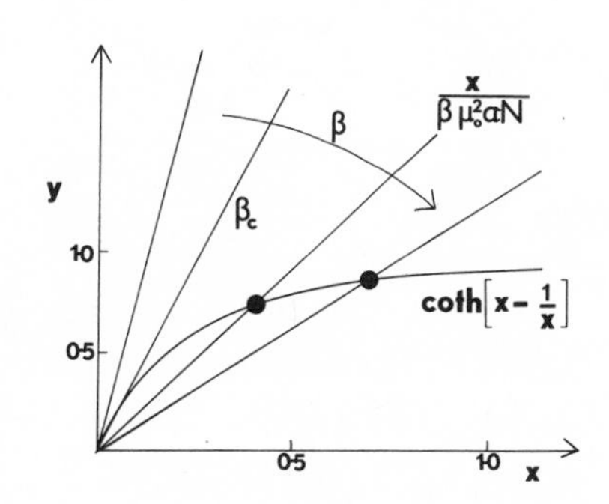

FIGURE 12.5. Graphical solution of Eq. (12.60) showing the appearance of a nonzero solution for β values above a critical value.

3. The solution to the two equations
   a. order parameter = some function of driving field
   b. driving field = some function of order parameter

   can be solved to give nontrivial results in the sense of a finite order parameter in some region determined by the variables (temperature, coupling parameter, etc.) in the problem.

As a quantum example consider the case of a one-dimensional solid with one conduction electron per atom (Figs. 12.6a and b) studied by Rice and Strassler. If we separate out the phonon and electronic coordinates, we may write as the basic Hamiltonian

$$\begin{aligned}\hat{H} &= \hat{H}_{\text{el}} + \hat{H}_{\text{ph}} + \hat{H}_{\text{INT}} \\ &= \sum_k E(k)\hat{c}_k^+\hat{c}_k + \sum_q \omega_q\left(\hat{a}_q^+\hat{a}_q + \frac{1}{2}\right) + \frac{1}{\sqrt{N}}\sum_{k,q}\gamma(q)\hat{c}_{k-q}^+\hat{c}_k(\hat{a}_q^+ + \hat{a}_{-q})\end{aligned} \tag{12.60}$$

where ω_q, $E(k)$ are the bare phonon and electronic energies (i.e., each considered without the effect of the other) and $\gamma(q)$ is the electron–phonon coupling matrix element.

Physically the situation is as follows. Consider first the effect of the

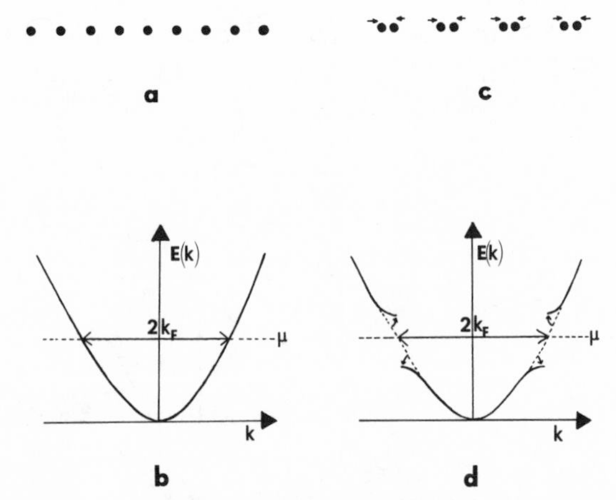

FIGURE 12.6. (a) Linear chain of atoms. (b) Energy–momentum relationship for the electrons in the linear chain. (c) Distorted chain. (d) Energy–momentum relationship for the electrons in the distorted linear chain.

electrons upon the phonon. In Chap. 11 we saw that the renormalization of the phonon energy (to Ω_q) was basically due to the screening effect of the electron gas, i.e.

$$\Omega_q \approx \frac{\omega_q}{[\omega(q,0)]^{1/2}} \qquad [12.61]$$

In three dimensions this produces some changes in the energy, but in one dimension, the dielectric response has an infinite singularity at twice the Fermi wavevector, which comes basically from a zero in the denominator of the polarization, i.e.

$$P(q,0) \approx \sum_k \frac{n_{(k)} - n_{(k+q)}}{E(k) - E(k+q)} \qquad [12.62]$$

So

$$P(2k_F,0) \approx \infty \qquad [12.63]$$

Since (Fig. 12.6b)

$$E(k_F) = E(k_F - 2k_F) \qquad [12.64]$$

From Eq. (12.61) it follows that the renormalized energy for the phonon may be zero at these wave vectors ($\pm 2k_F$). At finite temperatures the singularity is softened, so that the phonon frequency may be significantly reduced without going to zero. If the energy does become zero there is no obvious limitation on the number of phonons at these wave vectors that can be introduced into the system. Physically this corresponds to a *macroscopic* distortion of the atoms (Fig. 12.6c) to produce a new periodicity which effectively halves the Brillouin zone. The presence of the new distortion changes the system so that the number of phonons remains large but finite.

Turning now to the effect of the distortion on the electron distribution, simple one-electron physics tells us that the new periodicity will result in a gap appearing at the Fermi surface and hence a lowering of the energy of the valence electron states (Fig. 12.6d). This results in a net lowering in energy of the electronic system since only the unoccupied states are raised in energy.

The new equilibrium is reached when the lowering in energy of the electron system balances the rise in energy of the atomic system due to the added strain energy. At finite temperatures, states above and below the gap will be occupied so the net lowering of energy will be reduced. We would expect then that the distorted state will be stable, only below a critical temperature, if at all.

This simple physical argument which was described by Peierls many years ago, allows us to identify the new parameter which must be included, the distortion of the regular lattice due to a macroscopic occupancy of the phonon modes at $\pm 2k_F$ ($= q_0$). Let us apply mean-field theory to this system. If the energy of the phonon is reduced so that the expectation value of the number operators becomes macroscopic we have

$$\langle \hat{a}^+_{q0}\hat{a}_{q0}\rangle = M \qquad [12.65]$$

where M, the occupation number, is large. In the Hamiltonian [Eq. (12.60)] the electron–phonon interaction coupling terms will give rise to matrix elements of the form

$$\begin{aligned}
&\langle \text{electron states; } P \\
&\quad + 1 \text{ phonons in } q_0 | \hat{c}^+_{k-q0}\hat{c}_k\hat{a}^+_{q0} | \text{electron states; } P \text{ phonons in } q_0\rangle = \\
&(P+1)^{1/2}\langle \text{electron states; } P+1 \text{ phonons in } q_0 | \hat{c}^+_{k-q0}\hat{c}_k | \text{electron states;} \\
&\quad P+1 \text{ phonons in } q_0\rangle \approx \sqrt{M}\langle \text{electron states;} \\
&\quad P+1 \text{ phonons in } q_0 | \hat{c}^+_{k-q0}\hat{c}_k | \text{electron states; } P+1 \text{ phonons in } q_0\rangle
\end{aligned}$$

since all of the important P values will be around the value of M. So we replace a^+_{q0}, a_{q0} by their "expectation values"

$$\langle a_{q0}\rangle = \langle a^+_{q0}\rangle = \sqrt{M} \qquad [12.66]$$

The Hamiltonian is now, neglecting the microscopically occupied modes ($q \neq \pm q_0$)

$$\hat{H} = 2M\hbar\omega_{q0} + \sum_k E(k)\hat{c}^+_k\hat{c}_k + \frac{\sqrt{M}}{\sqrt{N}}\sum_k \gamma(q_0)(\hat{c}^+_{k-q0}\hat{c}_k + \hat{c}^+_{k+q0}\hat{c}_k) \qquad [12.67]$$

The first term is simply a constant (the strain energy in fact) while the second and third can be combined by a simple canonical transformation to give

$$\hat{H} = 2M\hbar\omega_{q0} + \sum_k \mathscr{E}(k)\, \hat{d}_k^+ \hat{d}_k \qquad [12.68]$$

where

$$\mathscr{E}(k) = \tfrac{1}{2}\{E(k) + E(k - q_0) \pm [(E(k) - E(k - q_0))^2 + 4\Delta^2]^{1/2} \qquad [12.69]$$

and

$$\Delta^2 = \frac{M\gamma^2(q_0)}{N} \qquad [12.70]$$

Essentially Eqs. (12.68) and (12.69) state the obvious fact that the introduction of a periodic distortion produces a mixing of the electronic states and a new set of single-particle excitations defined by the $\hat{d}$, $\hat{d}^+$ creation and annihilation operators. The Hamiltonian now contains both the strain energy and the new excitation spectrum for the electrons in terms of the single variable M (or Δ), which is the obvious order parameter for this problem. The electron spectrum at the Fermi level is given by

$$\mathscr{E}(k_F) = E(k_F) \pm \Delta \qquad [12.71]$$

so the gap is simply

$$2\Delta = \frac{2M\gamma^2(q_0)}{N} \qquad [12.72]$$

The solution can now be obtained for any temperature by minimizing the resulting free energy as a function of the order parameter; i.e., we require

$$\frac{\delta F}{\delta \Delta} = 0 \qquad [12.73]$$

where

$$F = 2\hbar N\, \Delta\omega_{q0} - \frac{1}{\beta} \int d\mathscr{E}\; N(\mathscr{E}) \ln\left(e^{-\beta\mathscr{E}} + 1\right) \qquad [12.74]$$

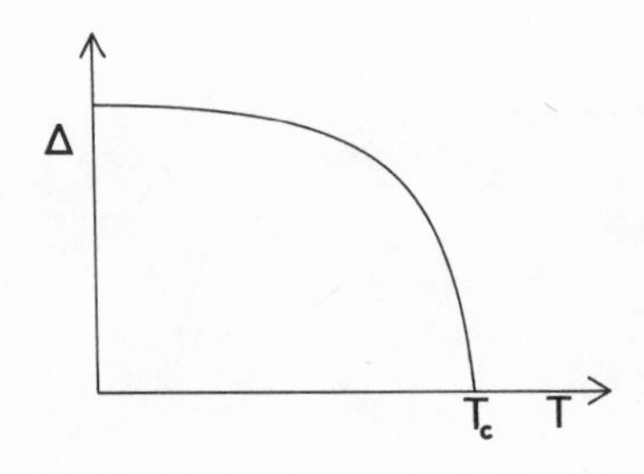

FIGURE 12.7. Schematic representation of variation of the order parameter (band gap) with temperature.

And $N(\mathcal{E})$ is the density of states of energy $\mathcal{E}$ in the distorted system. The algebra for this is complicated and results in an equation for the gap (or order parameter) as a function of temperature. This is shown schematically in Fig. 12.7. At a critical temperature the gap vanishes, and above this temperature the system is normal (i.e., as in Fig. 12.4a, b). Below this critical temperature there is a finite distortion of the lattice and a gap in the electron spectrum.

Obviously the mathematical principles are very much the same as in the magnetic case. The internal field is replaced by the lattice distortion and the self-consistent equation for the mean dipole [Eq. (12.59)] is replaced by minimization of the free energy (which says nothing more, after all, than the system should be in thermodynamic equilibrium). But in both cases it is important to connect the new order parameter (mean dipole moment, energy gap) to the driving force (internal field, distortion) self-consistently so that the possibility of a nonzero order parameter is not excluded. In all cases of mean-field theory, it is of vital importance to understand the physical input into the problem in order to reach a satisfactory solution. In other words, in this case the reduction of Eq. (12.60) to (12.67) contains the important aspects of the physics, the rest is algebra.

BIBLIOGRAPHY

Local Density Method

HOHENBERG, P., and KOHN, W., *Phys. Rev.*, **136,** 864, 1964.
KOHN, W., and SHAM, L. J., *Phys. Rev.*, **140,** 1133, 1965.

Localized Orbitals

HUBBARD, J., *Proc. Roy. Soc.*, **A276,** 238, 1963.
ANDERSON, P. W., *Phys. Rev.*, **124,** 49, 1961.

ADLER, D., *Solid State Phys.*, **22,** 1, 1968.
HALDANE, F. M. W., and ANDERSON, P. W., *Phys. Rev. B* **13,** 2553.
KONDO, J., *Solid State Phys.*, **23,** 183, 1969.

Mean-Field Theory

RICE, M. J., and STRASSLER, S., *Solid State Commun.*, **13,** 125, 1973.
PEIERLS, R. E., *Quantum Theory of Solids*, Oxford University Press, London, 1955.

PROBLEMS

1. Use the canonical transformation method to obtain the new energy spectrum and creation and annihilation operators for the Hamiltonian

$$H = \hbar\omega(\hat{a}^+\hat{a} + \tfrac{1}{2}) + \alpha(\hat{a}^+\hat{a}^+ + \hat{a}\hat{a}), \qquad \alpha \ll \hbar\omega$$

2. Show that the Hamiltonian

$$H = \sum_k E(k)\hat{a}_k^+\hat{a}_k + V(\hat{a}_{k-G}^+\hat{a}_k + \hat{a}_{k+G}^+\hat{a}_k)$$

may be transformed to first order in V to a Hamiltonian

$$H = \sum_k \mathscr{E}(k)\hat{b}_k^+\hat{b}_k$$

which exhibits a gap in that

$$\mathscr{E}\left(\frac{G}{2}\right) = E\left(\frac{G}{2}\right) \pm V$$

3. Derive Eq. (12.74).

CHAPTER THIRTEEN

SUPER-CONDUCTIVITY

The effects of the electron–electron and electron–phonon interaction on the normal solid are remarkably small. The results we have described simply amount to replacing a set of noninteracting one-particle states by a set of quasiparticles whose properties are only slight modifications of the original ones. In mathematical terms the perturbation series we have considered for the self energies have converged rapidly. This need not always be the case and is, in fact, highly dependent upon the way that the initial system is chosen and how the perturbation series is constructed.

Consider, as an example, the electron gas. The initial system was taken to be a set of one-electron states mainly because the available evidence pointed to the interacting system having many of the properties associated with such a system. The self-energy expansion in the bare interaction, however, gave, in first order, a result (Hartree–Fock) which did not accord with experiment and an expansion parameter which was large. The solution to the problem was to introduce the essentially classical concept of screening, which produced reasonable first-order results and a small expansion parameter. The concept of screening was, in turn, connected with the appearance of a new quasiparticle, the plasmon, which mathematically required the solution of the screened interaction equation

$$\mathsf{W} = \mathsf{v} + \mathsf{vPW} \qquad [13.1]$$

by a *nonperturbative* technique. Because this equation was "buried" in the self-energy expansion, no perturbative approach could give the correct result. Another example we have had was in Sec. 2.3 where we saw that an attractive scattering potential led to a new pole in the noninteracting Green's function representing a bound state. The appearance of the new solution required a nonperturbative approach, since the perturbation expansion failed to converge in the relevant energy range.

The lesson of these two examples is clear: the essential requirement of applying many-body theory is that the initial starting point should contain the basic physical properties of the interacting system. Otherwise no perturbative approach, which in any real system is the only one available, is going to work. In simple terms, unless the initial assumption contains within it the seeds of the solution, there is no way in which small changes (i.e., perturbations) from that assumption will lead to the correct answer.

The decision as to which physical ingredients to include in the problem is often difficult and is best answered either by considering simple models or, more normally, from the results of experiment. The problem of superconductivity arising at low temperatures is a spectacular example of how the introduction of a physical concept into the normal equations of motion for the electron Green's function can completely change the character of the solution.

13.1. COOPER PAIRS

The presence of both electron–electron and electron–ion interactions may, as we have seen, lead to an attractive interaction over a limited energy range. It is worthwhile considering the implications of this for the perturbation series. The self-energy contains a set of diagrams of the form

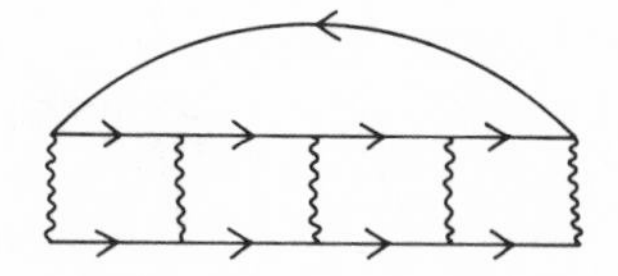

The "ladder" part of this diagram corresponds to two electrons repeatedly scattering by way of the effective interaction. If this interaction is attractive, a bound state may be formed leading to a failure of the self-energy perturbation series. The solution to this problem is to solve "exactly" the ladder series and then insert the solution into the appropriate place in the self-energy expansion. (This is equivalent to solving for the screened interaction in the normal system before developing the self-energy perturbation series). Thus, we require the t matrix for electron–electron scattering, which is the sum of all the repeated scattering events. Diagrammatically we can write

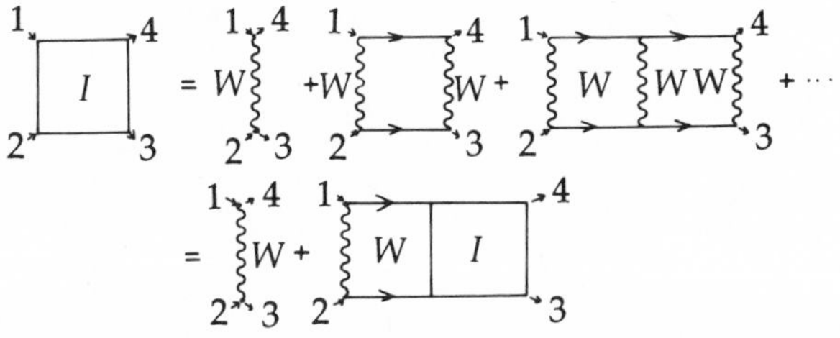

[13.2]

where I is the sum of the repeated scatterings and W the basic electron–electron interaction. Algebraically this is equivalent to

$$I(1,2,3,4) = W(1,2)\delta(1,4)\delta(2,3) + \int W(1,2)G(2,6)G(1,5)I(5,6,3,4)\, d[5]\, d[6] \tag{13.3}$$

This integral equation, similar in form to a Dyson equation, is known as a Bethe–Salpeter equation. The solution is dependent upon W and is in general extremely difficult. The physics of the solution is, however, quite general, and it is sufficient to illustrate it with a very simple interaction. Let us take a contact interaction approximation:

$$W(1,2) = \lambda\delta(1,2) \tag{13.4}$$

since the screened interaction has a short range in any case. The series for the interaction degenerates to a series of "bubbles"

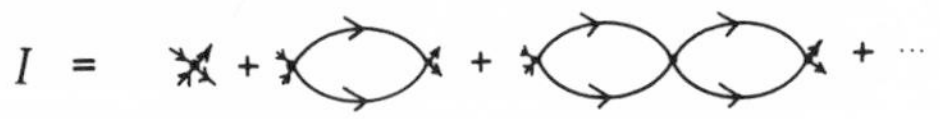

[13.5]

from which we see that

$$I(1,2,3,4) \rightarrow I(1,3)\delta(1,2)\delta(3,4) \tag{13.6}$$

and

$$I(1,3) = \lambda\delta(1,3) + \lambda\int G(1,2)G(1,2)I(2,3)\, d[2] \tag{13.7}$$

Defining

$$G(1,2)G(1,2) = A(1,2) \qquad [13.8]$$

we can write formally

$$\mathsf{I} = \lambda 1 + \lambda \mathsf{A} \mathsf{I} \qquad [13.9]$$

or

$$\mathsf{I} = \lambda 1(1 - \lambda \mathsf{A})^{-1} \qquad [13.10]$$

The failure of the perturbation series for the interaction is associated with the zeros of the function $1 - \lambda \mathsf{A}$. If we consider a uniform system, we can work in energy–momentum space and then

$$A(\mathbf{p},\Omega) = \frac{i}{(2\pi)^4} \int d\mathbf{q}\, d\omega\, G(\mathbf{p} - \mathbf{q},\Omega - \omega)G(\mathbf{q},\omega) \qquad [13.11]$$

an expression very similar to the polarization propagator of Eq. (9.3). As in that case we can perform, using the single-particle Green's functions, the energy integration, picking up the poles of the Green's functions so that

$$A(\mathbf{p},\Omega) = -\frac{1}{(2\pi)^3} \int d\mathbf{q} \frac{(n(\mathbf{q}) + n(\mathbf{p} - \mathbf{q}) - 1)}{\Omega - (E(\mathbf{q}) + E(\mathbf{p} - \mathbf{q}))} \qquad [13.12]$$

The Fermi factors $n(\mathbf{q})$, $n(\mathbf{p} - \mathbf{q})$ decide the basic contributions to the integral.

(i) $\quad |\mathbf{q}|, |\mathbf{p} - \mathbf{q}| > k_F;\ n(\mathbf{q}) = n(\mathbf{p} - \mathbf{q}) = 0$

$A(\mathbf{p},\Omega)$ corresponds to the motion of two added electrons.

(ii) $\quad |\mathbf{q}|, |\mathbf{p} - \mathbf{q}| < k_F;\ n(\mathbf{q}) = n(\mathbf{p} - \mathbf{q}) = 1$

$A(\mathbf{p},\omega)$ corresponds to the motion of two holes.

The denominator zeros are given by

$$\Omega = E(\mathbf{q}) + E(\mathbf{p} - \mathbf{q}) \qquad [13.13]$$

These are the excitation energies required to produce the two electrons or holes. Like the polarization propagator, $A(\mathbf{p},\Omega)$ is a two-particle propagator, the difference being the type of particles allowed.

Considering only the electron–electron interaction (since the character of the particles involved in A are determined by the external lines to the four vertices of I)

$$A_{ee}(\mathbf{p},\Omega) = \frac{1}{(2\pi)^3} \int_{\substack{|q|>k_F \\ |p-q|>k_F}} \frac{d\mathbf{q}}{\Omega - (E(q)+E(\mathbf{p} - \mathbf{q}))} \qquad [13.14]$$

The poles of this function are given by Eq. (13.13). This is best illustrated by plotting $A(\mathbf{p},\Omega)$ as if $\mathbf{q}$ is a discrete variable, as in Fig. 13.1. We can now solve Eq. 13.10 graphically for the new roots of $I(\mathbf{p},\Omega)$ (the Fourier transform of I) since we have a pole whenever

$$1 - \lambda A(\mathbf{p},\Omega) = 0 \qquad [13.15]$$

or

$$A(\mathbf{p},\Omega) = \lambda^{-1} \qquad [13.16]$$

Figure 13.2 illustrates the two cases for λ positive and negative. The repulsive potential produces no new roots—the poles change smoothly from those of the noninteracting (i.e., $\lambda^{-1} \rightarrow \infty$) to the interacting one

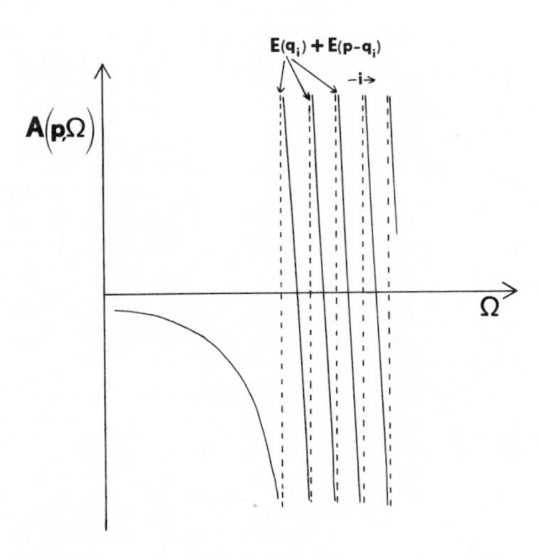

FIGURE 13.1. Energy structure of $A(\mathbf{p},\Omega)$ with the integration variable $\mathbf{q}$ considered as a discrete variable $\mathbf{q}_i$.

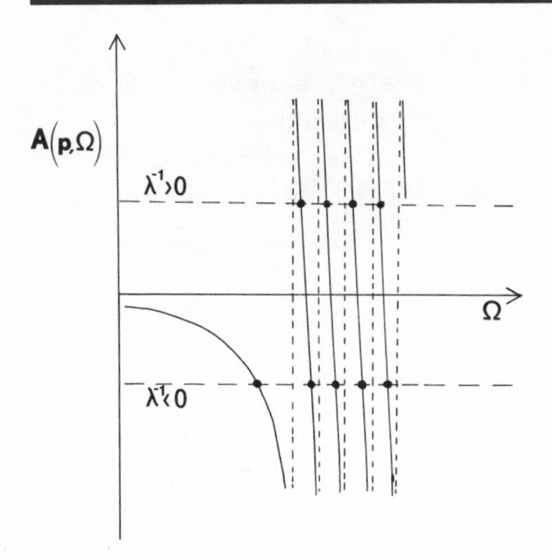

FIGURE 13.2. Graphical solutions for the poles of the interaction $I(\mathbf{p},\Omega)$.

and in addition are trapped close to the unperturbed value. For the attractive interaction, however, a new pole appears below the normal spectrum of states and furthermore does not disappear in the limit of $\lambda \to 0$; i.e., no *perturbative* approach can give this extra root.

The introduction of an attractive interaction, no matter how weak, gives, then, a bound state consisting of a pair of electrons. For a *given* strength of interaction λ we see that the maximum binding energy is obtained for the largest possible value of $A(\mathbf{p},\Omega)$. For a simple spherical Fermi surface this is obtained for $\mathbf{p} = 0$ as illustrated in Fig. 13.3. This makes physical sense, since it corresponds to the center of mass momentum being zero. The most energetically favorable bound state thus corresponds to pairing of electrons on the Fermi surface with equal and opposite momentum. We have neglected spin in this discussion, but it is fairly apparent that since particles of the same spin tend to "avoid" each other, the attractive interaction will be effectively weaker for these particles. Thus, paired particles of equal and opposite momentum and spin are most energetically favored.

More important than the simple consideration of the pairing of electrons into bound states is the consequence that the Fermi surface itself has become unstable. The Fermi surface is the boundary between empty and filled single-particle states. Once one can couple the particles and produce a bound pair with a lower net energy, the system in which the particles occupy single-particle levels is no longer the one of lowest energy and a transition to a new equilibrium state must take place.

The pairing process described above is, however, too simple to apply

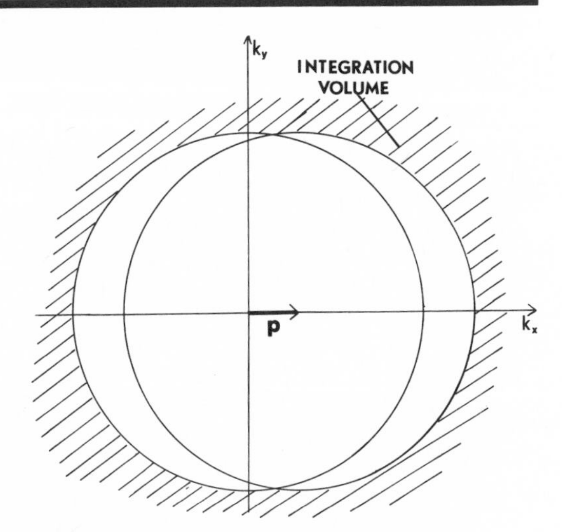

FIGURE 13.3. Integration volume for $A(\mathbf{p},\Omega)$ [compare with Fig. 9.1 for $\epsilon(\mathbf{q},\omega)$].

to obtain either the properties of the transition or the form of the new equilibrium state, which will necessarily involve large numbers of paired particles. The important point about this analysis is that it gives very clearly the new parameter that must be included in the initial approximation in order to develop continuously to the correct solution. That is, any attempted solution must include a parameter allowing for the presence of a macroscopic number of paired electron states. We have two ways of proceeding from this point, either by the canonical transformation method or mean-field theory. In either case we must be careful to include the pairing possibility from the beginning.

13.2. CANONICAL TRANSFORMATIONS

If perturbative approaches fail to solve a problem, the use of the canonical transformation is always a possible alternative. We will treat a model system in which there is an attractive interaction between electrons. To make the algebra simpler we will consider only the interaction between particles of equal and opposite momentum and spin.

These pairs are expected to be the most important in any transition. Since they are most energetically favorable, this pairing is most likely to appear first and so dominate the subsequent development of the new state. Because the Fermi level is not necessarily constant, it is better to work with the *grand canonical hamiltonian* $\hat{K}$, so that we write

$$\hat{K} = \sum_{\mathbf{k},\lambda} \hat{a}^{+}_{\mathbf{k}\lambda}\hat{a}_{\mathbf{k}\lambda}(\epsilon(\mathbf{k}) - \mu) + \hat{H}_{\text{INT}} \quad [13.17]$$

where k, λ are momentum and spin indices and

$$\hat{H}_{\text{INT}} = -\frac{1}{2}\sum_{\substack{\mathbf{k},\mathbf{k}_1\\ \lambda}} \langle \mathbf{k},-\mathbf{k}|V|\mathbf{k}_1,-\mathbf{k}_1\rangle \hat{a}^{+}_{-\mathbf{k}_1,-\lambda}\hat{a}^{+}_{\mathbf{k}_1,\lambda}\hat{a}_{\mathbf{k}\lambda}\hat{a}_{-\mathbf{k}-\lambda} \quad [13.18]$$

The creation and annihilation operators correspond to a single-particle representation, but we expect special properties in the new state—the pairing of the states of opposite momentum—which must be included explicitly in the transformation. Thus, for the canonical transformation, we try

$$\hat{\alpha}_{\mathbf{k}} = u_{\mathbf{k}}\hat{a}_{\mathbf{k}\lambda} - v_{\mathbf{k}}\hat{a}^{+}_{-\mathbf{k},-\lambda} \quad [13.19a]$$

$$\hat{\beta}^{+}_{-\mathbf{k}} = u_{\mathbf{k}}\hat{a}^{+}_{-\mathbf{k},-\lambda} + v_{\mathbf{k}}\hat{a}_{\mathbf{k},\lambda} \quad [13.19b]$$

where $v_{\mathbf{k}}$, $u_{\mathbf{k}}$ are real coefficients to be determined. The anticommutator relationships of the canonical transformation

$$\{\hat{\alpha}_{\mathbf{k}},\hat{\alpha}_{\mathbf{k}'}\} = \{\hat{\beta}_{\mathbf{k}},\hat{\beta}^{+}_{\mathbf{k}'}\} = \delta_{\mathbf{k},\mathbf{k}'} \quad [13.20]$$

serves to give one relationship between the $u_{\mathbf{k}}$ and $v_{\mathbf{k}}$, i.e.,

$$u^2_{\mathbf{k}} + v^2_{\mathbf{k}} = 1 \quad [13.21]$$

but this still leaves one more relationship required to complete the transformation. Equations (13.19) serve to define what is known as the *Bogoliubov transformation.* In his original work Bogoliubov considered a general attractive interaction ($\hat{H}_{\text{INT}}$), but the simpler form we have taken includes all of the important physics.

Because of the importance of the pairing, we write

$$\sum_{\mathbf{k},\lambda} (\epsilon(k) - \mu)\hat{a}^{+}_{\mathbf{k},\lambda}\hat{a}_{\mathbf{k},\lambda} = \sum_{\mathbf{k}} (\epsilon(k) - \mu)(\hat{a}^{+}_{\mathbf{k},\lambda}\hat{a}_{\mathbf{k},\lambda} + \hat{a}^{+}_{-\mathbf{k},-\lambda}\hat{a}_{-\mathbf{k},-\lambda}) \quad [13.22]$$

where we have simply coupled together the opposite spin and momentum terms together and assumed that the energy $\epsilon(\mathbf{k})$ depends only upon the magnitude of the momentum and is independent of the spin. Equation (13.22) can now be rewritten in terms of the new creation and annihilation operators to give

$$\sum_{\mathbf{k}} (\epsilon(k) - \mu)(\hat{a}^{+}_{\mathbf{k}\lambda}\hat{a}_{\mathbf{k}\lambda} + \hat{a}^{+}_{-\mathbf{k},-\lambda}\hat{a}_{-\mathbf{k},-\lambda}) = \sum_{\mathbf{k}} (\epsilon(k) - \mu)[2v^2_{\mathbf{k}} + (u^2_{\mathbf{k}} - v^2_{\mathbf{k}})(\hat{\alpha}^{+}_{\mathbf{k}}\hat{\alpha}_{\mathbf{k}} + \hat{\beta}^{+}_{-\mathbf{k}}\hat{\beta}_{-\mathbf{k}}) + 2u_{\mathbf{k}}v_{\mathbf{k}}(\hat{\beta}_{-\mathbf{k}}\hat{\alpha}_{\mathbf{k}} + \hat{\alpha}^{+}_{\mathbf{k}}\hat{\beta}^{+}_{-\mathbf{k}})] \quad [13.23]$$

In this expression, the first term is simply a constant, the second is in the correct form for the canonical transformation, while the last will need to be removed in some way.

The interaction term is quartic in the creation and annihilation operators. If we substitute for the $\hat{\alpha}$ and $\hat{\beta}$ operators we obtain a series of terms

$$a^{+}_{-\mathbf{k}_1,-\lambda}a^{+}_{\mathbf{k},\lambda}\hat{a}_{\mathbf{k},\lambda}\hat{a}_{-\mathbf{k},-\lambda} \Rightarrow \text{(i)} + \text{(ii)a} + \text{(ii)b} + \text{(iii)} \quad [13.24]$$

where

(i) is a constant $u_{\mathbf{k}}v_{\mathbf{k}}u_{\mathbf{k}_1}v_{\mathbf{k}_1}$,

(ii) are quadratic terms

(a) $u_{\mathbf{k}}v_{\mathbf{k}}u_{\mathbf{k}_1}v_{\mathbf{k}_1}(\hat{\alpha}^{+}_{\mathbf{k}}\hat{\alpha}_{\mathbf{k}} + \hat{\beta}^{+}_{-\mathbf{k}}\hat{\beta}_{-\mathbf{k}} + \hat{\alpha}^{+}_{\mathbf{k}_1}\hat{\alpha}_{\mathbf{k}_1} + \hat{\beta}^{+}_{-\mathbf{k}_1}\hat{\beta}_{-\mathbf{k}_1})$

of the required form

(b) $u_{\mathbf{k}}v_{\mathbf{k}}(u^2_{\mathbf{k}_1}\hat{\beta}^{+}_{-\mathbf{k}_1}\hat{\alpha}^{+}_{\mathbf{k}_1} - v^2_{\mathbf{k}_1}\hat{\alpha}_{\mathbf{k}_1}\hat{\beta}_{-\mathbf{k}_1}) + u_{\mathbf{k}_1}v_{\mathbf{k}_1}(u^2_{\mathbf{k}}\alpha_{\mathbf{k}}\beta_{-\mathbf{k}} + v^2_{\mathbf{k}}\hat{\beta}^{+}_{-\mathbf{k}}\hat{\alpha}^{+}_{\mathbf{k}})$

which need to be removed, and

(iii) quartic terms. These are numerous *but if the transformation is to have any meaning* their net effect must be small and correspond to interactions between the new quasiparticles.

If we combine these with the terms in Eq. (13.23) after summing the spin variable and using the symmetries inherent in the interaction potential in a uniform medium

$$\langle \mathbf{k},-\mathbf{k}|V|\mathbf{k}_1,-\mathbf{k}_1\rangle = \langle \mathbf{k}_1,-\mathbf{k}_1|V|\mathbf{k},-\mathbf{k}\rangle \cdots \quad [13.25]$$

we obtain

$$\hat{H} = \sum_{k} (\hat{\alpha}^{+}_{\mathbf{k}}\hat{\alpha}_{\mathbf{k}} + \hat{\beta}^{+}_{-\mathbf{k}}\hat{\beta}_{-\mathbf{k}})[u^2_{\mathbf{k}} - v^2_{\mathbf{k}})(\epsilon(k) - \mu) + 2u_{\mathbf{k}}v_{\mathbf{k}}\,\Delta(\mathbf{k})] + \sum_{k} (\hat{\alpha}^{+}_{k}\hat{\beta}^{+}_{-k} + \hat{\beta}_{-k}\hat{\alpha}_{k})[(\epsilon(k) - \mu)2u_{\mathbf{k}}v_{\mathbf{k}} - (u^2_{\mathbf{k}} - v^2_{\mathbf{k}})\,\Delta(\mathbf{k})] + \sum_{k} [2v^2_{\mathbf{k}}(\epsilon(k) - \mu) + u_{\mathbf{k}}v_{\mathbf{k}}\,\Delta(\mathbf{k})] + \text{(quartic terms)} \quad [13.26]$$

where

$$\Delta(\mathbf{k}) = -\sum_{\mathbf{k}_1} \langle \mathbf{k},-\mathbf{k}|V|\mathbf{k}_1 - \mathbf{k}_1\rangle u_{\mathbf{k}_1}v_{\mathbf{k}_1} \tag{13.27}$$

Of these constituents, the first is of the correct form, the third is simply a constant and need not be considered further, while the second term must be zero in order for the transformation to work. We must therefore have

$$2(\epsilon(k) - \mu)u_{\mathbf{k}}v_{\mathbf{k}} = (u_{\mathbf{k}}^2 - v_{\mathbf{k}}^2)\,\Delta(\mathbf{k}) \tag{13.28}$$

for all **k.** From Eq. (13.21) we already have that

$$u_{\mathbf{k}}^2 + v_{\mathbf{k}}^2 = 1 \tag{13.21}$$

It is normal to use this to define a function $\chi_{\mathbf{k}}$ such that

$$u_{\mathbf{k}} = \sin \chi_{\mathbf{k}} \tag{13.29a}$$

$$v_{\mathbf{k}} = \cos \chi_{\mathbf{k}} \tag{13.29b}$$

automatically satisfying Eq. (13.21). Then we have from Eq. (13.28)

$$\tan 2\chi_{\mathbf{k}} = \frac{\Delta(\mathbf{k})}{(\epsilon(k) - \mu)} \tag{13.30}$$

$$\sin 2\chi_{\mathbf{k}} = \frac{\Delta(\mathbf{k})}{[(\epsilon(k) - \mu)^2 + \Delta^2(\mathbf{k})]^{1/2}} \tag{13.31}$$

$$\cos 2\chi_{\mathbf{k}} = \frac{(\epsilon(k) - \mu)}{[(\epsilon(k) - \mu)^2 + \Delta^2(\mathbf{k})]^{1/2}} \tag{13.32}$$

If we define

$$E(\mathbf{k}) = [\Delta^2(\mathbf{k}) + (\epsilon(k) - \mu)^2]^{1/2} \tag{13.33}$$

Eq. (13.26) now becomes

$$\hat{K} = \sum_{\mathbf{k}} E(\mathbf{k})(\alpha_{\mathbf{k}}^{+}\alpha_{\mathbf{k}} + \beta_{-\mathbf{k}}^{+}\beta_{-\mathbf{k}}) + (\text{consts}) + (\text{quadratic terms}) \tag{13.34}$$

$E(\mathbf{k})$ is the new eigenvalue spectrum for the $\alpha_{\mathbf{k}}$ and $\beta_{-\mathbf{k}}$ quasiparticles. From Eq. (13.33), we see that there is a minimum value for that energy so that

$$E(\mathbf{k}) \geq \Delta(\mathbf{k}) \tag{13.35}$$

The term $\Delta(\mathbf{k})$ is the minimum energy required to produce a quasiparticle and, as such, plays a crucial role in the properties of the new state. From the definition of $\Delta(\mathbf{k})$ we can develop a self-consistent integral equation. Using Eqs. (13.27) and (13.29)

$$\Delta(\mathbf{k}) = -\frac{1}{2}\sum_{\mathbf{k}1} \langle \mathbf{k} - \mathbf{k}|V|\mathbf{k}_1, -\mathbf{k}_1\rangle \sin\frac{1}{2}\chi_{\mathbf{k}1} \tag{13.36}$$

Then from (13.31) and (13.33)

$$\Delta(\mathbf{k}) = -\frac{1}{2}\sum_{\mathbf{k}1} \langle \mathbf{k}, -\mathbf{k}|V|\mathbf{k}_1 - \mathbf{k}_1\rangle \frac{\Delta(\mathbf{k}_1)}{E(\mathbf{k}_1)} \tag{13.37}$$

This is the gap equation, and the solution for a given interaction will largely determine the properties of the new paired system. Taking the simplest, physically realistic form for the interaction we can write

$$\langle k, -k|V|k_1, -k_1\rangle = -\frac{g}{\Omega}\,\theta[\omega_0 - (\epsilon(k) - \mu)]\theta[\omega_0 + (\epsilon(k_1) - \mu)] \tag{13.38}$$

where g is the strength of the interaction, Ω the volume of the crystal, and the two θ functions restrict the interaction to electron states within ω_0 of the Fermi surface. This mimics the electron–phonon interaction of Sec. 11.3. Assuming that Δ is not pathological, we write

$$\Delta(\mathbf{k}) \approx \Delta(\mathbf{k}_F) = \Delta_0 \tag{13.39}$$

and then

$$\Delta_0 = -\frac{1}{2}\frac{g}{\Omega}\sum_{\mathbf{k}} \frac{\Delta_0}{[\Delta_0^2 + (\epsilon(k) - \mu)^2]^{1/2}} \tag{13.40}$$

where the integral over k is restricted to the energy range of the interaction. This gives

$$\Delta_0 = 2\omega_0 e^{-1/N(\mu)g} \qquad [13.41]$$

where $N(\mu)$ is the density of states at the Fermi level. The physics of the system is now described by a vacuum state $|0\}$ plus the creation and annihilation operators $\hat{\alpha}^+, \hat{\alpha}, \hat{\beta}^+, \hat{\beta}$ so that any state of the system can be described in the form

$$\Psi = \prod_{\mathbf{k}} [\alpha_{\mathbf{k}}^+)^{A_k} + (\beta_{-\mathbf{k}}^+)^{B_k}]|0\} \qquad [13.42]$$

where A_k, B_k are the occupation numbers of the quasiparticles.

Any experiment performed on such a system would include within the original Hamiltonian [Eq. (13.17)] a further interaction term to describe the probe used. Such a Hamiltonian could be very complicated, but generally used probes, such as light, sound, fast electrons, may be described in terms of an effective single-particle term of the form (describing the scattering produced)

$$\delta\hat{H}_{\text{exp}} = \sum_{\substack{\mathbf{k},\mathbf{k}' \\ \lambda,\lambda}} V_{\lambda,\lambda'}(\mathbf{k},\mathbf{k}')\hat{a}_{\mathbf{k},\lambda}^+\hat{a}_{\mathbf{k}',\lambda'} \qquad [13.43]$$

Since the true quasiparticles are the paired states, Eq. (13.43) is transformed by substitution from Eq. (13.19) to the form

$$\delta\hat{H}_{\text{exp}} = \sum_{\mathbf{k},\mathbf{k}'} [Q(\mathbf{k},\mathbf{k}')\hat{\alpha}_{\mathbf{k}}^+\hat{\alpha}_{\mathbf{k}'} + R(\mathbf{k},\mathbf{k}')\hat{\beta}_{-\mathbf{k}}^+\hat{\beta}_{-\mathbf{k}'} + S(\mathbf{k},\mathbf{k}')\hat{\alpha}_{\mathbf{k}}^+\hat{\beta}_{-\mathbf{k}'}^+ + T(\mathbf{k},\mathbf{k}')\hat{\alpha}_{\mathbf{k}}\hat{\beta}_{-\mathbf{k}'}] \qquad [13.44]$$

The *transformed* matrix elements $Q(\mathbf{k},\mathbf{k}') \cdots T(\mathbf{k},\mathbf{k}')$ determine the physical effect of the probe. Consider, for instance, the spin invariant case

$$\delta\hat{H}_{\text{exp}} = \sum_{\substack{\mathbf{k},\mathbf{k}' \\ \lambda}} V(\mathbf{k},\mathbf{k}')\hat{a}_{\mathbf{k}\lambda}^+\hat{a}_{\mathbf{k}'\lambda} \qquad [13.45]$$

Substituting for $\hat{a}^+_{\mathbf{k}\lambda}\hat{a}_{\mathbf{k}'\lambda'}$ we have

$$a^+_{\mathbf{k}\lambda}\hat{a}_{\mathbf{k}'\lambda} = u_{\mathbf{k}}u_{\mathbf{k}'}\hat{\alpha}^+_{\mathbf{k}}\hat{\alpha}_{\mathbf{k}'} + v_{\mathbf{k}}v_{\mathbf{k}'}\hat{\beta}_{-\mathbf{k}}\hat{\beta}^+_{-\mathbf{k}'} + u_{\mathbf{k}}v_{\mathbf{k}'}\hat{\alpha}^+_{\mathbf{k}}\hat{\beta}^+_{-\mathbf{k}'} + v_{\mathbf{k}}u_{\mathbf{k}'}\hat{\beta}_{-\mathbf{k}}\hat{\alpha}_{\mathbf{k}'} \quad [13.46]$$

But similar terms are obtained from the pair

$$\hat{a}^+_{-\mathbf{k}'-\lambda}\hat{a}_{-\mathbf{k}-\lambda} = v_{\mathbf{k}}v_{\mathbf{k}'}\hat{\alpha}_{\mathbf{k}'}\hat{\alpha}^+_{\mathbf{k}} + u_{\mathbf{k}}u_{\mathbf{k}'}\hat{\beta}^+_{-\mathbf{k}'}\hat{\beta}_{-\mathbf{k}} - v_{\mathbf{k}'}u_{\mathbf{k}}\hat{\alpha}_{\mathbf{k}'}\hat{\beta}_{-\mathbf{k}} - u_{\mathbf{k}'}v_{\mathbf{k}}\hat{\beta}^+_{-\mathbf{k}'}\hat{\alpha}^+_{-\mathbf{k}} \quad [13.47]$$

Because of the anticommutator relationships, these two terms may be added in the Hamiltonian to give

$$\delta\hat{H}_{\text{exp}} = \frac{1}{2}\sum_{\mathbf{k},\mathbf{k}'}[u_{\mathbf{k}}u_{\mathbf{k}'}V(\mathbf{k},\mathbf{k}') - v_{\mathbf{k}}v_{\mathbf{k}'}V(\mathbf{k}',\mathbf{k})](\hat{\alpha}^+_{\mathbf{k}}\hat{\alpha}_{\mathbf{k}'} + \hat{\beta}^+_{-\mathbf{k}}\hat{\beta}_{-\mathbf{k}'}) + [u_{\mathbf{k}}v_{\mathbf{k}'}V(\mathbf{k},\mathbf{k}') + u_{\mathbf{k}'}v_{\mathbf{k}}V(\mathbf{k}',\mathbf{k})](\hat{\alpha}^+_{\mathbf{k}}\hat{\beta}^+_{-\mathbf{k}'} - \hat{\alpha}_{\mathbf{k}'}\hat{\beta}_{-\mathbf{k}}) \quad [13.48]$$

The experimental results for transitions may be calculated from the Fermi golden rule and will depend upon the *modulus* of the matrix elements between the initial and final states of the system and the density of final states. For a *normal* system, with an interaction term δH_{exp}, this means a typical term will be $|V(\mathbf{k},\mathbf{k}')|^2$. For the superconducting state, however, the equivalent term is $|u_{\mathbf{k}}u_{\mathbf{k}'}V(\mathbf{k},\mathbf{k}') + v_k v_{k'}V(\mathbf{k}',\mathbf{k})|^2$. The factors $u_{\mathbf{k}}$, $v_{\mathbf{k}}$ which couple the paired states produce a change in the matrix elements and hence a *change in the experimental result*. It is this "coherent" change in the matrix elements which makes the pairing description of the superconducting state so important.

In addition to the matrix elements the density of states is also changed, of course, being zero within Δ of the ground-state energy. The range of applications of the pairing theory is very large and is the subject of a comprehensive literature, so we will not attempt to cover it here (see, for instance, Schrieffer's book in the bibliography).

13.3. PROPAGATOR APPROACH

The canonical transformation and the concept of pairing, useful though it is, does not really have the generality and flexibility to cope rigorously with temperature effects or the many and varied phenomena encountered in the superconducting phase. In order to treat this system on a rational basis, the equivalent of the Green's function approach of the normal system is required.

As in the previous section, the Hamiltonian is divided into two parts:

$$H = H_0 + H_{\text{INT}} \qquad [13.49]$$

H_0 is the "normal" Hamiltonian in the sense that it produces an interacting Green's function that may be developed perturbatively from the noninteracting system by some well-behaved series expansion. We assume that H_0 presents a soluble problem in principle. The term H_{INT} is that part of the interaction which produces the nonanalyticity and so needs to be treated explicitly. We are concerned here with a pairing interaction, and the solution will obviously depend upon the exact nature of that interaction. Following Bardeen, Cooper, and Schrieffer we use the simplest interaction, an attractive contact potential [cf. Eq. (13.4)] and hope that this contains the basic physics. In terms of the field operators this gives

$$H_{\text{INT}} = -\tfrac{1}{2} g \int d\mathbf{r}\; \hat{\psi}_{\lambda}^{+}(\mathbf{r}) \hat{\psi}_{\lambda'}^{+}(\mathbf{r}) \hat{\psi}_{\lambda'}(\mathbf{r}) \hat{\psi}_{\lambda}(\mathbf{r}) \qquad [13.50]$$

where g is a measure of the strength of the interaction. Even with this interaction, the appearance of pairing is not obvious but it must be included explicitly in the form of the solution we attempt.

Consider the interaction term in the equation of motion for the normal Green's function. From Eq. (6.57) this is given by

$$i \int v(x_1 - x_3) \langle N | T[\hat{\psi}^{+}(x_3,t_1) \hat{\psi}(x_3,t_1) \hat{\psi}(x_1,t_1) \hat{\psi}^{+}(x_2,t_2)] | N \rangle \, dx_3$$

and contains a two-particle Green's function. With this interaction it was possible to develop the Hartree and Hartree–Fock solutions by approximating the two-particle Green's function by two noninteracting Green's functions (Sec. 6.7). That is, we wrote

$$G(1,2,3,3^{+}) = G(1,2)G(3,3^{+}) + G(1,3^{+})G(3,2) \qquad [13.51]$$

corresponding to the diagrammatic expansion

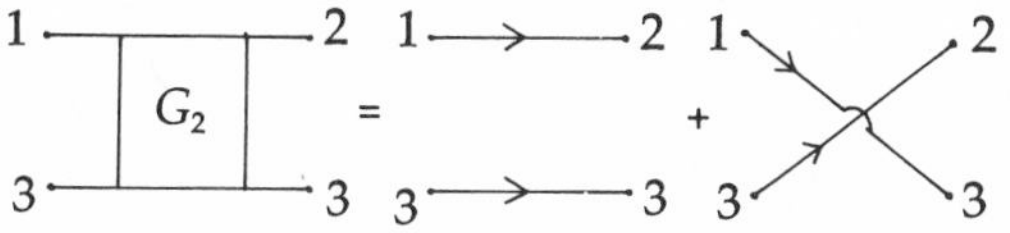

1 —[G_2]— 2 = 1 ⟶ 2 + 1 ↘ ↗ 2

3 —[G_2]— 3 = 3 ⟶ 3 + 3 ↗ ↘ 3 [13.52]

In effect we were taking out what we considered to be the most important terms in the motion of two particles. In terms of the interaction part of the Hamiltonian, this is equivalent to the simplification

$$
\begin{aligned}
&\int v(x_1,x_3)\hat{\psi}^+(x_1,t_1)\hat{\psi}^+(x_3,t_1)\hat{\psi}(x_3,t_1)\hat{\psi}(x_1,t_1)\;dx_1\;dx_3\\
&\Rightarrow \int v(x_1,x_3)\langle N|\hat{\psi}^+(x_3,t_1)\hat{\psi}(x_3,t_1)|N\rangle\hat{\psi}^+(x_1,t_1)\hat{\psi}(x_1,t_1)\;dx_1\;dx_3\\
&+ \int v(x_1,x_3)\langle N|\hat{\psi}^+(x_3,t_1)\hat{\psi}(x,t_1)|N\rangle\hat{\psi}^+(x_1,t_1)\hat{\psi}(x_3,t_1)\;dx_1\;dx_3\\
&+ (\text{neglected terms})
\end{aligned}
$$

The replacement of the field operators by their expectation values corresponds to averaging the interaction to produce a self-consistent, if approximate, one-particle potential.

In the present system we are required to explicitly allow the pairing of particles. Therefore, it seems best to write

$$G_2 = G_2^0 + (\text{pairing amplitude terms}) \qquad [13.53]$$

where G_2^0 is the best normal Green's function and the pairing amplitude terms will take the form

$$\langle\hat{\psi}_\downarrow^+(\mathbf{r})\hat{\psi}_\uparrow^+(\mathbf{r})\rangle$$

for the pairing of electrons, and

$$\langle\hat{\psi}_\uparrow(\mathbf{r})\hat{\psi}_\downarrow(\mathbf{r})\rangle$$

for holes. Thus, we split the interaction term into two parts.

$$
\begin{aligned}
H_{\text{INT}} = -g\int d^3r\,[&\langle\hat{\psi}_\downarrow^+(\mathbf{r})\hat{\psi}_\uparrow^+(\mathbf{r})\rangle\hat{\psi}_\uparrow(\mathbf{r})\hat{\psi}_\downarrow(\mathbf{r})\\
&+\hat{\psi}_\downarrow^+(\mathbf{r})\hat{\psi}_\uparrow^+(\mathbf{r})\langle\hat{\psi}_\uparrow(\mathbf{r})\hat{\psi}_\downarrow(\mathbf{r})\rangle] + (\text{“ordinary” terms}) \qquad [13.54]
\end{aligned}
$$

The $\langle\ \rangle$ brackets indicate a thermal average as in Eq. (10.45). The interaction is now a consistent potential ($\approx g\langle\psi_\downarrow\psi_\uparrow\rangle$) produced by the paired states plus other terms amenable to the normal analysis (we will incorporate these in H_0 from now on). This separation is only useful if there is a large density of such pairs; in a normal system such a separation would only give an infinitesimal contribution to the interaction. We have now as our Hamiltonian

$$H = H_0 - g \int d^3r\, [\langle\hat{\psi}_\downarrow^+(\mathbf{r})\hat{\psi}_\uparrow^+(\mathbf{r})\rangle\hat{\psi}_\uparrow(\mathbf{r})\hat{\psi}_\downarrow(\mathbf{r}) + \hat{\psi}_\downarrow^+(\mathbf{r})\hat{\psi}_\uparrow^+(\mathbf{r})\langle\hat{\psi}_\uparrow(\mathbf{r})\hat{\psi}_\downarrow(\mathbf{r})\rangle] \tag{13.55}$$

We can now use the thermal Heisenberg equation of motion for the field operators

$$\hbar \frac{\partial}{\partial\tau}\hat{\psi}_k = [\hat{K},\hat{\psi}_k] \tag{13.56}$$

to obtain the equation of motion for the Green's function. Because of the particular form of the interaction, however, there are a number of equations which may be obtained. From (13.56) we have

$$\left(-\hbar\frac{\partial}{\partial\tau} - K_0(\mathbf{r}_1)\right)\hat{\psi}_\uparrow(\mathbf{r}_1,\tau_1) + g\langle\hat{\psi}_\downarrow(\mathbf{r}_1,\tau_1)\hat{\psi}_\uparrow(\mathbf{r}_1,\tau_1)\rangle\hat{\psi}_\downarrow^+(\mathbf{r}_1,\tau_1) = 0 \tag{13.57}$$

We may also write

$$\left(-\hbar\frac{\partial}{\partial\tau_1} + K_0(\mathbf{r}_1)\right)\psi_\downarrow^+(\mathbf{r}_1,\tau_1) + g\langle\psi_\downarrow^+(\mathbf{r}_1,\tau_1)\psi_\uparrow^+(\mathbf{r}_1,\tau_1)\rangle\psi_\uparrow(\mathbf{r}_1,\tau_1) = 0 \tag{13.58}$$

In terms of Green's functions, Eq. (13.57) gives

$$\left(-\hbar\frac{\partial}{\partial\tau_1} + K_0(\mathbf{r}_1)\right)\mathcal{G}_{\uparrow\uparrow}(\mathbf{r}_1,\tau_1,\mathbf{r}_2,\tau_2) + g\langle\hat{\psi}_\downarrow(\mathbf{r}_1,\tau_1)\hat{\psi}_\uparrow(\mathbf{r}_1,\tau_1)\rangle\langle T[\hat{\psi}_\downarrow^+(\mathbf{r}_1,\tau_1)\hat{\psi}_\uparrow^+(\mathbf{r}_2,\tau_2)]\rangle = \hbar\delta(\mathbf{r}_1-\mathbf{r}_2)\delta(\tau_1-\tau_2) \tag{13.59}$$

In this equation the term $\langle T[\hat{\psi}_{\downarrow}^{+}(\mathbf{r}_1,\tau_1)\hat{\psi}_{\uparrow}(\mathbf{r}_2,\tau_2)]\rangle$ is obviously a Green's function but of a particular kind. The presence of two creation field operators means that in a normal system it must automatically vanish, since it consists of terms like

$$\langle N|a^{+}a^{+}|N\rangle = \langle N|N+2\rangle = 0$$

In the grand canonical ensemble there is no such requirement of course. What it really describes is the effect of the pairing upon the one-electron states and is the anomalous Green's function in the sense of Sec. 10. We write

$$F^{+}(\mathbf{r}_1,\tau_1,\mathbf{r}_2,\tau_2) = -\langle T[\psi_{\downarrow}^{+}(\mathbf{r}_1,\tau_1)\hat{\psi}_{\uparrow}^{+}(\mathbf{r}_2,\tau_2)]\rangle \qquad [13.60]$$

and

$$F(\mathbf{r}_1,\tau_1,\mathbf{r}_2,\tau_2) = -\langle T[\hat{\psi}_{\uparrow}(\mathbf{r}_1,\tau_1)\hat{\psi}_{\downarrow}(\mathbf{r}_2,\tau_2)]\rangle \qquad [13.61]$$

which appears in the equation of motion

$$\left[-\hbar\frac{\partial}{\partial\tau_1} + K_0(\mathbf{r}_1)\right]\mathcal{G}_{\downarrow\downarrow}(\mathbf{r}_2,\tau_2,\mathbf{r}_1,\tau_1)$$
$$+ g\langle\hat{\psi}_{\downarrow}^{+}(\mathbf{r}_1,\tau_1)\hat{\psi}_{\uparrow}^{+}(\mathbf{r}_2,\tau_2)\rangle\langle T[\hat{\psi}_{\uparrow}(\mathbf{r}_1,\tau_1)\hat{\psi}_{\downarrow}(\mathbf{r}_2,\tau_2)]\rangle = \hbar\delta(\mathbf{r}_1-\mathbf{r}_2)\delta(\tau_1-\tau_2)$$

[13.62]

In just the same way that the Hartree potential could be written in terms of the Green's function [Eq. (6.64), we can define a mean pairing potential

$$\begin{aligned}\Delta(\mathbf{r}_1,\tau_1) &= -g\langle\hat{\psi}_{\uparrow}(\mathbf{r}_1,\tau_1)\hat{\psi}_{\downarrow}(\mathbf{r}_1,\tau_1)\rangle \\ &= g\langle\hat{\psi}_{\downarrow}(\mathbf{r}_1,\tau_1)\hat{\psi}_{\uparrow}(\mathbf{r}_1\tau_1)\rangle \\ &= gF(\mathbf{r}_1,\tau_1,\mathbf{r}_1,\tau_1^{+})\end{aligned} \qquad [13.63]$$

which can be thought of loosely as $g \times$ (density of pairs). There are two other equations to be derived from Eqs. (13.57) and (13.58):

$$\left(-\hbar\frac{\partial}{\partial\tau_1} - K_0(\mathbf{r}_1)\right)F(\mathbf{r}_1\tau_1,\mathbf{r}_2\tau_2) = \Delta(\mathbf{r}_1,\tau_1)\mathcal{G}_{\downarrow\downarrow}(\mathbf{r}_2\tau_2,\mathbf{r}_1,\tau_1) \qquad [13.64]$$

$$\left(-\hbar \frac{\partial}{\partial \tau_1} + K_0(\mathbf{r}_1)\right) F^+(\mathbf{r}_1\tau_1\mathbf{r}_2\tau_2) = \Delta^+(\mathbf{r}_1,\tau_1)\mathcal{G}_{\uparrow\uparrow}(\mathbf{r}_1,\tau_1,\mathbf{r}_2,\tau_2) \quad [13.65]$$

The four equations (13.59), (13.62), (13.64), and (13.65) serve to define a solution to the problem. They can, however, be put in a more concise form by defining a matrix formulation. Thus we write

$$\mathsf{G} = \begin{bmatrix} \mathcal{G}_{\uparrow\uparrow}(\mathbf{r}_1,\tau_1,\mathbf{r}_2,\tau_2) & F(\mathbf{r}_1,\tau_1,\mathbf{r}_2,\tau_2) \\ F(\mathbf{r}_1,\tau_1,\mathbf{r}_2,\tau_2) & -\mathcal{G}_{\downarrow\downarrow}(\mathbf{r}_2,\tau_2,\mathbf{r}_1,\tau_1) \end{bmatrix} \quad [13.66]$$

and

$$\mathsf{K} = \begin{bmatrix} -\hbar \dfrac{\partial}{\partial \tau_1} - K_0(\mathbf{r}_1,\tau_1) & \Delta(\mathbf{r}_1,\tau_1) \\ -\Delta^*(\mathbf{r}_1,\tau_1) & -\hbar \dfrac{\partial}{\partial \tau_1} + K_0(\mathbf{r}_1,\tau_1) \end{bmatrix} \quad [13.67]$$

so that the equations can be written in the formal form

$$\mathsf{KG} = \hbar \mathsf{1} \quad [13.68]$$

This is the *Nambu formulation of superconductivity.*

Suppose we consider the application to the simplest situation, where the normal Hamiltonian is both time independent and corresponds to a uniform medium. Thus

$$\mathcal{G}(\mathbf{r}_1,\tau_1,\mathbf{r}_2,\tau_2) \rightarrow \mathcal{G}(\mathbf{r}_1 - \mathbf{r}_2,\tau_1 - \tau_2) \rightarrow \mathrm{G}(\mathbf{k},\omega_n)$$

and similarly for the "anomalous" Green's function F. The pair potential Δ becomes a constant, and from Eqs. (13.59) and (13.65) we have

$$[i\omega_n - (\epsilon(k) - \mu)]\mathcal{G}(\mathbf{k},\omega_n) + \Delta F^+(\mathbf{k},\omega_n) = 1 \quad [13.69]$$

$$[-i\omega_n - (\epsilon(k) - \mu)]F^+(\mathbf{k},\omega_n) - \Delta^+\mathcal{G}(\mathbf{k},\omega_n) = 0 \quad [13.70]$$

where $(\epsilon(k) - \mu)$ is the eigenvalue for the normal system. These can be simply solved to give

$$\mathcal{G}(\mathbf{k},\omega_n) = \frac{-[i\hbar\omega_n + (\epsilon(k) - \mu)]}{[\omega_n^2 + (\epsilon(k) - \mu)^2 + |\Delta|^2]} \quad [13.71]$$

and

$$F^{+}(\mathbf{k},\omega_n) = \frac{\Delta^*}{\omega_n^2 + (\epsilon(k) - \mu)^2 + |\Delta|^2} \tag{13.72}$$

If we put

$$E^2(k) = (\epsilon(k) - \mu)^2 + |\Delta|^2 \tag{13.73}$$

and

$$v_k^2 = \frac{1}{2}\left(1 - \frac{(\epsilon(k) - \mu)}{E(k)}\right) \tag{13.74}$$

we can write these in the form

$$\mathcal{G}(\mathbf{k},\omega_n) = \frac{u_k^2}{i\omega_n - E(k)} + \frac{v_k^2}{i\omega_n + E(k)} \tag{13.75}$$

$$F(\mathbf{k},\omega_n) = u_k v_k \left[\frac{1}{i\omega_n - E(k)} - \frac{1}{i\omega_n + E(k)}\right] \tag{13.76}$$

The poles of the Green's functions now relate directly to the pairing hypothesis with excitation energies appropriate to the new quasiparticles. The resulting equations in both formulations give essentially the same information in the zero-temperature limit but now the extension to more complicated systems and finite temperatures is more straightforward. The gap equation (now temperature dependent) comes from Eq. (13.63). For the uniform material we have

$$\begin{aligned}\Delta &= gF(\mathbf{r}_1 - \mathbf{r}_1,\tau^+ - \tau_1)\\ &= gF(0,0^+)\\ &= \frac{g}{(2\pi)^3\beta}\sum_n \int \frac{\Delta}{\omega_n^2 + (\epsilon(\mathbf{k}) - \mu)^2 + \Delta^2}\, d\mathbf{k}\end{aligned} \tag{13.77}$$

From which we have

$$1 = \frac{g}{(2\pi)^3\beta}\sum_n \int \frac{d\mathbf{k}}{\omega_n^2 + (\epsilon(k) - \mu)^2 + \Delta^2} \tag{13.78}$$

By evaluating the energy sum this can be rewritten as

$$1 = \frac{g}{(2\pi)^3} \int d\mathbf{k} \, \frac{\tanh\{[(\epsilon(k) - \mu)^2 + \Delta^2]^{1/2}\beta/2\}}{2[(\epsilon(k) - \mu)^2 + \Delta^2]^{1/2}} \qquad [13.79]$$

This integral equation for Δ can be solved in a number of limits.

The approximation for the interaction given in Eq. (13.38) is normally used to determine the integral over $\mathbf{k}$ and introduce a *physical* limitation of the scattering to the region of the Fermi surface. This gives the integral equation

$$1 \approx gN(\mu) \int_{-\omega_0}^{+\omega_0} d\mathscr{E} \, \frac{\tanh[(\mathscr{E}^2 + \Delta^2)^{1/2}\beta/2]}{[\mathscr{E}^2 + \Delta^2)^{1/2}} \qquad [13.80]$$

The numerical solution to this equation is shown in Fig. 13.4. The gap vanishes for temperatures greater than a critical value (T_c). The two limits of low T ($\ll T_c$) and high T ($\sim T_c$) temperature can be treated analytically, the former reducing to Eq. (13.41), while the latter may be obtained from an expansion of the integral in powers of the (now-small) quantity Δ to obtain

$$\Delta \approx 3.06 k_B T_c (1 - T/T_c)^{1/2} \qquad [13.81]$$

where

$$T_c = 1.13 \, \frac{\omega_0}{k_B} \, e^{-1/N(\mu)g} \qquad [13.82]$$

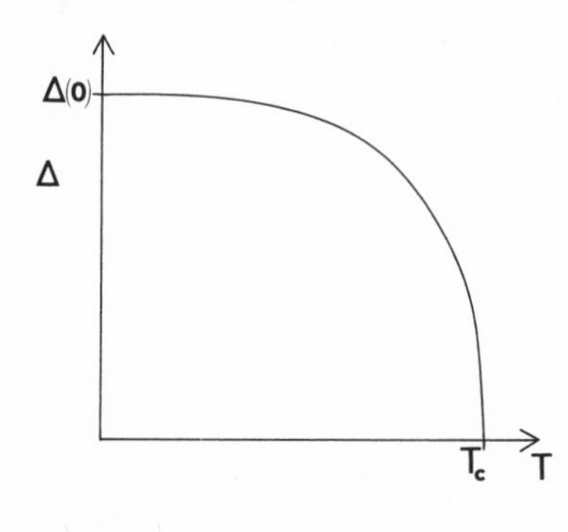

FIGURE 13.4. Schematic representation of the variation of the superconductor gap as a function of temperature.

The appearance of the gap and the phase transition from a normal to a superconducting state at the critical temperature is the major result of the theory and is contained in the initial reduction of the interaction into a self-consistent pairing field. The feedback mechanism implicit in the equations of motion, in which the pairing potential (Δ) is itself dependent upon the density of pairs present (through F^{+}), serves to accentuate the temperature variation of the gap, as can be seen in Fig. 13.4. As the temperature rises the attractive interaction completes with the disruptive thermal motion until the possibility of a macroscopic density of pairs disappears quite suddenly.

The applications of the Nambu formalism are widespread. Of particular importance is the inclusion of electromagnetic phenomena into the Hamiltonian. This is necessary to describe the very important physical properties of the condensed state but is outside the scope of this text.

BIBLIOGRAPHY

PARKS, R. D., Ed., *Superconductivity*, Marcel Dekker, New York, 1969.
SCHRIEFFER, J. R., *Theory of Superconductivity*, Benjamin, Menlo Park: CA, 1965.
TINKHAM, D., *Introduction to Superconductivity*, McGraw-Hill, New York, 1975.
BARDEEN, J., COOPER, L. N., and SCHRIEFFER, J. R., *Phys. Rev.*, **108,** 1175, 1957.
BOGOLIUBOV, N. N., *Sov. Phys. JETP*, **34,** 41, 1958.
COOPER, L. N., *Phys. Rev.*, **104,** 1189, 1956.
NAMBU, Y., *Phys. Rev.*, **117,** 648, 1960.

PROBLEMS

1. The interaction may be averaged over the Fermi surface in order to obtain an effective interaction for superconductivity purposes by using the average

$$\lambda(\omega) = \frac{1}{4\pi k_F^2} \int_{\text{Fermi surface}} W(q,\omega - E(q))\, dq$$

This corresponds to the interaction taking an electron from a state with energy ω to the Fermi surface. Show that for this average the interaction parameter is positive for all energies if the electronic response is used corresponding to

$$\epsilon(q,\omega) = 1 + \frac{1}{(q^2/\gamma^2) - \omega^2/\omega_p^2}$$

where ω_p is the plasmon energy ($\approx E_F$) and γ is the Thomas–Fermi wave vector ($\approx k_F$).

2. Derive Eqs. (13.23) and (13.26).

3. Show that the Hartree–Fock approximation is indeed equivalent to that in (13.53).

4. Derive Eqs. (13.57) and (13.58).

APPENDIX

LIST OF SYMBOLS

In the following table of symbols, the many transitory symbols that have no application beyond their immediate relevance to the derivation involved have not been included for the sake of clarity. Even so, many symbols have had to be used for more than one purpose. Where possible I have kept to the standard usage, but many of the symbols still need to be taken in the context of the equation. For example, Z is used for both the ionic charge number and the renormalization constant, but since the former always appears with the electronic charge e, the danger of confusion is small. As an indication the chapter number has been given if a symbol has a definition restricted to that chapter. Throughout, the following standard notation has been used to indicate the possible meanings of a symbol (0).

$\hat{O}$	An operator.
O	The matrix form of a function.
$\mathbf{O}$	A vector.
O^+	The Hermitian conjugate.
O^{-1}	The inverse.
$\dot{O}$	The time derivative.
$\langle O \rangle$	The expectation value.
O_{lm}	The matrix element $\langle l \vert \hat{O} \vert m \rangle$.

$\hat{A}$	Heisenberg to Schrödinger representation transformation operator.
$A(\mathbf{k},\omega)$	Spectral weight function.
$A(1,2)$	The two-particle propagator $[G(1,2)G(1,2)]$.
$A_n(x)$	A known function of x (Chap. 5).
$A_\lambda(t)$	Fourier coefficient of the electromagnetic potential for the wave vector $\mathbf{k}_\lambda$ and polarization λ.
$\mathbf{A}(\mathbf{r},t)$	Electromagnetic vector potential.
a	Interatomic separation of atoms on the linear chain.

a_l, b_l — lth components of the expansion of the Landau parameters in Legendre polynomials.

$\hat{a}_k^+, \hat{a}_k$ — Creation and annihilation (boson) operators for the normal mode k.

$\hat{a}_p^+(\tau), \hat{a}_p(\tau)$ — Creation and annihilation operators in the "thermal" Heisenberg representation.

$\hat{a}_0, \hat{a}_0^+$ — Annihilation and creation operators corresponding to the condensate state.

$a_{\mathbf{G},n}(\mathbf{k})$ — Coefficient of the quasiparticle state of wave vector $\mathbf{k}$, band index n.

$a(k_F,\theta), b(k_F,\theta)$ — Landau interaction parameters.

$B(\mathbf{k},\omega)$ — Spectral weight function.

$\hat{b}_j, \hat{b}_j^+$ — Annihilation and creation operators for the state j.

$\hat{b}_\mathbf{k}, \hat{b}_\mathbf{k}^+$ — Annihilation and creation operators for the plasmon excitation of wave vector $\mathbf{k}$.

c — Velocity of light.

$\hat{c}_m, \hat{c}_m^+$ — Fermion annihilation and creation operators for the state m.

$D(\mathbf{r}_1,t_1,\mathbf{r}_2,t_2)$ — Phonon Green's function (propagator).

$D_0(\mathbf{r}_1,t_1,\mathbf{r}_2,t_2)$ — Bare phonon Green's function.

E — Energy of a system.

E_M^j — Energy of the jth state of the M-particle system ($j = 0$ corresponds to the ground state).

E_n — Energy of the state n.

E_n^i — Eigenvalue of the particle i in the state $\psi_n^i(\mathbf{r})$ (Chap. 1).

E_R, E_i — Real and imaginary parts of the energy (Chap. 5).

E_ϕ — Energy of the N-particle system for a given external potential $\phi(\mathbf{r})$. (Chap. 12).

E_0 — Ground-state energy of the interacting system (Chap. 5).

$E(\mathbf{k})$ — Energy of the isolated Landau quasiparticle state $\mathbf{k}$ in the presence of other quasiparticle (Chap. 5).

$E(\mathbf{k})$ — Energy of the state $\mathbf{k}$.

$E(\mathbf{k})$	Energy eigenvalue for the state corresponding to the transformed operators $\hat{\alpha}_{\mathbf{k}}\hat{\beta}_{\mathbf{k}}$ (Chap. 13).
$E_0(\mathbf{k})$	Energy of the isolated Landau quasiparticle state $\mathbf{k}$ (Chaps. 5 and 9).
$E(\mathbf{v})$	Energy of the N-particle system as a function of the velocity of the observer (Chap. 5).
$\mathbf{E}(\mathbf{r},t)$	Electric field.
$E[n]$	The energy as a functional of the electron density.
$E_{xc}[n]$	The exchange and correlation energy functional.
e	Electronic charge.
$\mathbf{e}$	Polarization vector.
F	Number of closed fermion loops in a diagram (Chap. 8).
F	Free energy (Chaps. 10, 12).
$F(\mathbf{r}_1,\tau_1,\mathbf{r}_2,\tau_2)$	Anomalous Green's function (Chap. 13).
$f(\mathbf{k},\mathbf{k}')$	Landau quasiparticle interaction function.
$f^d(k_F,\theta),f^e(k_F,\theta)$	The direct and exchange parts of the quasiparticle interaction.
$f(\mathbf{r})$	Arbitrary function (Chap. 2).
$\mathbf{G}$	Reciprocal lattice vector (Chap. 9).
$G_{nn'}$	Expansion coefficients for the Green's function (Chap. 2).
G'_{12},G'_{21}	Anomalous Green's functions.
G_2^0	The normal (i.e., nonpairing) part of the two-particle Green's function (Chap. 13).
$G(\mathbf{k},\omega)$	Momentum–energy dependent Green's function.
$G(\mathbf{r},\mathbf{r}',t,t')$	Time-dependent Green's function (Chap. 2).
$G(\mathbf{r},\mathbf{r}',E)$	Energy-dependent Green's function (Chap. 2).
$G(x,x',\omega)$	Energy-dependent Green's function.
$G[n]$	The functional which contains the kinetic, exchange, and correlation contributions to the total energy (Chap. 12).
$G_c(x_1,t_1,x_2,t_2)$	Condensate part of the Green's function.
$G_\beta(x_1,t_1,x_2,t_2)$	Real-time thermal Green's function.
$G_\beta^A(x_1,t_1,x_2,t_2)$	Advanced real-time thermal Green's function.
$G_\beta^R(x_1,t_1,x_2,t_2)$	Retarded real-time thermal Green's function.
$G_0(x_1,t_1,x_2,t_2)$	Noninteracting Green's function.

$G_{00}(x_1,t_1,x_2,t_2)$ — Green's function corresponding to the single-particle Hamiltonian without the Hartree potential (Chap. 7).

$G_{(n)}(x_1,t_1,x_2,t_2)$ — nth order approximation to the single-particle Green's function.

$G_2(x_1,t_1,x_2,t_2,x_3,t_3,x_4,t_4)$ — Two-particle Green's function.

$G_{\pm}(\mathbf{r},\mathbf{r}',t - t')$ — Advanced (+) and retarded (−) Green's function (Chap. 2).

$G'(x_1,t_1,x_2,t_2)$ — Noncondensate part of the Green's function (Chap. 11).

$G^{-1}(x_1,t_1,x_2,t_2)$ — Inverse of the single-particle Green's function.

$\mathrm{G}'(x_1,t_1,x_2,t_2)$ — The matrix Green's function for a system with a condensate (Chap. 11).

$G(k,\omega_n)$ — Momentum-energy dependent thermal Green's function.

$\mathcal{G}(x,\tau,x',\tau')$ — Thermal Green's function.

$\mathcal{G}_0(x,\tau,x',\tau')$ — Noninteracting thermal Green's function.

$\mathcal{G}_0(x_1,x_2,\omega_n)$ — Noninteracting energy-dependent thermal Green's function.

$\mathcal{G}_{\uparrow\uparrow}(\mathbf{r}_1,t_1,\mathbf{r}_2,t_2)$ — The spin-up electron propagator (Chap. 13).

g — The strength of the effective electron–electron interaction (Chap. 13).

$g_i(n(r))$ — Energy functions (not functionals) of the electron density.

$g_{\mathbf{r}}[n]$ — The energy density functional.

H — Hamiltonian.

$\hat{H}_b$ — Hamiltonian describing the background neutralizing charge (Chap. 11).

$\hat{H}_{\mathrm{CORR}}$ — Hamiltonian containing the coulomb repulsion (Chap. 12).

$\hat{H}_{\mathrm{el}}$ — Electron Hamiltonian (Chaps. 11 and 12).

$\hat{H}_{\mathrm{el-ion}}$ — Electron–ion interaction Hamiltonian (Chap. 11).

$\hat{H}_{\mathrm{INT}}$ — Hamiltonian describing the interaction between the N-particle system and the added electron (Chap. 11).

$\hat{H}_{\mathrm{INT}}$ — Electron–phonon interaction Hamiltonian (Chap. 12).

$\hat{H}_{\mathrm{ION}}$ — Ion Hamiltonian.

$H_{l,f}$	Coupling Hamiltonian between localized and delocalized states (Chap. 12).
$\hat{H}_N$	Hamiltonian for the N-particle system (Chap. 11).
$\hat{H}_{0,f}$	Hamiltonian for the delocalized states (Chap. 12).
$\hat{H}_{\mathrm{ph}}$	Phonon Hamiltonian (Chap. 12).
$\hat{H}_0$	Single-particle Hamiltonian.
$\hat{H}_1$	Hamiltonian for one particle (Chap. 11).
$\hat{H}_0(x_1)$	Single-particle Hamiltonian.
$\mathbf{H}(\mathbf{r},t)$	Magnetic field (Chap. 3).
$H[n]$	Hamiltonian as a functional of the electron density.
$H[\phi]$	Hamiltonian as a function of the external potential.
$\mathcal{H}$	Hamiltonian density.
$\mathcal{H}$	Magnetic field (Chap. 12).
$\mathcal{H}_I$	An effective internal magnetic field (Chap. 12).
$\mathcal{H}_0$	External magnetic field (Chap. 12).
$\hbar$	Plank's constant.
$I(1,2,3,4)$	T-matrix for electron–electron scattering.
$\hat{J}(x)$	Single-particle operator.
$J_0(\mathbf{r})$	Zero-order Bessel function (Chap. 9).
$\hat{\mathcal{J}}(x)$	Operator density.
K	Effective Spring constant for internuclear forces (Chap. 3).
$\hat{K}$	Grand canonical Hamiltonian.
$\hat{K}_0$	Soluble part of the grand canonical hamiltonian.
K_n	Eigenvalue of the grand canonical hamiltonian for the state n.
$\hat{K}_0(\mathbf{r}_1)$	The normal (i.e., nonpairing) part of the grand canonical Hamiltonian (Chap. 13).
k	Variable conjugate to x [$=(\mathbf{k}$, spin)].
k_B	Boltzmann constant.
k_F	Fermi wave vector, i.e., wave vector of state at the Fermi level.
$\mathbf{k}$	Wave vector of a plane-wave state.
L	Lagrangian.
$\mathcal{L}$	Lagrangian density.

M	Atomic mass.
M	Occupation number of the $q = 2k_F$ phonon in the one-dimensional solid (Chap. 12).
$M(\mathbf{k},\mathbf{j},\mathbf{q})$	Matrix element appearing in the response function of an insulator.
m	Electron mass.
m^*	Effective mass of an electron.
N	Number of atoms in the linear chain (Chap. 3).
N	Electron density (Chap. 11).
$N(E)$	Density of states at energy E.
N_n	Number eigenvalue of the state $\|n\rangle$ (Chap. 10).
n	Number of particles *not* in the condensate (Chap. 11).
n	Order number of a diagram (Chap. 8).
$n(\mathbf{k})$	Number of particles in the eigenstate **k.**
$n(\mathbf{r})$	Density of electrons.
$\hat{n}(\mathbf{r})$	Density operator.
$n[\phi]$	The electron density as a functional of the external potential.
$\hat{n}_k$	Number operator for the normal mode of wave vector k.
$\hat{n}_{\mathbf{l},\sigma}$	Number operator for the state $\mathbf{l},\sigma$.
$n_0(\mathbf{k})$	Ground-state occupation factor (Chap. 5).
$\hat{O}$	General operator.
$\hat{O}_H(q,t)$	Heisenberg representation operator.
$\hat{O}_I(q,t)$	Interaction representation operator.
$\hat{O}_K(q,\tau)$	Operator in the "thermal" Heisenberg representation.
$\hat{O}_S(q,t)$	Schrödinger representation operator.
$P(q,\omega)$	Energy–momentum dependent polarization propagator.
$P(x_1,t_1,x_2,t_2)$	Polarization propagator.
$P(x_1,\tau_1,x_2,\tau_2)$	Thermal response function.
$P\{\ \}$	Permutation operator which produces all possible permutation of the labels; thus, $P\{\psi_\alpha\psi_\beta\psi_\gamma\} = \psi_\alpha\psi_\beta\psi_\gamma + \psi_\alpha\psi_\gamma\psi_\beta + \psi_\gamma\psi_\alpha\psi_\beta + \psi_\gamma\psi_\beta\psi_\alpha + \psi_\beta\psi_\gamma\psi_\alpha + \psi_\beta\psi_\alpha\psi_\gamma$.
P_n	Probability of finding a system in the state $\|n\}$ (Chap. 10).

$P_0(x_1,t_1,x_2,t_2)$	First-order polarization propagator.
$\mathbf{P}$	Momentum operator.
$\hat{P}$	Momentum operator (single-particle).
p_i	Momentum of *i*th atom (Chap. 3).
p_k	Fourier transform of p_i and momentum coordinate of the normal mode of wave vector k (Chap. 3).
$\mathbf{p}$	Particle momentum.
$\mathbf{p}_i$	Momentum of the *i*th atom (Chap. 11).
$\mathbf{p}_{\mathbf{k},\lambda}$	Momentum coordinate for the normal mode $(\mathbf{k},\lambda)$.
$Q(\mathbf{k},\mathbf{k}')$	The transformed interaction matrix element in a paired electron system (Chap. 13).
q	Generalized position variable.
q_i	Shift of *i*th atom from its equilibrium position (Chap. 3).
q_k	Position variable of the normal mode of wave vector k. The Fourier transform of q_i (Chap. 3).
$q_{\mathbf{k},\lambda}$	Position coordinate for the normal mode $(\mathbf{k},\lambda)$ (Chap. 11).
$\mathbf{q}$	Wave vector variable.
$R(\mathbf{k},\mathbf{k}')$	As $Q(\mathbf{k},\mathbf{k}')$ (Chap. 13).
$R(\mathbf{r}_1,t_1,\mathbf{r}_2,t_2)$	The plasmon Green's function (or propagator).
$\hat{R}^\lambda$	Operator occurring in the electron electromagnetic field interaction (Chap. 4).
$\mathbf{R}_i$	Atomic position vector.
$\mathbf{R}_{i0}$	Position vector describing the equilibrium position of the *i*th atom in a lattice.
$\mathbf{r}_n$	Electron position vector for the *n*th electron.
S	Entropy of a system (Chap. 10).
$\hat{S}$	Operator used to explicitly remove the effects of a perturbation from the system variables (Chap. 7).
$S(\mathbf{k},\mathbf{k}')$	As $Q(\mathbf{k},\mathbf{k}')$ Chap 13).
$S(\mathbf{p},\omega)$	Denominator of the anomalous Green's function (Chap. 11).
$S(1,2,3,4)$	Summation of the effects of repeated electron–hole scattering (Chap. 8).
S_k	A canonical transformation variable (Chap. 12).

$\hat{S}^{\lambda,j}$	Operator occurring in the electron electromagnetic field interaction (Chap. 4).
$S_0(1,2,3,4)$	Lowest-order electron–hole scattering function (Chap. 8).
T	Temperature.
$T(\mathbf{k},\mathbf{k}')$	As $Q(\mathbf{k},\mathbf{k}')$ (Chap. 13).
$T[\ ]$	Time ordering operator.
$T[n]$	Kinetic energy functional (Chap. 12).
$\mathrm{Tr}[\ ]$	Trace operator.
T_c	Critical temperature.
$T_\tau[\ ]$	Ordering operator.
t	Time variable.
$U(\mathbf{r})$	Potential function.
$U_0(\mathbf{r})$	Applied potential (Chap. 1).
$U(\mathbf{R}_i - \mathbf{R}_j)$	Two-body interaction (Chap. 11).
$U(\mathbf{r}_1,t_1,\mathbf{r}_2,t_2)$	Effective two-body interaction including phonon effects (Chap. 11).
U_l	Contribution to the energy of a state from multiple occupancy (Chap. 12).
u_k	Canonical (Bogoliubov) transformation coefficient (Chap. 13).
V	Volume of the system (Chap. 11).
$V(\mathbf{r})$	The potential energy of a particle at a position $\mathbf{r}$.
V_I	Imaginary part of the potential (Chap. 5).
$V_{k,l}$	Matrix element for the coupling between localized l and delocalized k states (Chap. 12).
$V^{P}_{\mathbf{k},\lambda}$	Electron–phonon scattering matrix element (Chap. 11).
V_R	Real part of the potential (Chap. 5).
$V_{\mathrm{ex}}(\mathbf{k})$	Exchange potential of state $\mathbf{k}$ (Chap. 5).
$V^{\alpha}_{\mathrm{ex}}(\mathbf{r})$	Local Hartree–Fock exchange potential for the state (Chap. 1).
$V_{\mathrm{ex}}(\mathbf{r},\mathbf{r}')$	Hartree–Fock exchange potential.
$V_H(\mathbf{r},t)$	Hartree potential.
$V_H^{(n)}(x,t)$	nth order approximation to the Hartree potential (Chap. 7).
$V_{\mathrm{INT}}(\mathbf{r}_1 \cdots \mathbf{r}_N)$	The interaction caused by the coulomb potential between the electrons situated at position vectors $\mathbf{r}_1,\mathbf{r}_2,\ldots,\mathbf{r}_N$.

$V_{\mathbf{k}}(x)$	Effective potential for state **k** (Chap. 9).
$V_{\lambda\lambda'}(\mathbf{k},\mathbf{k}')$	The effective scattering matrix element produced by an experimental probe (Chap. 13).
$V_{1S}(\mathbf{r}), V_{2S}(\mathbf{r})$	Atomic Hartree potentials (Chap. 1).
$v(q)$	Momentum-dependent coulomb interaction.
$v(x,x')$	Coulomb interaction.
v_k	Canonical transformation coefficient (Chap. 13).
v_0	Fermi velocity.
$\mathbf{v}$	Velocity variable.
$\mathbf{v}_k$	Velocity contribution of state k (Chap. 5).
$W(q,\omega)$	Energy- and momentum-dependent screened interaction.
$W(x_1,t_1,x_2,t_2)$	Screened coulomb interaction.
$W(x_1,\tau_1,x_2,\tau_2)$	Thermal screened interaction.
$W_0(x_1,t_1,x_2,t_2)$	First-order screened coulomb interaction.
X	Total number of systems in a grand canonical ensemble (Chap. 10).
X_n	Number of systems in a grand canonical ensemble which are in the state n.
x	A general spatial + spin variable [$\equiv (\mathbf{r},\alpha)$].
Z	Partition function (Chap. 10).
Z	Spectral weight or renormalization constant.
Z	Ionic charge.
α	Spin index.
α	Magnetic field coupling parameter (Chap. 12).
α_k	A perturbation matrix element (Chap. 12).
$\hat{\alpha}_k, \hat{\alpha}_k^+$	Annihilation and creation operators for the paired states in the Bogoliubov transformation (Chap. 13).
α_m	Expansion coefficient for a general wave function in terms of the eigenstates (Chaps. 4, 5).
$\alpha_{\lambda j}^{lm}$	A function containing the details of the electron photon interaction arising from the operator $\hat{R}^{\lambda}$ (Chap. 4).
β	Spin index.
β	Thermodynamic variable ($\equiv 1/k_BT$).
$\hat{\beta}_k, \hat{\beta}_k^+$	Annihilation and creation operators for the

	paired states in the Bogoliubov transformation.
$\beta_{\lambda j}^{lm}$	A function containing implicitly the details of the electron–phonon interaction arising from the operator $\hat{S}^{\lambda,j}$.
$\Gamma(\mathbf{k},\omega,\mathbf{q},\Omega)$	Energy–momentum dependent vertex function.
$\Gamma(x_1,t_1,x_2,t_2,x_3,t_3)$	Vertex function.
$\Gamma(x_1,\tau_1,x_2,\tau_2,x_3,\tau_3)$	Thermal vertex function.
$\Gamma_s[\ ]$	Subset of the complete set of eigenstates of the system symmetrized to conform to the Pauli principle.
γ	Electron–phonon coupling constant (Chap. 11).
$\gamma(\mathbf{q})$	Electron–phonon coupling matrix element (Chap. 12).
$\gamma_{k,n}$	An expansion coefficient (Chap. 12).
$\gamma(k,l,m)$	Matrix element in the anharmonic linear chain expansion (Chap. 3).
$\gamma(\mathbf{r})$	Electron–phonon coupling function (Chap. 11).
$\gamma'(\mathbf{r}_1,t_1,\mathbf{r}_2,t_2)$	Screened electron–phonon coupling function (Chap. 11).
γ_p	Electron–plasmon coupling constant.
Δ	Gap parameter (or order parameter) (Chap. 12).
Δ	An infinitesimal (Chap. 9).
ΔE	Change in energy from the ground-state energy due to changes in the distribution function (Chap. 5).
$\Delta n(k)$	Deviation from the ground-state occupation for the quasiparticle state k (Chap. 5).
$\Delta E(\mathbf{k})$	Contribution to the energy of the quasiparticle state **k** from the self-energy (Chap. 9).
$\Delta(\mathbf{k})$	Superconducting gap parameter (Chap. 13).
$\Delta(\mathbf{r}_1,\tau_1)$	The mean pairing potential (Chap. 13).
Δ_0	The Fermi level approximation to the gap parameter (Chap. 13).
δH_{EXP}	The contribution to the Hamiltonian of a system arising from an experimental probe (Chap. 13).
$\delta(\mathbf{r} - \mathbf{r}')$	Dirac delta function.
$\delta_{n,n'}$	Kronecker delta function.
ϵ	Infinitesimal energy (Chap. 2).

ϵ	Coefficient for the first anharmonic term in the interatomic potential (Chap. 3).
$\mathcal{E}(\mathbf{k})$	Quasiparticle energy (Chap. 9).
$\epsilon(\mathbf{q})$	Dielectric response function (Chap. 1).
$\epsilon(q,\omega)$	Energy- and momentum-dependent dielectric response function.
$\epsilon(x_1,t_1,x_2,t_2)$	Dielectric response function.
$\epsilon(0)$	Static long-range dielectric response function (classical dielectric constant) (Chap. 9).
$\epsilon_{\text{eff}}(q,\omega)$	An approximation to the Lindhard response function for the electron gas (Chap. 9).
ϵ_l	Energy of the state l.
$\epsilon_M(q,\omega)$	Metallic response function (Chap. 9).
$\epsilon_M(j,k)$	jth excitation energy of M-particle state with momentum $\hbar\mathbf{k}$ (Chap. 6).
$\epsilon_{N\pm1}(j)$	Excitation of the $(N \pm 1)$-particle system (Chap. 6).
$\epsilon_{\text{xc}}(n(r))$	The local exchange and correlation energy density function (Chap. 12).
ϵ_0	Permittivity of free space.
$\epsilon_1(q_1\omega),\epsilon_2(q,\omega)$	Real and imaginary parts of the dielectric response function (Chap. 9).
$\epsilon^{-1}(x_1,t_1,x_2,t_2)$	Inverse of the dielectic response function.
$\epsilon_{\mathbf{k},\lambda}$	Polarization vector for the normal mode $\mathbf{k}\lambda$ (Chap. 11).
η	An infinitesimal.
θ	Angle between quasiparticles interacting new Fermi surface (Chap. 5).
$\theta(q)$	Step function: $\theta(q) = 0,\ q < 0;\ \theta(q) = 1,\ q > 0$.
λ	Normal-mode label (includes wave vector and polarization) (Chap. 3).
λ	Spin index.
λ	Strength of the contact interaction potential (Chap. 13).
λ	Thomas–Fermi wave vector (Chaps. 1 and 9).
μ	Chemical potential (Fermi energy).
μ_0	An atomic magnetic moment (Chap. 12).
μ_0	Permeability of free space.

$\xi(x,t)$	Part of the condensate Green's function (Chap. 11).
$\hat{\xi}(x,t)$	Condensate part of the field operator (Chap. 11).
$\xi_0(x)$	Condensate wave function (Chap. 11).
$\pi(\mathbf{r},t)$	Generalized momentum function.
$\hat{\rho}$	Statistical density matrix operator (Chap. 10).
$\hat{\rho}(\mathbf{r})$	Density operator (Chap. 11).
$\rho(x,t)$	Charge density (Chap. 7).
$\hat{\rho}_{\mathbf{k}}$	Fourier transform of the density operator (Chap. 11).
$\rho_{\mathbf{k}}(E)$	Density of state of wave vector **k** (Chap. 5).
$\hat{\rho}_n$	Density matrix operator for state $\|n\}$ (Chap. 10).
$\Sigma(q,\omega)$	Energy- and momentum-dependent self-energy.
$\Sigma(x_1,x_2,\omega)$	Energy-dependent self-energy.
$\Sigma(x_1,t_1,x_2,t_2)$	The self-energy.
$\Sigma(x_1,\tau_1,x_2,\tau_2)$	Thermal self-energy.
$\Sigma_{\text{C.H.}}(\mathbf{k},\omega)$	The coulomb hole (or coulomb correlation) part of the self-energy.
$\Sigma_{\text{H-F}}(\mathbf{k},\omega)$	The Hartree–Fock approximation to the self-energy.
$\Sigma^{(n)}(x_1,t_1,x_2,t_2)$	nth order approximation to the self-energy (Chap. 7).
$\Sigma_{\text{S.E.}}(\mathbf{k},\omega)$	The screened exchange part of the self-energy.
$\Sigma_1(q,\omega),\Sigma_2(q,\omega)$	Real and imaginary parts of the self-energy (Chap. 7).
$\Sigma_{11},\Sigma_{12},\Sigma_{21},\Sigma_{22}$	Components of the matrix self-energy used in a Bose system with a condensate (Chap. 11).
$\Sigma(x_1,t_1,x_2,t_2)$	Matrix self-energy for a system with a condensate (Chap. 11).
σ	Spin variable.
$\hat{\sigma}_k$	Fourier transform of the momentum operator for a set of particles (Chap. 11).
τ	"Thermal" Heisenberg representation variable (imaginary time variable) (Chap. 10).
τ	Time difference variable (zero-temperature formulation).
Φ	General basis state for the n-particle system (Chap. 4).

$\hat{\Phi}(\mathbf{r},t)$	Plasmon field operator.
$\Phi_l(r)$	A localized state (Chap. 12).
ϕ_m	Eigenfunction of the single-particle Hamiltonian.
$\hat{\phi}(\mathbf{r})$	Phonon field operator.
$\phi(\mathbf{r},t)$	Generalized position function [conjugate to $\pi(\mathbf{r},t)$] (Chap. 3).
$\phi(x,t)$	External potential applied to the N-particle system, used in deriving the iterative solution to the Green's function equation of motion.
$\phi(x,\tau)$	Perturbation used in deriving the iterative solution to the thermal Green's function equations of motion.
$\hat{\phi}_H(\mathbf{r},t)$	Phonon field operator in the Heisenberg representation.
$\phi_P(x)$	A general eigenstate (Chap. 10).
$\hat{\chi}(x,t)$	Noncondensate part of the field operator (Chap. 11).
$\chi_\mathbf{k}$	A Bogoliubov transformation variable (Chap. 13).
$\Psi(\mathbf{r}_1,\mathbf{r}_2, \ldots ,\mathbf{r}_N)$	The total wave function for the N-particle system.
$\Psi_S(q,t)$	Schrödinger representation wave function (Chap. 4).
$\Psi_H(q,t)$	Heisenberg representation wave function (Chap. 4).
$\Psi_I(q,t)$	Interaction representation wave function (Chap. 4).
$\psi_\uparrow(\mathbf{r}),\psi_\downarrow(\mathbf{r})$	The field operators corresponding to spin-up and spin-down electrons (Chap. 13).
$\psi_l(\mathbf{r}_i)$	Single-particle wave function. The subscript denotes the state concerned.
$\hat{\psi}_K(x,\tau)$	Field operator in the "thermal" Heisenberg representation.
$\hat{\psi}(q,\tau)$	Field operator.
Ω	An energy or frequency variable.
Ω	Probability of a configuration of the grand canonical ensemble (Chap. 10).
Ω	Volume of the system (Chaps. 1, 13).

Ω_q — Renormalized phonon energy (Chap. 12).

ω — An energy variable.

ω_k — Oscillation frequency of the normal mode of wave vector k (Chap. 3).

$\omega_{\mathbf{k},\lambda}$ — Energy of the normal mode $\mathbf{k},\lambda$ (Chap. 11).

ω_n — Energy variable for the thermal functions conjugate to τ (Chap. 10).

ω_p — Plasmon energy (long-wavelength limit).

$\omega_P(\mathbf{q})$ — Plasmon energy at wave vector $\mathbf{q}$ (Chap. 11).

ω_{ph} — The "bare" phonon energy, i.e., the phonon energy of the ion system in which the electrons do not respond (Chap. 11).

$\omega_{\text{ph}}(\mathbf{q})$ — The energy of a phonon with wave vector $\mathbf{q}$ (Chap. 11).

ω_0 — Energy interval defining the energy range of the electron–electron interaction (Chap. 13).

$|\cdots\rangle$ — A wave function in the Dirac notation. Details of the wave functions are contained inside the bracket.

$|M,j\rangle$ — jth state of the M-particle system (Chaps. 6, 11).

$|M,j,\mathbf{k}\rangle$ — jth state of the M-particle system with momentum $\hbar\mathbf{k}$ (Chap. 6).

$|N\rangle$ — Ground state of the N-particle system.

$|N,\phi\rangle$ — The N-particle eigenstate for the applied external potential (Chap. 12).

$|N_I,\pm\infty\rangle$ — Ground state of the N-particle system at $t = \pm\infty$ (Chap. 7).

$|\mathbf{k}_1,\mathbf{k}_2\rangle$ — Two-particle eigenstate with particle in $\mathbf{k}_1$, $\mathbf{k}_2$ (Chap. 13).

$|n_1n_2,\ldots,n_k,\ldots\rangle$ — Eigenstate of the system specified by the number operator eignevalues for the normal modes (Chap. 3).

$|\mathbf{k},n\rangle$ — The nth eigenstate of the normal mode $\mathbf{k}$ (Chap. 12).

$|\xi\rangle$ — Eigenfunction with energy ξ.

$|0\rangle$ — Vacuum state of the phonon system (Chap. 11).

$|n\}$ — General state of a system (Chap. 10).

$|0\}$ — Vacuum state (Chap. 12).

$\langle\mu_O\rangle$ — Residual magnetic dipole moment (Chap. 12).

INDEX